MECHANICS

MECHANICS

N. C. Barford
Department of Physics, Imperial College, London

JOHN WILEY & SONS
London · New York · Sydney · Toronto

Library of Congress catalog card number 72–2639

ISBN 0 471 04840 2 Cloth bound
ISBN 0 471 04841 0 Paper bound

Set on Monophoto Filmsetter and printed by
J. W. Arrowsmith Ltd., Bristol, England

To
ALMA MAY

Preface

Classical mechanics is well established, and rigorous treatments of its many applications and extensions were made well before this century, and in many cases much earlier still. As a result it suffers, as far as undergraduate teaching is concerned, from several disadvantages.

Firstly, because it is not 'modern physics' it tends to be thought of as out-of-date. Yet even in the most modern fields of experimental physics and engineering the research worker or designer will constantly find himself looking for solutions to 'classical' problems.

Secondly, its long existence has produced many powerful formal treatments of the theory covering very wide fields of application. The difficulty for the student is that their logical mathematical framework touches his everyday experience only, if at all, when problems at the ends of chapters are solved. Yet some of the manifestations of classical mechanics form our earliest experience of physics. Our ability to judge the bounce of a ball from a wall, to ride a bicycle round a corner, to use a spanner to tighten a nut, represents a considerable understanding and application of mechanical principles, unaware as we may be of the depths of our skill! Wherever possible therefore, particularly in the earlier chapters, common experience and easily understood observations are used as particular, and not necessarily precisely stated, starting points to lead to the general and exact formulation of concepts and the laws relating them.

Thirdly, because classical mechanics can start with the statement of a few simple laws and by the end of two or three hundred pages, using derivations of great exactitude and certainty, describe all the ramifications arising and the different viewpoints which are possible, a student may conclude that it is a closed book. If he knows already of the existence of relativity and quantum mechanics he may wonder how classical mechanics can be so precisely correct, yet apparently 'wrong'. In this book emphasis is put on recognizing that physical theory is 'true' only for certain ranges of the quantities described and for certain classes of observer. The student should find the necessity for quantum mechanics and relativity no surprise and he is shown in the last chapter how the basic ideas of special relativity follow from principles already inherent in classical mechanics.

Given the wide spread of abilities and interests among students, how should one present a subject, such as mechanics, with which they may already have some familiarity? I have attempted to deal with this problem partly by establishing most of the concepts and laws in precise but simple form in the first few chapters. They may be regarded as an elementary course by themselves, or as an introduction to the more general and detailed topics of the later chapters for those who have already covered some of the ground.

In addition comments and worked examples of two types are given at the end of each chapter. Some are revisions of ideas and techniques which the reader may have forgotten. (Even in the more advanced chapters there are repetitions of earlier, simple work—it would be rash to assume that they are read only when the preceding chapters have been thoroughly understood and remembered.) The others are intended to encourage further thinking, and caution, among those whose familiarity with the subject might breed contempt!

The other difficulty I have tried to meet is that of introducing the mathematics convincingly and effectively. Undergraduate physics and engineering are often accompanied by separate mathematics courses in which students show great competence. It is a sad but common experience, however, to find that this is unaccompanied by any great ability to deal with physical problems in a coherent and meaningful mixture of prose and mathematics. A deadly distinction seems to become established between what is 'mathematics' and what is 'physics'.

Although a concurrent mathematics course would be very useful for the reader, a great deal of the mathematics that occurs is discussed also at the ends of chapters. The topics are dealt with as they became necessary to describe the physics. This is a departure from the usual method of collecting together the mathematics as complete appendices to the book. While this tidy practice has the advantage of conciseness it would conflict with the attempt here to present the physics and mathematics as a whole.

Another aspect of this viewpoint is the emphasis put on describing mechanics in a way that accords directly with physical measurements. In this respect the use of vector notation, while extremely illuminating in some ways, can obscure the fundamental fact that any experiment concerned with the motion of a particle in space must be based upon *three* coordinate measurements (or four if we include time). Vectors are, of course, used extensively in the book, but wherever it is the *measurement* that is of primary importance, for example in transforming between different frames of reference, the mathematical descriptions are formulated directly in terms of such measurements.

This does not necessitate a return to the old-fashioned tedious triplication of every equation, provided that matrix methods are used. The notation is then as concise as that for vectors, and the effort of mastering the few laws of matrix algebra that are required here is only on the same scale as that needed for learning the various vector operations. One great advantage of this approach is the comparitive ease with which one can move from three-dimensional classical mechanics to four-dimensional special relativity.

Courses on mathematical methods for physicists or engineers usually present the elements of matrix algebra, but all too often these are learned (and quickly forgotten) simply as mathematical exercises because rather few physicists and engineers have the courage to use this very powerful mode of description in their undergraduate lectures. It should also be remembered that as the influence of the 'new mathematics' spreads in the schools, students will have had experience of transformation theory which has been quite unknown until now.

For this reason, too, I feel that there is much to be gained if students acquire an early facility for using matrix methods in a subject whose physical content concerns the familiar macroscopic world, and is relatively easily understood, rather than to wait until late in their studies when the difficulties of a new or forgotten technique are compounded with those of the unfamiliar and rather abstract regions of physics that belong to quantum theory.

Comments from many of my colleagues have played a very useful part in the preparation of this book; in particular, I should like to thank Dr. K. Ruddock who read the first manuscript and made many valuable suggestions, and Professor M. Blackman with whom I had several enlightening discussions about the first chapter. I am especially indebted to Mr. R. M. Hill whose most careful and constructive critique of the whole book led to many changes in its final form.

My thanks must also go to Jill Farmer who undertook the main burden of the typing with great patience and skill, and to my daughter, Sarah Barford, who spent much time with me reading the proofs.

Acknowledgements are due to the University of London for permission to use material from University examinations set for students of Imperial College.

Imperial College, N. C. BARFORD
July, 1972

Contents

Chapter 1

Basic Concepts; Newton's Laws

1.1. Introduction

The universe presents a great variety of objects in motion. The most distant stars move away from us with speeds approaching one hundred thousand kilometers per second. The planets orbit the sun with periods ranging from less than a year to more than a century. Our own earth spins with a daily rotation, and the moon encircles it once approximately every four weeks. Satellites and space probes, driven by rockets, leave the earth with speeds of over ten kilometers per second. On or near the earth's surface ships, aircraft, cars, trains and lifts, powered by internal combustion engines, jet turbines and electric motors, travel with more modest speeds. Clocks and gramophones rotate; footballs, tennis-balls and golf-balls fly through the air.

All around us we see this motion, aided perhaps by telescopes and microscopes. But, beyond the range of these aids, unseen motion exists. The molecules of a gas are in constant motion, colliding with each other and with the walls of the vessel containing them; a crystal, apparently rigid and at rest, is really vibrating in many ways.

Can we fit some sense of order into this bewildering scene? Are there common concepts which describe apparently dissimilar phenomena? Are there 'laws' which link these concepts?

1.2. Natural motion

One of the first ideas we need to establish is how to distinguish between the motion, if any, that an object would have if left alone (what we might call its *natural* motion) from that which it shows in response to outside influences (its *forced* motion).

It is not at all easy to free a body from all outside influences. We can avoid obvious things such as pushing it with our hands or pulling it with strings, and ensure that it does not collide with other objects. However, if we tried to carry out a complete programme of 'non-influence' in an earth laboratory we should have to remove any floor or table on which the body could be supported, and use for the laboratory an unobstructed mine or lift shaft. Even then the body would move towards the centre of the earth because of the influence of the earth's gravitation. There might also be electrical charges which could similarly affect the behaviour of the body although it was not in obvious contact with anything else.

The best we can do is to suppose that all influences between objects—interpreting 'objects' in a very wide sense, ranging from neutrinos to galaxies—diminish with the separation between them, and that it is possible to remove an object so far away from all others that they have no appreciable effect upon it. Is there then any feature common to the motion of objects when they are very distant from all others?

At first sight there is no common feature, since if we examine closely such isolated objects we may find that one is a spinning sphere, another an oscillating spring, another a hot gas with all its molecules moving randomly at high speed, and so on. However, if we view these objects from a very great distance they all appear as points, and the detailed local motions of rotation, vibration, and so on, become unobservable. (See Section 1.11.1, page 10.)

A very simple result then emerges:

The points representing the objects all either remain at rest or move with a constant velocity, relative to each other.

We shall call these points *inertial points.*

To use the phrase 'constant velocity' must, to an experimentalist, imply a method of measuring this, in principle at least. For this we shall use a three-dimensional, right-handed Cartesian frame of reference (see Figure 1.1). If we

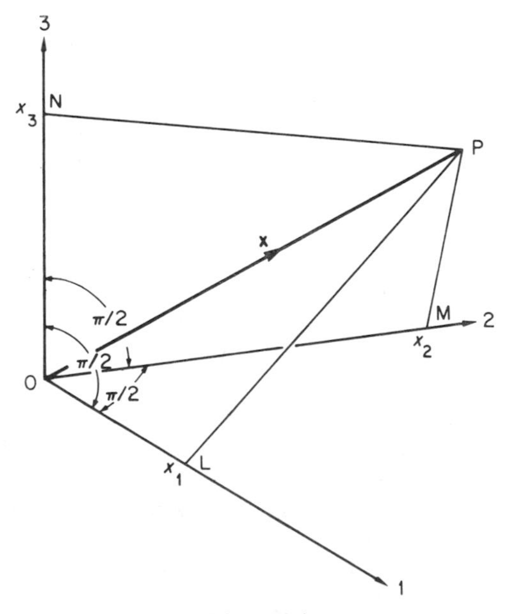

Figure 1.1

take an inertial point, O, as the origin of such a frame, an observer using this frame must still decide how to fix the direction of the axes. This he can do by choosing three of the most distant points he can observe. It does not matter if he cannot find three whose directions are all exactly perpendicular to each other. If he can, the Cartesian frame is immediately defined, but any three distinct directions are sufficient to enable him to construct the frame O1, O2, O3.

What is much more important is that the directions of the three points should not change relative to each other. Is it possible to find such points? Indeed it is. If the stars are observed for many years the great majority show no detectable difference in their *relative* directions. These are the so-called *fixed stars*, and it is these which we must use to fix the directions of the Cartesian axes.

Notice that we are not arguing in a circle here. We do not assume that the fixed stars themselves are inertial points, but simply use them to fix the axes of a frame of reference centred on one of a group of inertial points. Once this is done the position of any other point, P, of the group is given by measuring the three components x_1, x_2, x_3 (thus defining its position *vector* $\mathbf{x}$). Its velocity is then the vector $\mathbf{v}$ formed from the *measured* components dx_1/dt, dx_2/dt, dx_3/dt. A constant velocity vector then means one for which the three components show no variation with time. (See Section 1.11.2, page 10.)

1.3. Newton's first law

The process just outlined gives a frame of reference for every inertial point. These are the *inertial frames of reference*. We can now restate the observation made in the previous section:

In any inertial frame of reference all other inertial points are at rest or move with a constant velocity.

This is the form we shall take for Newton's first law. It may be written

$$\frac{d\mathbf{v}}{dt} = 0 \quad \text{or} \quad \frac{dv_1}{dt} = \frac{dv_2}{dt} = \frac{dv_3}{dt} = 0. \tag{1.1}$$

Note that since all inertial frames of reference are based upon inertial points, then all inertial frames of reference move with constant velocities or are at rest, relative to each other. The axes need not be parallel, but will remain at constant angles with respect to each other.

1.4. Forced motion

When objects are sufficiently close, the points representing them may no longer behave as inertial points, either remaining at rest or moving with constant velocity. This deviation from inertial or 'natural' motion we describe as 'forced' motion.

1.4.1. Force

Forced motion may be considered to arise from some influence or *force* acting on the object. One type of forced motion, in which objects tend always to move towards each other, we ascribe to *gravitational force*. Another type, in which either attraction or repulsion may occur, is only observed when electrical charges are present, and is ascribed to an *electrical force*.

These are the two fundamental forces that govern the motion of most macroscopic objects insofar as it differs from the natural motion expressed in Newton's first law. (See Section 1.11.3, page 10.) It is possible to develop an elegant and precise description of these forces by considering them solely in terms of the deviation from inertial motion that they produce. However, here we shall discuss force in a more experimentally practicable way by considering the effect of a small coil spring.

Suppose such a spring is held in a compressed condition by a simple trigger. If one end is fixed relative to some inertial frame of reference and the other end touches an object, then when the trigger is released, the spring expands, pushing the object in front of it. If the object is initially at rest it is thereby set in motion, or if it is already moving its velocity will be changed (see Figure 1.2).

Figure 1.2

This influence that the spring has upon an object we call the *force* exerted by the spring. Starting with one such spring we can show that forces can arise from other sources, such as a human muscle, a heated gas, or electrical charges in proximity. We can show that force can be measured in units of the initial spring force and that it is a *vector quantity*. We can therefore attach a definite quantitative meaning to the components F_1, F_2, F_3 of a force vector $\mathbf{F}$. (See Sections 1.11.4, page 11, and 1.11.5, page 13.)

1.4.2. Inertial mass

We must now try to relate the forced motion that occurs when a force is allowed to act on a body to the force itself. In detail this is a very complicated matter. If the force acts roughly at the centre of an object it will move without rotating, whereas if it acts near the edge it will be set spinning. Some objects, if they are sufficiently frail, may even be distorted by bending or oscillating as the force sets them in motion. Both the point of action of the force and the shape of the object, then, affect the motion.

However, there is one general observation that we are all familiar with. If the same force acts in turn on a variety of objects, some will respond more easily than others. It is, for example, very easy to push a bicycle along but very difficult to move a large motor car. Objects have a certain reluctance to respond to an applied force, which is called *inertia*, and which generally increases with the amount of material in the object.

Let us, therefore, restrict our observations to what happens when the same force acts on a variety of objects all made from the same volume of the same material. We might, for example, take equal volumes of iron and form one volume into a sphere, and others into a cube, a long rod and a thin sheet. We shall still find that their motion depends in a complicated way upon their shape and upon where the force acts.

However, if, as in Section 1.2, we view them from such a distance that all the objects appear as points, and the detailed motion becomes invisible, we obtain a remarkable result. *All the objects show the same motion.* The motion is in the direction of the force and increases with the magnitude of the force.

Motion is rather a vague word. What experiment shows us precisely is that, if we use any inertial frame of reference, the *acceleration* of the points is the significant observation, and that this is *proportional to the force*, the constant of proportionality being independent of time and of the direction of the force. Thus for any of the objects made from the same volume of the same material, there is a single constant of proportionality, m, in terms of which we may write the experimental results as

$$m\frac{\mathrm{d}^2x_1}{\mathrm{d}t^2} = F_1, \qquad m\frac{\mathrm{d}^2x_2}{\mathrm{d}t^2} = F_2, \qquad m\frac{\mathrm{d}^2x_3}{\mathrm{d}t^2} = F_3. \tag{1.2}$$

(See Section 1.11.6, page 17.) The equations (1.2) may be condensed into the vector equation

$$m\mathbf{a} = \mathbf{F}. \tag{1.3}$$

Because the acceleration $\mathbf{a}$ is linked to the force vector $\mathbf{F}$ in this way it also is a vector. m is a property only of the type and amount of material in a body, and since the larger it is, the smaller the acceleration produced by a given force, m is a measure of the reluctance to respond, or inertia, that we remarked upon earlier. We therefore call m the *inertial mass*. If we use equations (1.2) to define it numerically, its value would depend upon the units used to measure a and F, the magnitude of the vectors $\mathbf{a}$ and $\mathbf{F}$.

However, we can use mass itself as a basis of measurement in the following way. Suppose we carry out acceleration measurements for a group of objects

(1), all made from the same volume of the same material; repeat them for another group (2), all made from the same volume of the same material (but differing from those of group 1); and so on. If the same force **F** acts throughout, we shall find

$$m_{(1)}\mathbf{a}_{(1)} = \mathbf{F}, \qquad m_{(2)}\mathbf{a}_{(2)} = \mathbf{F}, \qquad m_{(3)}\mathbf{a}_{(3)} = \mathbf{F}, \ldots,$$

or, in terms of magnitudes,

$$F = m_{(1)}a_{(1)} = m_{(2)}a_{(2)} = m_{(3)}a_{(3)} = \ldots,$$

where $m_{(1)}$ and $\mathbf{a}_{(1)}$ are the inertial mass and acceleration common to all the objects of group (1); $m_{(2)}$ and $\mathbf{a}_{(2)}$ are those common to group (2); and so on.

Thus the ratios

$$\frac{m_{(2)}}{m_{(1)}} = \frac{a_{(1)}}{a_{(2)}}, \qquad \frac{m_{(3)}}{m_{(1)}} = \frac{a_{(1)}}{a_{(3)}}, \ldots$$

do not depend upon the units in which acceleration (or force) is measured. If we therefore arbitrarily choose $m_{(1)}$ to be our basic unit, all the other masses may be defined and measured by the relationships

$$m_{(2)} = \frac{a_{(1)}}{a_{(2)}} m_{(1)}, \qquad m_{(3)} = \frac{a_{(1)}}{a_{(3)}} m_{(1)}, \ldots. \tag{1.4}$$

(See Section 1.11.7, page 17.)

In practice a particular piece of metal (platinum, because of its resistance to chemical and physical change) is defined as having a mass of one kilogramme, and the inertial mass of any other object is determined from this directly or indirectly through relationships of the type (1.4). Equations (1.2) can now be used to measure F, not in units of a particular coil spring force but in units of the dimensions MLT^{-2} of the left-hand sides of these equations.

The internationally agreed standard system of units is the Système Internationale d'Unités, or SI, which is an extended version of the metre–kilogramme–second, or MKS, system. In this system inertial mass is measured in kilogrammes (kg), length in metres (m) and time in seconds (s). The unit of force, $1\ \mathrm{kg\ m\ s^{-2}}$, is then called the *Newton*.

Since, as we shall see in Chapter 2, experiment shows that inertial mass is equivalent to the gravitational mass that we measure on a balance, we shall normally simply use the term 'mass'. However, it must be remembered that it is the inertial, and not the gravitational, property of mass that is fundamental to our present discussion.

1.5. The point particle

The results of our experiments enable us to describe the response of objects to an applied force primarily in terms of the motion of a point possessing only one physical property—mass. This we shall call a *point particle*. Note that it is not merely one property that is involved but a property that is described by a single quantity, m. Thus, not only is the point particle a point in the sense that it is infinitesimally small, but it has no directional properties—equations (1.1) show that it looks the same in all directions. It is a much simpler concept than, for example, an elementary magnet (a magnetic dipole). This is also a 'point' as far as size is concerned, but the magnetic property it possesses involves a direction (the dipole axis) as well as the strength of the dipole.

To be realistic as far as laboratory experiments are concerned we ought to qualify some of the statements made above. Firstly, it would not always be true, even at a great distance, that all objects with the same mass show the same motion. A very thin sheet of material would be affected by air resistance much more than a compact lump. The experiments would have to be carried out in a vacuum or the variety of shapes and densities of materials would have to be considerably restricted. Secondly, with objects having masses of, say, a tenth to one kilogramme, it would not be easy to make measurements at a distance that would reduce them to point particles. In practice, the best approximation to point particle behaviour would be to use highly polished small circular steel discs floating on a horizontal frictionless (air cushion) surface, and to observe the motion of their centres.

1.6. Newton's second law

This simplification of the behaviour of an object to the motion and mass of a point particle enables us first to define the *momentum* as

$$\mathbf{p} = m\mathbf{v} = m\frac{\mathrm{d}\mathbf{x}}{\mathrm{d}t}, \tag{1.5}$$

and then to state the observations (1.2) or (1.3) as Newton's second law:

In any inertial frame of reference the rate of change of momentum of a point particle is equal to the force acting on it.

Written as

$$\mathbf{F} = \frac{\mathrm{d}\mathbf{p}}{\mathrm{d}t} = m\frac{\mathrm{d}\mathbf{v}}{\mathrm{d}t} = m\mathbf{a} \tag{1.6}$$

it can be seen to be equivalent to equation (1.3), and to include Newton's first law (equations 1.1) as a special case. (See Section 1.11.8, page 17.)

1.7. Forces between objects

When we discussed the action of a spring force in Section 1.4.1 we considered one end to be fixed in an inertial frame. Now we can only fix the end to some object. Thus we do not really have one end of the spring fixed in an absolute sense while the other end moves one object, but rather a spring attached to two objects, one at each end. A force is exerted by both ends of the spring when the trigger is released, and *both objects will move.*

We can obtain a very good approximation to the situation envisaged in Section 1.4.1 by ensuring that one object has very much more material in it and therefore possesses a much greater inertia than the other. For example, if the spring is attached at one end to a steel ball a few millimetres in diameter and at the other to a steel ball one metre in diameter, both initially at rest in an inertial frame, the larger ball will show no appreciable motion when the spring trigger is released. (See Section 1.11.9, page 18.)

However, we often have to deal with a situation in which both objects exhibit forced motion and in which we can therefore say that each object exerts a force upon the other. Experiments demonstrate a very simple result in this case: *the forces are equal and opposite.*

1.8. Newton's third law

By considering the equivalent point particles of the two objects we can state the preceding observation precisely as the following version of Newton's third law:

The forces that two point particles exert upon each other are directed along the line joining them and are equal in magnitude but opposite in direction.

(See Section 1.11.10, page 18.)

Thus, if particle 1 experiences a force $\mathbf{f}_{12}$ arising from the presence of particle 2, no other particles being involved, then particle 2 will experience a force $\mathbf{f}_{21}$, where

$$\mathbf{f}_{12} = -\mathbf{f}_{21},$$

and where the directions of both force vectors are along the line 12.

It is easy to demonstrate this law when the force between the particles is transmitted by some material object such as a spring or rod. However, it is true also for the fundamental electrostatic and gravitational forces which can act across a vacuum and may be considered as transmitted by the appropriate 'field'.

Compared with Newton's first two laws, the third is a very incomplete statement of the way in which objects interact with each other. For example, the rotational effect that one permanent magnet has upon another cannot be described in terms of a single force acting between them, however small (or point-like) the magnets may be. The interaction must take account of the basic directional properties of the magnets themselves.

Even when the objects do not have an intrinsic direction the law may fail. A uniformly charged sphere is in this category. It is true that two such spheres at rest, or moving only slowly, exert electrostatic forces upon each other which are equal but opposite and lie along the line joining their centres. However, at high speeds, the moving charges constitute currents which exert magnetic forces upon each other, and these additional forces do not obey the third law.

Thus, we must take care to apply this law only to cases where the forces depend neither upon any intrinsic directional properties of the objects nor upon their velocities. (See Sections 1.11.11 and 1.11.12, page 18.)

1.9. Practicability of Newton's laws

We have begun by establishing the concept of an inertial frame of reference and later asserted that the basic laws of motion were true for any such frame. This gives great flexibility in using the most convenient frame for analysing a dynamical problem.

Now it is clearly impracticable to test the laws for *all* inertial frames, and what we have really done so far is to overstate the experimental situation by assuming that what is true for some or many cases is true for all. However, we shall see in Chapter 8 that if the laws are true for only one frame of reference then they will be true for any other frame travelling with a constant velocity relative to the first.

Thus there is no sense in which one inertial frame has a special significance compared with any other as far as mechanical phenomena are concerned, and it is just this universality of Newton's laws among all inertial frames that gives them the status of fundamental physical laws.

Even establishing one inertial frame and experimentally testing the laws in it is a formidable undertaking, and much of our understanding of mechanics necessarily comes from experiments performed in earth laboratories. What is remarkable is how accurately the laws describe such experiments. In an earth laboratory no object will be far from any other. Moreover, the earth is orbiting the sun and at the same time spinning on its own axis, so that it is clear that a laboratory-based frame of reference has a very complicated motion relative to a frame based upon the fixed stars.

However, gravitational and electrical forces are weak, nuclear forces have a very short range and the effects of earth motion, as we shall also see in Chapter 8, are slight. The fact that the simple basic laws are so little affected by these complications has proved a most fortunate accident for our understanding of mechanics and, in consequence, of the whole of physics and engineering.

1.10. Limitations of Newton's laws

Although classical mechanics, based on Newton's three laws, covers phenomena over a wide scale, it is important to realize that most experience and verification of them extends over only a finite range and that extensions greatly beyond this may not be valid.

Roughly speaking, the range extends from the microscopic scale to that of the solar system; that is,

lengths from 10^{-6} m to 10^{10} m,
times from 10^{-9} s to 10^{10} s,
masses from 10^{-15} kg to 10^{30} kg,
speeds from 0 to 10^{5} m s^{-1}.

These are not sharp demarcations. By itself one limit may be exceeded by many orders of magnitude, but when two or more are passed classical predictions may be suspect. For example, the motion of an electron (whose mass is only $9{\cdot}1 \times 10^{-31}$ kg) subject to comparatively weak electric forces in a cathode ray tube is accurately described by classical mechanics, but not when it is accelerated in a betratron to speeds of more than 10^{8} m s^{-1} or when it is confined within an atom to distances of less than 10^{-9} m.

1.11. Comments and worked examples

1.11.1. See Section 1.2

This is how we normally see the stars and planets—all as points of light, moving smoothly or at rest with respect to each other. In fact, they have a great variety of shapes, sizes and local motions, some spinning, some exploding, some with their own moons encircling them, and so on.

1.11.2. See Section 1.2

Cartesian axes. Cartesian axes are orthogonal (i.e. each axis is perpendicular to the other two) and form a right-handed or positive system if a rotation from O1 to O2 as we look along O3 is in the clockwise or right-handed sense. The coordinates x_1, x_2, x_3 are the (algebraic) lengths of the *projections* OL, OM, ON of OP on the axes (see Figure 1.1).

If P remains stationary in the frame then it does not matter how slowly we find, experimentally, the values x_1, x_2, x_3. However, when P is moving, the statement that its velocity $\mathbf{v}$ is given by the components dx_1/dt, dx_2/dt, dx_3/dt assumes that the feet of the perpendiculars L, M, N are exactly following the motion of P. Now, even in an optical method a finite time will be required to establish, by means of light signals, the positions of L, M and N corresponding to any position of P. Thus the assumption is justified only to the extent that P is sufficiently near to the axes and moving sufficiently slowly for light signals to be considered as instantaneous.

1.11.3. See Section 1.4.1

Gravitational forces are normally observed directly, being principally that of the earth on terrestial objects and that of the sun on the planets. Electrical, particularly electromagnetic, forces may also be observed directly, but all 'mechanical' forces, such as elastic stresses and the pressure of gases, arise from interatomic forces and are therefore also electrical in origin.

In nuclear phenomena two additional types of force are known—the 'weak' and 'strong' interactions. Although these are of fundamental importance in determining the distribution of energy in the universe, their influence does not extend beyond nuclear dimensions (10^{-15} m) and they do not enter explicitly into any dynamical calculations on an atomic or larger scale.

1.11.4. See Section 1.4.1

Measurement of force. We can set about measuring force in the following way, starting with equality. If we wish to decide whether force A is equal to force B we can take another spring C and adjust its angle and compression so that its force just counteracts that of A. We test this by seeing whether an object remains stationary when the triggers of A and C are released simultaneously so that their forces act on it together (see Figure 1.3a).

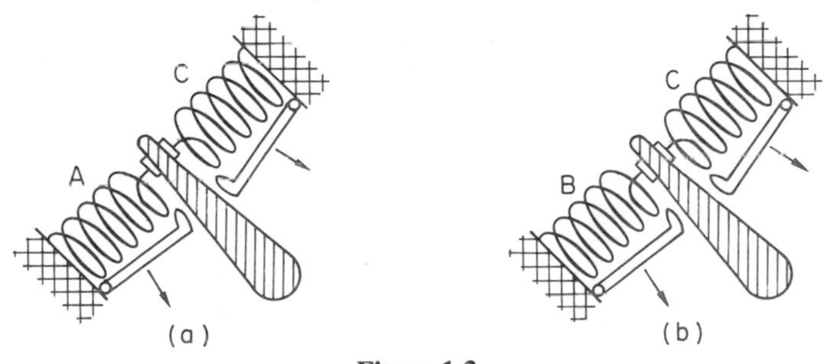

Figure 1.3

It does not matter exactly where force A acts on the object, as long as, when making the test, C acts at the same point. We can then say that A and C exert equal but *opposite* forces. Experiments will show us that once C is adjusted to be equal and opposite to A by ensuring that no motion of the object arises from their simultaneous application at one point of an object, this will be true for any other point of the object. It will also be true for any other object, so that in practice we do not need to use an 'object' specially for the purpose; we simply let the two compressed springs touch and ensure that neither moves when they are simultaneously released. The 'object', if you still wish to think of one, is made up of the molecules of dirt between the springs where they touch!

Once we have C equal and opposite to A we can replace A by B and see whether C is still equal and opposite to B, as in Figure 1.3b. If it is, then A and B exert *equal* forces. One important point that arises in such an experimental test of equality is that A and B do not both have to be coil springs.

B could be the push exerted by a human arm or the repulsion between two positive charges. We have thus discovered a way of testing a property—*force*—which is common to these and any other methods by which objects are set in motion.

Another important aspect of our test is that equal forces implies equality *both in magnitude and direction.* The amount of compression in the spring C *and* its direction must remain fixed while we substitute B for A. The direction is easily specified; it lies along the axis of the spring. From this we can find the direction of other types of force. For example the force exerted on one electrically charged metal sphere by another can be shown to lie along the line joining their centres.

To discuss the magnitude we could prepare several equal 'coil spring' forces, A, A′, A″, . . . , tested for equality as we have just described. We then couple A and A′ together as in Figure 1.4a. (We have assumed that the springs have different diameters so that they can fit inside each other and thus act at the same point; the springs do not need to have the same dimensions in order to exert equal forces.)

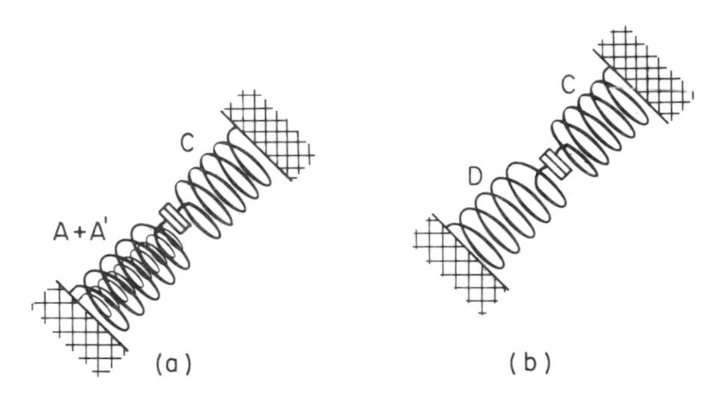

Figure 1.4

To balance A plus A′ by C we find that the direction of C remains unchanged but it has to be compressed further. In this condition a force D substituted for A plus A′ (Figure 1.4b) which just counteracts C would be defined as twice the force A (direction the same as A but twice the magnitude).

In a similar way we could define forces E, F, . . . equal to three times A, four times A, and so on. This appears to give us a scale of force only in steps of A. However, each of these steps corresponds to a definite compression of the test spring C which we could plot as in Figure 1.5. If we draw a smooth curve through these points we can then interpolate between them so that C could be set to any value (say 4·5 A as indicated in Figure 1.5), by choosing the corresponding compression from the graph. In practice, provided the compression is not too great, the relationship between force and compression is linear and the interpolation is very simple. What we have now done, of course, is to turn C into a spring balance, calibrated in units of the force A with which we started.

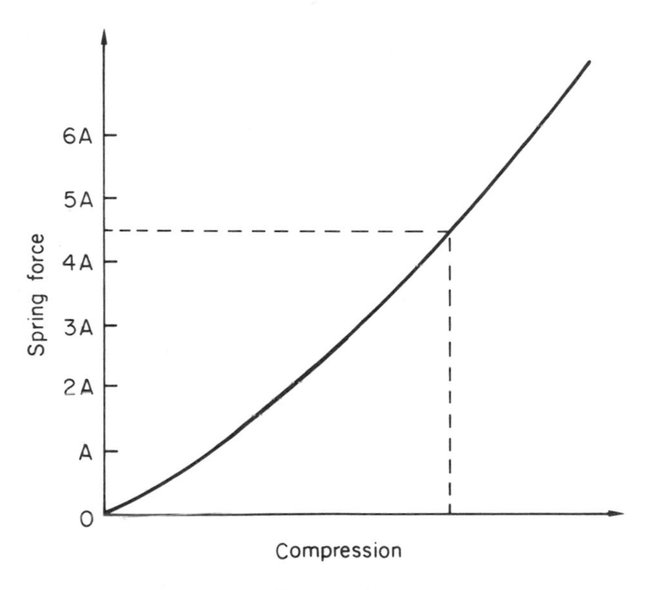

Figure 1.5

1.11.5. See Section 1.4.1

Compound forces, vectors. Very often two or more forces act on an object. For example, in the laboratory the gravitational attraction of the earth is always present, in addition to any forces applied by springs, electric fields, and so on. We can measure any such force in magnitude and direction; now we must find *experimentally* how forces are compounded. The basic experimental result is as follows: if two forces, **H** and **K**, act at a common point, O, they are equal to (where equality is defined as on page 11) a single or resultant force, **F**, acting at O, the magnitude and direction of **F** being given by the *parallelogram of forces* illustrated in Figure 1.6. In this, OL has the direction of **H** and a length proportional to the magnitude of **H**; OL′ the direction of **K** and length proportional to the magnitude of **K**; and P completes the parallelogram OLPL′. Then OP has the direction and a length proportional to the magnitude of the resultant force **F**.

We can see from Figure 1.6 that it is not necessary to construct the complete parallelogram. The triangle OLP alone is sufficient to determine the resultant force **F**.

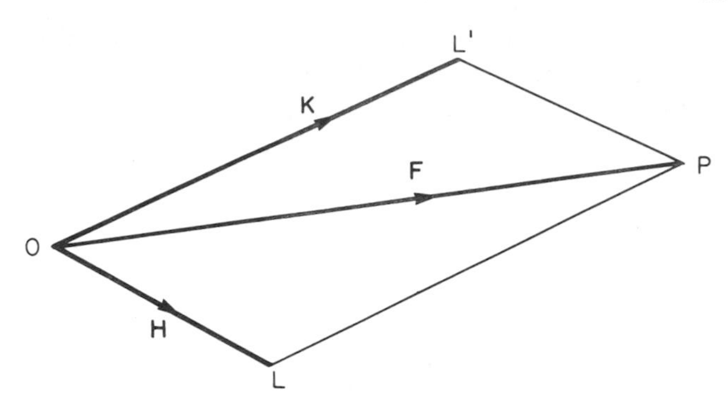

Figure 1.6

The experiment works both ways: we can either start with two forces and show that they are equivalent to a single resultant, obtainable by the rule just described, or we can start with a single force and show that it is equivalent to two forces acting in two chosen directions. For the latter we use the same construction as in Figure 1.6. OP is now the line, initially known in length and direction, that represents the single force; L and L′ are chosen so that OL, OL′ lie along the chosen directions, forming the parallelogram OLPL′.

There are important restrictions on the two chosen directions: they must be different, and the two directions and the given force must all lie in one plane. Unless these two conditions are satisfied it is clearly impossible to construct the parallelogram OLPL′ to have the required properties. If we demand an arbitrary choice of directions then we must choose *three*, although they must all be different from each other and not all in the same plane.

Suppose the three directions are O1, O2, O3, and OP represents the force (Figure 1.7). We choose a point L on O1 such that LP is parallel with the plane O23 (i.e. the plane in which the directions O2 and O3 both lie). We construct in the plane O23 a line OR parallel with LP. OLPR is now a parallelogram with the directions OL, OR different from each other and coplanar with OP. We can therefore assert that the force **F** (represented by OP) is equivalent to a force $\mathbf{F}_1$ (represented by OL) and a force **H** (represented by OR). Now O2 and O3 are different directions and are coplanar with OR. Hence, if we choose points M on O2 and N on O3 to form a parallelogram in the plane O23 we can split the force **H** (represented by OR) into equivalent forces $\mathbf{F}_2$ (represented by OM) and $\mathbf{F}_3$ (represented by ON). Thus, overall, we have shown that the force **F** is equivalent to the forces $\mathbf{F}_1$, $\mathbf{F}_2$, $\mathbf{F}_3$ acting along the chosen directions O1, O2, O3.

The experimentally observed properties of force—(a) that it has magnitude and direction and (b) that forces may be compounded and decomposed in the manner we have just described—are what is meant by calling force a *vector*.

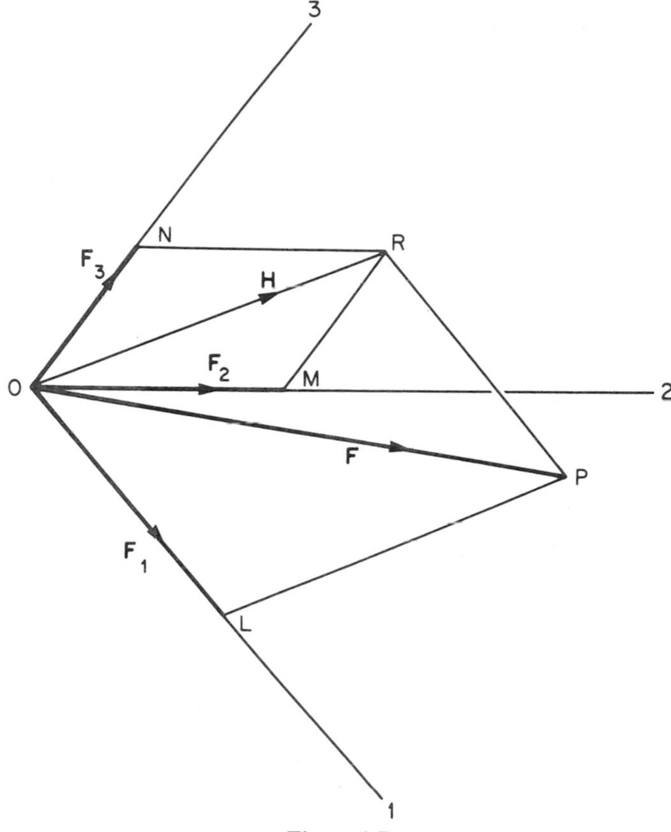

Figure 1.7

Cartesian components. In practice, the three directions O1, O2, O3 are usually chosen to be those of a right-handed set of *Cartesian* axes. This is shown in Figure 1.8. Since PL is parallel with the plane O23, which is at right angles to O1, the angle OLP is a right angle; this will also be true of the angles OMP and OLP. To decompose the force **F** (represented by OP), in this case we therefore simply construct perpendiculars from P to the three Cartesian axes. This gives the three forces $\mathbf{F}_1$, $\mathbf{F}_2$, $\mathbf{F}_3$ (represented by OL, OM, ON), which are called the *components* or *projections* of **F** on the three axes. Since OPL is a right-angled triangle and the lengths OP, OL are proportional to the magnitudes of the forces **F**, $\mathbf{F}_1$, then

$$\frac{F_1}{F} = \frac{\mathrm{OL}}{\mathrm{OP}} = \cos \mathrm{L\hat{O}P},$$

where F, F_1 are the magnitudes of **F**, $\mathbf{F}_1$. Thus,

$$F_1 = F \cos \mathrm{L\hat{O}P},$$

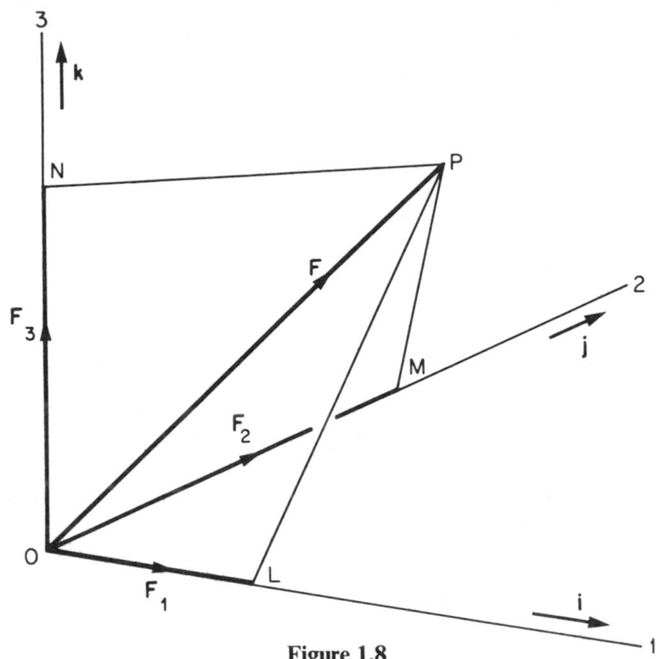

Figure 1.8

and by similar reasoning

$$F_2 = F \cos \mathrm{M\hat{O}P}$$

and

$$F_3 = F \cos \mathrm{N\hat{O}P}.$$

If we define **i**, **j**, **k** as vectors of unit length lying along the three axes, the decomposition of the force **F** that we have just carried out can be written as

$$\mathbf{F} = F_1\mathbf{i} + F_2\mathbf{j} + F_3\mathbf{k}.$$

We may write the components in yet another way,

$$F_1 = \mathbf{F}\,.\,\mathbf{i}, \qquad F_2 = \mathbf{F}\,.\,\mathbf{j}, \qquad F_3 = \mathbf{F}\,.\,\mathbf{k},$$

by using the concept of the *scalar product*. For any two vectors **A**, **B**, this is defined as

$$\mathbf{A}\,.\,\mathbf{B} = AB \cos \theta_{\mathbf{AB}},$$

where A, B are the magnitudes of the vectors and $\theta_{\mathbf{AB}}$ is the angle between them. It is a scalar quantity (i.e. a single number with no implied direction). Thus projecting the force **F** on to the axis 1 defined by the unit vector **i** is equivalent to taking the scalar product,

$$\mathbf{F}\,.\,\mathbf{i} = F \cos \theta_{\mathbf{Fi}} = F \cos \mathrm{L\hat{O}P} = F_1.$$

Any vector can then be expressed in the form

$$\mathbf{F} = \mathbf{F}.\mathbf{i}\mathbf{i} + \mathbf{F}.\mathbf{j}\mathbf{j} + \mathbf{F}.\mathbf{k}\mathbf{k}.$$

1.11.6. See Section 1.4.2

As the spring of Figure 1.2 expands, the force it exerts will diminish and the direction of its axis may change too. Thus if F_1, F_2, F_3 are the components of the spring force as measured by the static process outlined in Section 1.11.4, the observations (1.2) will only be true instantaneously, when the force is first applied. If the force is to be kept constant over a finite time and distance, either the 'fixed' end of the spring must be moved in step with the moving end to keep its compression and direction constant, or the spring must be sufficiently long for the motion of one end to have a negligible effect upon the compression and direction.

1.11.7. See Section 1.4.2

We can see from the relationships (1.4) that it is not necessary to measure force, or even to introduce the concept explicitly, in order to define and measure inertial mass. All we need to observe is that certain physical processes, for example the release of a compressed spring, are accompanied by the acceleration of a body, and that if the same process is repeated with two different bodies the ratio of their accelerations may be used to give the ratio of their inertial masses and thereby the mass of any object in terms of a standard mass.

1.11.8. See Section 1.6

It may seem rather artificial to introduce the concept of momentum and to use this to express the second law when mass is an intrinsic *constant* characteristic of a body, and equation (1.6) therefore only restates the observations (1.3). The chief reason for this is that the concept of momentum turns out to be extremely important in many branches of physics (for example, electromagnetism and high energy nuclear physics) where the simple definition (1.5) is no longer valid or even meaningful.

Even in the restricted field we consider here it is a most useful concept to use, as we shall see in Chapters 2, 4 and 5. However, it is important not to interpret equation (1.6) as

$$F = \frac{d\mathbf{p}}{dt} = m\frac{d\mathbf{v}}{dt} + \frac{dm}{dt}\mathbf{v}. \tag{1.8}$$

In classical physics the mass of any system, or of any particle of the system, is constant throughout any processes that take place. It is true that when the particles of a system are separated into groups, there may be an increase of particles in one group at the expense of those in another. For example, when a rocket is firing, the group of molecules that comprise the rocket case and the unburnt fuel is diminishing while the group of molecules making up the exhaust gases is increasing. When this happens, we can define a non-zero $\mathrm{d}m/\mathrm{d}t$, when m is the total mass of one such group. However, it could be very misleading to assume, without further thought, that the $\mathrm{d}m/\mathrm{d}t$ so defined may be used in equation (1.8) to give the motion of the group (see problem 5.14).

1.11.9. See Section 1.7

It is not strictly necessary to fix one end of the spring as suggested in Section 1.4.1 or even to approximate to this as suggested in Section 1.7. Even when objects are moved by both ends of the spring, the forces exerted when the trigger is first released will be the same, and the initial acceleration of either body will be the same as if the other end were fixed. The spring will, of course, expand more rapidly when both ends can move, and the subsequent forces and motion will therefore be different.

1.11.10. See Section 1.8

At first sight the third law and its vector formulation (1.7) may not seem to depend upon inertial frames of reference. However, it must be remembered that if equality of forces is based, as in Section 1.11.4, upon observing that point particles are not accelerated, then when such an observation is found to be true in an inertial frame, it will not be true for frames which are not inertial. For the present we shall consider the law as formulated only in inertial frames, and consider in Chapter 8 the problems that arise when non-inertial frames are used.

1.11.11. See Section 1.8

If two bodies are represented by point particles, characterized only by their positions $\mathbf{x}_1, \mathbf{x}_2$, and by scalar (non-directional) quantities such as mass and electric charge, the *only* direction which they alone can define is that of the line joining them, $\mathbf{x}_{12}$. Thus if there is no absolute frame of reference to define any other directions, or to give absolute positions to the particles, and the forces between them do not depend upon their velocities, then any such force *must* be a function only of $\mathbf{x}_{12}$ (the distance between them) and its direction *must* lie along $\mathbf{x}_{12}$.

1.11.12. See Section 1.8

Example 1

A particle 1 approaches the free end of a spring, the other end of which is attached to a particle 2 at rest in an inertial frame. Initially the velocity of 1 is directed towards 2 along the axis of the spring.

What qualitative differences would you expect *in practice* between the forces and motions of the two particles when the initial speed of 1 is (a) very small, (b) very great? What implications does your answer have concerning Newton's third law?

If x is the natural length of any part of a spring and Δx is the amount by which it is compressed, then the force exerted at the ends of that part is $k\Delta x/x$, k being a constant for the particular material and form of the spring. Thus if particle 1 compresses the spring sufficiently slowly for the compression to be spread uniformly along the spring, the force will be the same at both ends of the spring. The masses and accelerations of the particles will therefore be related by

$$m_1\mathbf{a}_1 = \mathbf{f}_{12} = -\mathbf{f}_{21} = -m_2\mathbf{a}_2$$

in accordance with Newton's second and third laws.

However, if particle 1 strikes the spring rapidly it will not compress it uniformly, but instead send a compression wave down the spring. Until this reaches particle 2 the latter will experience no force, whereas particle 1 will have experienced a force from the initial compression and will have decelerated in consequence. There is no simple relationship, therefore, between the forces and the accelerations.

The essential point is that the spring itself has inertial properties which result in a force being transmitted with a *finite* velocity, not instantaneously. No force is known which is transmitted instantaneously over a finite distance. Thus restricting Newton's third law to forces which do not depend upon the velocities of the particles not only excludes such obviously velocity-dependent effects as the magnetic forces arising from moving charges, but also those arising from the finite velocity of propagation of the force. Strictly speaking, it applies only to particles which are stationary, at least relative to each other.

However, in many practical cases bodies move sufficiently slowly and are close enough for the forces between them to be indistinguishable from static forces obeying the third law.

1.12. Problems

1.1. Devise an experiment to show that if m_1, m_2 are the inertial masses of two bodies, then, when they are joined together, the inertial mass of the composite body will be $m_1 + m_2$.

1.2. A mass M_1 approaches a mass M_2 as shown in the following diagram. Between them are n masses m and $n + 1$ springs, each so short that it may be considered to transmit a force instantaneously and without change along its length.

M_2 — m — m … m — m — M_1

Show that if A_1, A_2 are the accelerations of M_1, M_2 and $a_1, a_2, \ldots, a_n$ are the accelerations of the masses m (all accelerations towards the right), then

$$M_1A_1 + m(a_1 + a_2 + \ldots + a_n) + M_2A_2 = 0.$$

Hence, show that if the n masses and the $n + 1$ springs are the means by which M_1 and M_2 exert forces on each other, Newton's third law will describe these forces only when m is zero.

1.3. Determine the lengths of the following vectors and the angles they make with each of the coordinate axes:

(a) $2\mathbf{i} + 3\mathbf{j} + 6\mathbf{k}$ (b) $6\mathbf{i} + 9\mathbf{k} + 18\mathbf{j}$ (c) $\frac{1}{3}\alpha\mathbf{i} + \frac{1}{2}\alpha\mathbf{j} + \alpha\mathbf{k}$
(d) $\mathbf{i} + \sqrt{3}\mathbf{j}$ (e) $2\mathbf{i} + 3\mathbf{j} - 6\mathbf{k}$ (f) $2\mathbf{i} - 3\mathbf{j} + 6\mathbf{k}$
(g) $-2\mathbf{i} + 3\mathbf{j} + 6\mathbf{k}$ (h) $2\mathbf{i} - 3\mathbf{j} - 6\mathbf{k}$ (i) $-2\mathbf{i} + 3\mathbf{j} + 6\mathbf{k}$

1.4. Find the angles between the following pairs of vectors:

(a) $2\mathbf{i} + 3\mathbf{j} + 6\mathbf{k}$, $2\mathbf{i} + 3\mathbf{j} - 6\mathbf{k}$
(b) $2\mathbf{i} - 3\mathbf{j} + 6\mathbf{k}$, $-2\mathbf{i} + 3\mathbf{j} + 6\mathbf{k}$
(c) $2\mathbf{i} + 3\mathbf{j} + 6\mathbf{k}$, $3\mathbf{i} + 2\mathbf{j} - 2\mathbf{k}$

1.5. A force $\mathbf{F}$ is represented in magnitude and direction by a line from the point (3, 5, 7) to the point (−1, 6, 9). If one unit of length represents 1 N, what is the magnitude of the force and of its components? A second force is represented by a line from the point (3, 5, 7) to a point on the axis 1. What are the coordinates of this point if the forces are to be at right angles, and what is the magnitude of the second force in this case?

1.6. Show that when a vector $\mathbf{x}$ is changed by adding to it a small vector $\mathbf{y}$ ($|\mathbf{y}| \ll |\mathbf{x}|$), then, to a first approximation, the length of $\mathbf{x}$ is increased to $|\mathbf{x}| + y_{\parallel}$, and its direction rotated through the angle $y_{\perp}/|\mathbf{x}|$, where $y_{\parallel}$ and $y_{\perp}$ are the components of $\mathbf{y}$ parallel with, and perpendicular to, $\mathbf{x}$.

1.7. Show that with proper choice of axes two vectors which are nearly parallel may be expressed as

$$\mathbf{x} = b_1\mathbf{i} + b_2\mathbf{j} + b_3\mathbf{k}, \qquad \mathbf{y} = \alpha b_1\mathbf{i} + \alpha b_2\mathbf{j} + \alpha(b_3 + \varepsilon)\mathbf{k},$$

where

$$\varepsilon \ll b_3,$$

and the angle, θ, between them then has the approximate value

$$\theta \simeq \frac{\varepsilon}{|\mathbf{x}|}\left[1 - \left(\frac{b_3}{|\mathbf{x}|}\right)^2\right]^{1/2}.$$

Chapter 2

One-dimensional Motion; Energy and Momentum

The whole universe forms one gigantic physical system. Within the range of times and distances available to us various parts of this, such as our own galaxy, appear to be freely moving. For nearly all purposes the much smaller part that constitutes only the solar system behaves as an isolated system, giving one inertial point very near the centre of the sun and an inertial frame with this as origin and the directions of its axes determined by the fixed stars.

All bodies within the solar system exert forces upon each other, all in different directions. If we represent the bodies by point particles obeying Newton's laws, our model of the solar system would be a three-dimensional, many-particle system, with all the forces and accelerations of the particles arising from internal effects. This is an impracticable approach when the motion of a particular object or small group of objects has to be found. Instead, we consider separately the particles of immediate interest from the rest of the system and call only the forces between these selected particles *internal forces*. Any others are taken as *external forces*.

Such a division is useful only if the sources of the external forces are clearly identifiable and are unaffected by the motion of the selected group of particles. In practice, this usually means that the sources of the external forces must have a much greater mass than that of the group.

Thus if we consider a single passenger travelling in a large motor coach, his motion will be controlled by the forces exerted by his own muscles, the forces exerted by the coach on the various parts of his body with which it is in contact and the gravitational force exerted by the earth. The mass of a coach is so large compared with that of a passenger that any muscular activity of his has no appreciable effect upon it; the motion of the coach is determined by the pull of the engine, gravity, the gradient, the curvature of the road, and so on. It is even more true that the earth's motion is unaffected by a single human being. It would therefore make good sense to consider the passenger as the separated 'group of particles' experiencing internal forces from his own muscles and external forces from the coach and the earth.

On the other hand, fifty passengers in the coach could have some effect upon it if they acted together, since their total mass would no longer be negligible compared with that of the coach. To deal properly with this situation we would have to consider passengers plus coach as the group, with forces between them now as internal forces. The external forces would then be gravity and the forces exerted by the road on the wheels. The source of both is the earth, still so massive compared with coach and passengers as to be unaffected by them. (See Section 2.9.1, page 42.)

The simplest situation is one in which we isolate a single object and represent it by a point particle. In this case any forces acting on it are necessarily external, and we need not explicitly name them as such.

2.1. One-dimensional motion

It is often unnecessary to deal with the full three-dimensional form of Newton's laws. A sphere near the earth's surface falling freely under gravity and a train moving along a straight horizontal track are two cases where the motion is linear and only one varying coordinate is required to give the position of the moving object. In the first case this might be the height above the earth's surface, and in the second the horizontal distance from some fixed point on the track.

If we take this coordinate to be x_1, the coordinates at right angles to this, x_2 and x_3, will be constant so that, from equation (1.2), the corresponding components of forces will be zero,

$$F_2 = F_3 = 0.$$

This does not necessarily mean that *no* forces act perpendicular to the motion. It is the total or resultant force acting which has no perpendicular component. For example, there will be a gravitational force acting vertically downwards on the train, but there will also be a reaction force exerted by the track on the train which will exactly balance this in the vertical direction.

We are thus left with the single equation of motion for a point particle,

$$m\frac{d^2x_1}{dt^2} = F_1. \qquad (2.1)$$

Since only one coordinate is involved we may now omit the suffix.

2.2. Constant force

The gravitational force on a particle is proportional to its inertial mass. (See Section 2.9.2, page 42.) For motion near the earth's surface it is constant and directly downwards. Thus if x is measured vertically *upwards* the force is

$$F = -mg,$$

where g has the value 9·81 m s^{-2}. The equation of motion (2.1) then becomes

$$\frac{d^2x}{dt^2} = -g.$$

This is the most familiar example of a constant force. For all such forces the equation of motion will be of the form:

$$\frac{d^2x}{dt^2} = \frac{F}{m} = C = \text{constant.} \tag{2.2}$$

Since it tells us only the rate at which the velocity is changing, the equation of motion by itself will not give the velocity or position of the particle in any particular case. This is why, when it is integrated to give

$$x(t) = A + Bt + \tfrac{1}{2}Ct^2, \tag{2.3}$$

there are two constants, A and B, which can be chosen to fit two other physical (boundary) conditions. For example, if the position and velocity at time zero are x_0 and v_0,

$$x_0 = x(0) = A, \qquad v_0 = \frac{dx(0)}{dt} = B,$$

and the general solution (2.3) is then restricted to

$$x(t) = x_0 + v_0 t + \tfrac{1}{2}Ct^2.$$

After a sufficiently long time the third term, $\frac{1}{2}Ct^2$, will dominate the solution. Since this has the same sign as F, the motion will then be in the direction of the force, whatever the initial conditions.

2.3. Impulse and momentum; work and kinetic energy

The longer the force acts on a particle the more it changes its motion. How we measure this change depends upon whether 'longer' refers to time or distance. Thus for a constant force F acting on the particle from time t_1 to time t_2 the simplest description of its 'time' effect is the *impulse*, I_{12}, it delivers:

$$I_{12} = F \times (t_2 - t_1)$$

—the product of the force and the time for which it acts.

If the particle moves from x_1 to x_2 while the force acts on it, the analogous 'distance' effect is the *work*, $\mathscr{W}_{12}$, done by the force:

$$\mathscr{W}_{12} = F \times (x_2 - x_1)$$

—the product of the force and the distance over which its point of application moves.

It is easy to show from equations (2.2) and (2.3) that

$$F(t_2 - t_1) = mv_2 - mv_1$$

and

$$F(x_2 - x_1) = \tfrac{1}{2}mv_2^2 - \tfrac{1}{2}mv_1^2.$$

We can therefore say that the impulse delivered by the force is the gain in *momentum*,

$$P_2 - P_1 = I_{12} = F(t_2 - t_1), \tag{2.4}$$

momentum having already been defined for a single particle in Chapter 1 as

$$p = mv = m\frac{dx}{dt}. \tag{2.5}$$

Similarly, the work done by the force is the gain in *kinetic energy*,

$$\mathcal{T}_2 - \mathcal{T}_1 = \mathcal{W}_{12} = F(x_2 - x_1), \tag{2.6}$$

kinetic energy being defined as

$$\mathcal{T} = \frac{1}{2}mv^2 = \frac{1}{2}m\left(\frac{dx}{dt}\right)^2. \tag{2.7}$$

Constant forces are rather rare in nature. A charged particle near a conductor whose electric potential is changing experiences a force that depends both upon the time and its position at that time. Its equation of motion would be of the form:

$$m\frac{d^2x}{dt^2} = F(x, t).$$

If magnetic fields were present the force would also depend upon the velocity of the charged particle,

$$m\frac{d^2x}{dt^2} = F\left(x, \frac{dx}{dt}, t\right). \tag{2.8}$$

The concepts we have just introduced will still apply, however. For if we write the last equation as

$$\frac{dp}{dt} = \frac{d}{dt}\left(m\frac{dx}{dt}\right) = F\left(x, \frac{dx}{dt}, t\right),$$

we may integrate this immediately to give

$$\int_{(1)}^{(2)} dp = \int_{(1)}^{(2)} F\left(x, \frac{dx}{dt}, t\right) dt. \tag{2.9}$$

Thus

$$p_2 - p_1 = I_{12}, \tag{2.10}$$

where the impulse, I_{12}, delivered by the force is defined by

$$I_{12} = \int_{(1)}^{(2)} F\left(x, \frac{dx}{dt}, t\right) dt. \tag{2.11}$$

Similarly, using equation (2.8),

$$\frac{d\mathscr{T}}{dx} = \frac{d\mathscr{T}}{dt}\frac{dt}{dx} = \frac{d}{dt}\left[\frac{1}{2}m\left(\frac{dx}{dt}\right)^2\right]\frac{dt}{dx}$$

$$= m\frac{dx}{dt}\frac{d^2x}{dt^2}\frac{dt}{dx} = m\frac{d^2x}{dt^2} = F\left(x, \frac{dx}{dt}, t\right),$$

which gives

$$\int_{(1)}^{(2)} d\mathscr{T} = \int_{(1)}^{(2)} F\left(x, \frac{dx}{dt}, t\right) dx. \tag{2.12}$$

Thus

$$\mathscr{T}_2 - \mathscr{T}_1 = \mathscr{W}_{12} \tag{2.13}$$

where the work, $\mathscr{W}_{12}$, done by the force is defined by

$$\mathscr{W}_{12} = \int_{(1)}^{(2)} F\left(x, \frac{dx}{dt}, t\right) dx. \tag{2.14}$$

Underlying both the results (2.4) and (2.6) or their more general forms (2.10) and (2.13) is the idea that the particle gains something from the force or the source of the force; in other words, a dynamical quantity is being transferred from one to the other while being *conserved* overall. The extent to which this is a useful concept depends upon how clearly the expressions for impulse and work can be given a physical basis and how readily they can be calculated.

2.4. Time-dependent forces

If the equation of motion (2.8) can be solved to give the position of the particle at any time, $x(t)$, then x and dx/dt, both now known as functions of t, can be substituted into equation (2.11) to calculate the impulse. This would be an empty exercise, since we can calculate changes in momentum directly once dx/dt is known.

However, when F depends *only upon time*, then the impulse,

$$I_{12} = \int_{(1)}^{(2)} F(t)\, dt, \tag{2.15}$$

can be found without solving the equation of motion. Thus if the momentum (or velocity) of the particle is known at one time, equations (2.10) and (2.15) enable it to be found at any other time without necessarily knowing where the particle is. Even when the force does depend upon the particle's position as well as the time, the impulse can be calculated provided that the force acts for so short a time that the position does not change appreciably. (See Section 2.9.3, page 43.)

The most common application of the impulse concept is to collision phenomena, such as the striking of a tennis-ball by a racket or the bouncing back of a gas molecule from the walls of its container. Here a very large force acts on the ball or molecule for a very short time, so that its velocity and therefore momentum appear to change instantaneously with no change of position. The force will have the appearance shown in Figure 2.1, and the change in

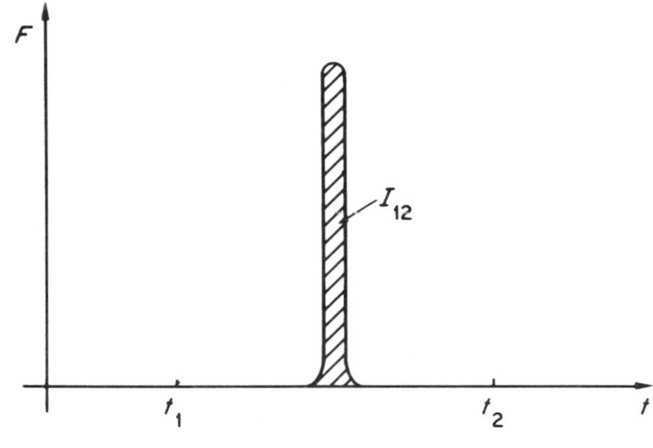

Figure 2.1

momentum from some time t_1 before the collision to some time t_2 afterwards will be

$$p_2 - p_1 = I_{12} = \int_{(1)}^{(2)} F(t)\,\mathrm{d}t \tag{2.16}$$

—the area under the $F(t)$ curve. The time average $\bar{F}$ of the force over the interval t_1 to t_2 will be given by

$$\int_{(1)}^{(2)} F(t)\,\mathrm{d}t = \bar{F}\int_{(1)}^{(2)} \mathrm{d}t = \bar{F}(t_2 - t_1).$$

Hence,

$$\bar{F} = \frac{I_{12}}{t_2 - t_1} = \frac{p_2 - p_1}{t_2 - t_1}.$$

(See Section 2.9.4, page 44.)

2.5. Position-dependent forces; conservation of energy

When the force acting on a particle depends only upon its *position*, the work done by the force,

$$\mathscr{W}_{12} = \int_{(1)}^{(2)} F(x)\,\mathrm{d}x, \tag{2.17}$$

can be found without solving the equation of motion. Thus if the kinetic energy is known for one position of the particle, equations (2.13) and (2.17) enable it to be found for any other position without necessarily knowing the corresponding time. (See Section 2.9.5, page 45.) Since kinetic energy depends upon the square of the velocity, only the magnitude of the velocity, or speed, can then be deduced, but not its direction.

Instead of using equations (2.13) and (2.17) a better alternative is to define the *potential energy* at any point x as the work which would be done by the force as the particle moves from x to some arbitary fixed point, X_0,

$$\mathscr{V}(x, X_0) = \int_{x}^{X_0} F(x')\,\mathrm{d}x'. \tag{2.18}$$

(See Section 2.9.6, page 46.) Then

$$\begin{aligned}\mathscr{W}_{12} &= \int_{x_1}^{x_2} F(x')\,\mathrm{d}x' = \int_{X_0}^{x_2} F(x')\,\mathrm{d}x' - \int_{X_0}^{x_1} F(x')\,\mathrm{d}x' \\ &= -\int_{x_2}^{X_0} F(x')\,\mathrm{d}x' + \int_{x_1}^{X_0} F(x')\,\mathrm{d}x' \\ &= -\mathscr{V}(x_2, X_0) + \mathscr{V}(x_1, X_0).\end{aligned} \tag{2.19}$$

Thus if we write, more concisely,

$$\mathscr{V}_1 = V(x_1, X_0), \qquad \mathscr{V}_2 = V(x_2, X_0),$$

equations (2.13) and (2.19) give

$$\mathscr{T}_2 + \mathscr{V}_2 = \mathscr{T}_1 + \mathscr{V}_1. \tag{2.20}$$

Since this will be true for all points of the path, the sum of the kinetic and potential energies is always a constant, which we call the total energy, $\mathscr{E}$:

$$\mathscr{T} + \mathscr{V} = \mathscr{E}. \tag{2.21}$$

This is the law of *conservation of energy*.

A force which conserves total energy is called a *conservative force*. We have just seen that *in one dimension* such a force may depend only upon the position of the particle and not, for example, upon the time when the particle is at that position or upon its velocity. The corresponding potential energy determines the force uniquely, since from the definition (2.18),

$$F(x) = -\frac{\partial \mathscr{V}}{\partial x}. \tag{2.22}$$

The converse is not true, for X_0 can be chosen arbitrarily and this choice will affect the value of the potential energy. Since

$$\mathscr{V}(X_0, X_0) = 0,$$

whatever the value of X_0, choosing X_0 is equivalent to choosing the zero of the potential energy scale. A convenient choice is often to take the potential energy as zero at the coordinate origin or at $+\infty$ or $-\infty$ which corresponds to

$$X_0 = 0 \quad \text{or} \quad X_0 = +\infty \quad \text{or} \quad X_0 = -\infty.$$

The choice of X_0 can never affect the physics of any problem since the conservation of energy equation (2.20) only involves the difference of two potential energies, which, from equation (2.19), is independent of X_0.

A great deal of qualitative information about the motion of a particle can be gained once a potential energy function is known. The conservation law (2.21) may be written

$$\mathscr{T}(x) = \mathscr{E} - \mathscr{V}(x) \tag{2.23}$$

where, for simplicity, X_0 is not explicitly mentioned. Thus if $\mathscr{V}(x)$ and the constant $\mathscr{E}$ are plotted as shown in Figure 2.2, $\mathscr{T}(x)$ will be represented by

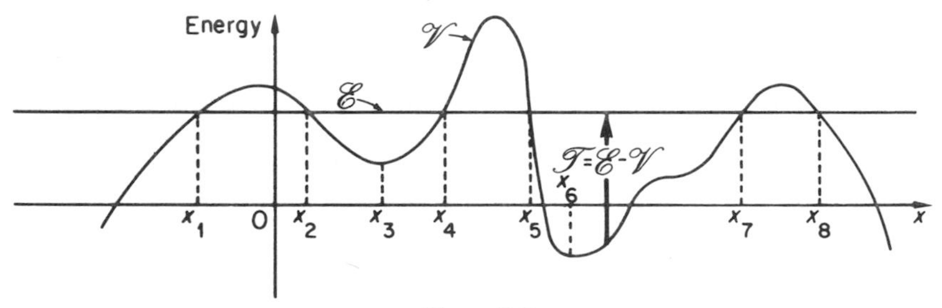

Figure 2.2

the difference between the two corresponding lines. However, since

$$\mathscr{T} = \frac{m}{2}\left(\frac{\mathrm{d}x}{\mathrm{d}t}\right)^2 \tag{2.24}$$

is necessarily positive or zero, the particle could only be observed in those regions such as

$$x \leqslant x_1, \qquad x_2 \leqslant x \leqslant x_4, \qquad x_5 \leqslant x \leqslant x_7, \qquad x_8 \leqslant x,$$

where

$$\mathscr{E} - \mathscr{V}(x) \geqslant 0.$$

2.5.1. Unbound motion

Let us consider the motion of the particle at a point x in the first region. Its velocity, from equations (2.23) and (2.24), will be

$$\frac{dx}{dt} = \pm\sqrt{\frac{2}{m}[\mathscr{E} - \mathscr{V}(x)]}, \tag{2.25}$$

while the force is

$$F(x) = -\frac{\partial \mathscr{V}}{\partial x}.$$

Since the potential energy gradient is positive, for $x \leqslant x_1$, the force will be negative (i.e. directed towards the left). Thus if the velocity is already negative the particle will be accelerated towards the left, gaining greater and greater speed as it moves towards $-\infty$.

If the positive velocity solution of equation (2.25) is appropriate the particle will travel to the right, but with steadily decreasing velocity since the left-ward force still gives it a *negative* acceleration. This continues until the particle just reaches x_1, by which position the velocity has been reduced to zero and the total energy is therefore completely in the form of potential energy,

$$\mathscr{T}(x_1) = 0, \qquad \mathscr{E} = \mathscr{V}(x_1).$$

At this point the (still negative) force,

$$F(x_1) = -\left(\frac{\partial \mathscr{V}}{\partial x}\right)_{x=x_1},$$

begins to move the particle leftwards, the kinetic energy increases at the expense of the potential energy and the particle accelerates towards $-\infty$ as before. Similar considerations would show that if the particle were in the last region, $x \geqslant x_8$, it would move off towards $+\infty$, either directly or after being turned back at x_8.

2.5.2. Bound motion

In the other two regions the motion is very different. Here the particle exists in a potential well, with a minimum at x_3 in the first region and at x_6 in the second. From equations (2.22) and (2.23) the velocity will have a maximum magnitude at a potential energy minimum. If at x_3 the velocity is positive then, using the arguments of the preceding section, the particle will move in the positive direction against a negative force with steadily decreasing velocity until it just reaches x_4, where all its kinetic energy will have been transformed into potential energy. It will then accelerate in the negative direction passing x_3 with a velocity equal in magnitude to, but opposite in sign from, what it possessed before. The reverse process now takes place. The force and therefore acceleration is positive, so the negative velocity diminishes in magnitude until x_2 is just reached, with zero velocity and kinetic energy. The particle then moves to the right again and the motion repeats itself indefinitely.

The essential feature of this motion is that it is *oscillatory* and *bound* between the fixed limits x_2 and x_4. There is a cyclic interchange of energy between the potential and kinetic forms. The amount of this interchange varies with the total energy and with the depth of the potential energy well. Thus at the minimum x_6 of the third region the kinetic energy and the magnitude of the velocity will be greater than at x_3 for a given value of $\mathscr{E}$.

It was pointed out in Section 2.5 that because kinetic energy is necessarily positive a particle could not exist in regions such as

$$x_4 < x < x_5, \qquad x_7 < x < x_8.$$

Thus a particle once bound in a potential hollow or well could never 'jump' across a potential hump into another well or into a region of unbound motion; all regions of possible motion are mutually exclusive. (See Section 2.9.7, page 47.)

2.5.3. The reality of potential energy

Since the kinetic energy $\frac{1}{2}mv^2$ is determined only by the mass and speed of the particle, existing independently of whether or not a force is acting, we may consider this energy as located at the particle. Potential energy, although dependent upon the position of the particle, cannot similarly be assigned simply to the particle since it depends also upon the force. Where, then, *is* the potential energy?

When the force arises from a spring and we measure the position x of the particle from the point at which the spring is unstrained, the force acting on the particle is, for small values of x,

$$F(x) = -Kx,$$

where K is the spring strength constant. The potential energy will then be

$$\mathscr{V}(x) = \int_x^0 F(x')\,\mathrm{d}x' = \tfrac{1}{2}Kx^2,$$

taking, in this case, $X_0 = 0$. This will be true whatever particle happens to be at the end of the spring, and depends simply upon the stretched or compressed length of the spring and its strength constant, K. We may therefore consider the potential energy to reside *in the spring*. The conservation law (2.21) then expresses the fact that the motion of the linked particle and spring is governed by an exact interchange of energy between the particle (with kinetic (dynamic) energy arising from its mass and velocity) and the spring (with potential (static) energy arising from its strength and length).

Although electric and gravitational forces do not have a tangible mechanical means of transmission like a spring, nevertheless the fact that a particle may experience them even in an evacuated space suggests that such a vacuum should not be considered as mere 'nothingness'. This is expressed by saying that a *field* exists in the space and that the particle experiences a force by being coupled (by its mass or charge) to the field, just as, in the mechanical case, it does by being attached to a hook at the end of the spring.

The potential energy may then be considered as a static energy stored in the field, arising from its strength and shape. (See Section 2.9.8, page 47.)

2.6. Two interacting particles; energy, momentum and impulse

So far the force acting on a particle has been assumed to arise from some outside source. In fact the source of the force will itself be a particle or a collection of particles. The general problem of describing in three dimensions the properties of groups of particles exerting forces upon each other is treated in Chapter 4. However, even when only two particles are concerned, some important principles can be established which will later be shown to be generally true.

We shall consider their motion to be constrained to a straight line. In fact, when the two point particles are initially at rest, or moving along the line joining them, the force between them will also act along this line, and subsequent motion arising from this force will be along the same line. Our 'constrained' motion, therefore, only excludes the case when the particles are moving initially at an angle to the line joining them.

Particle 1 has mass m_1 and position X_1; particle 2 has mass m_2 and position X_2. Their relative position is

$$x = X_1 - X_2,$$

and the force exerted by the particles on each other will be taken to depend only upon this. If f_{12} is the force exerted on particle 1 by particle 2 and f_{21} is the force exerted on particle 2 by particle 1, then these will be functions only of x, and, by the third of Newton's laws, will be equal and opposite:

$$f_{12} = -f_{21} = f(x) = f(X_1 - X_2). \tag{2.26}$$

Since we are concerned with only the two particles, there will be no external forces acting upon them.

The equations of motion of the two particles will therefore be

$$m_1 \frac{d^2 X_1}{dt^2} = f_{12} = f(x) = f(X_1 - X_2), \tag{2.27}$$

$$m_2 \frac{d^2 X_2}{dt^2} = f_{21} = -f(x) = -f(X_1 - X_2). \tag{2.28}$$

2.6.1. *Momentum and impulse*

When we integrate equation (2.27) with respect to time between conditions (1) and (2),

$$\int_{(1)}^{(2)} m_1 \frac{d^2 X_1}{dt^2} dt = \int_{(1)}^{(2)} f(x) \, dt,$$

we obtain, on the left-hand side, the change in momentum of particle 1,

$$m_1 \left(\frac{dX_1}{dt} \right)_{(1)} - m_1 \left(\frac{dX_1}{dt} \right)_{(2)} = P_{1(2)} - P_{1(1)}.$$

The right-hand side is the impulse delivered to the particle in the time interval $t_{(1)}$ to $t_{(2)}$,

$$\int_{(1)}^{(2)} f(x) \, dt = I_{12}. \tag{2.29}$$

Thus,

$$P_{1(2)} - P_{1(1)} = I_{12}. \tag{2.30}$$

For particle 2, in a similar way, the change in momentum will be equal to the impulse delivered to it,

$$P_{2(2)} - P_{2(1)} = I_{21}, \tag{2.31}$$

where

$$I_{21} = - \int_{(1)}^{(2)} f(x) \, dt. \tag{2.32}$$

From equations (2.29) and (2.32) the two impulses are equal and opposite, and therefore equations (2.30) and (2.31) give

$$P_{1(2)} + P_{2(2)} = P_{1(1)} + P_{2(1)}.$$

This may be expressed as a law of conservation of total momentum for the two particles,

$$P_1 + P_2 = P = \text{constant}. \tag{2.33}$$

2.6.2. Kinetic and potential energy

Multiplying equation (2.27) by $\mathrm{d}X_1/\mathrm{d}t$, equation (2.28) by $\mathrm{d}X_2/\mathrm{d}t$, adding and integrating between conditions (1) and (2),

$$m_1 \int_{(1)}^{(2)} \frac{\mathrm{d}^2 X_1}{\mathrm{d}t^2}\frac{\mathrm{d}X_1}{\mathrm{d}t}\,\mathrm{d}t + m_2 \int_{(1)}^{(2)} \frac{\mathrm{d}^2 X_2}{\mathrm{d}t^2}\frac{\mathrm{d}X_2}{\mathrm{d}t}\,\mathrm{d}t$$

$$= \int_{(1)}^{(2)} f(X_1 - X_2)\frac{\mathrm{d}(X_1 - X_2)}{\mathrm{d}t}\,\mathrm{d}t$$

or

$$\int_{(1)}^{(2)} \frac{\mathrm{d}}{\mathrm{d}t}\left[\frac{m_1}{2}\left(\frac{\mathrm{d}X_1}{\mathrm{d}t}\right)^2 + \frac{m_2}{2}\left(\frac{\mathrm{d}X_2}{\mathrm{d}t}\right)^2\right]\mathrm{d}t = \int_{(1)}^{(2)} f(x)\,\mathrm{d}x. \tag{2.34}$$

The first integral is the change in the sum of the kinetic energies of the two particles, $\mathscr{T}_{1(2)} + \mathscr{T}_{2(2)} - \mathscr{T}_{1(1)} - \mathscr{T}_{2(1)}$. The second is the work done by the internal force; it can be expressed in terms of an internal or mutual potential energy of the two particles,

$$\mathscr{V}(x, X_0) = \int_x^{X_0} f(x')\,\mathrm{d}x', \tag{2.35}$$

X_0 being the arbitary separation (in practice, usually infinite) at which this potential energy is taken to be zero. Then

$$\int_{(1)}^{(2)} f(x')\,\mathrm{d}x' = \int_{x(1)}^{X_0} f(x')\,\mathrm{d}x' - \int_{x(2)}^{X_0} f(x')\,\mathrm{d}x'$$

$$= \mathscr{V}(x_{(1)}, X_0) - \mathscr{V}(x_{(2)}, X_0).$$

Writing this change in potential energy more concisely as $\mathscr{V}_{(1)} - \mathscr{V}_{(2)}$, equation (2.34) becomes

$$\mathscr{T}_{1(2)} + \mathscr{T}_{2(2)} - \mathscr{T}_{1(1)} - \mathscr{T}_{2(1)} = \mathscr{V}_{(1)} - \mathscr{V}_{(2)}$$

or

$$(\mathscr{T}_1 + \mathscr{T}_2 + \mathscr{V})_{(1)} = (\mathscr{T}_1 + \mathscr{T}_2 + \mathscr{V})_{(2)}.$$

Since (1) and (2) represent any two conditions during the motion of the particles, this equation expresses the conservation of total energy $\mathscr{E}$:

$$\mathscr{T}_1 + \mathscr{T}_2 + \mathscr{V} = \mathscr{E} = \text{constant}. \tag{2.36}$$

(See Section 2.9.9, page 47.)

Note that the condition for this is that the force of interaction between the particles should depend only upon their relative distance, in which case it is known as a conservative force, in accordance with Section 2.5.

2.6.3. Elastic collisions

When the force between the two particles is conservative (depending only upon their separation x) any process in which they move together and then separate is called an *elastic collision.* Any force of repulsion between them will correspond to a potential energy of a form intermediate between those of Figure 2.3, where we have taken the zero of potential energy to be that at infinite separation. A slowly varying force such as that between like electric charges is illustrated by Figure 2.3a, whereas that between two steel balls, which is very large but only acts when they are in contact, corresponds to Figure 2.3b.

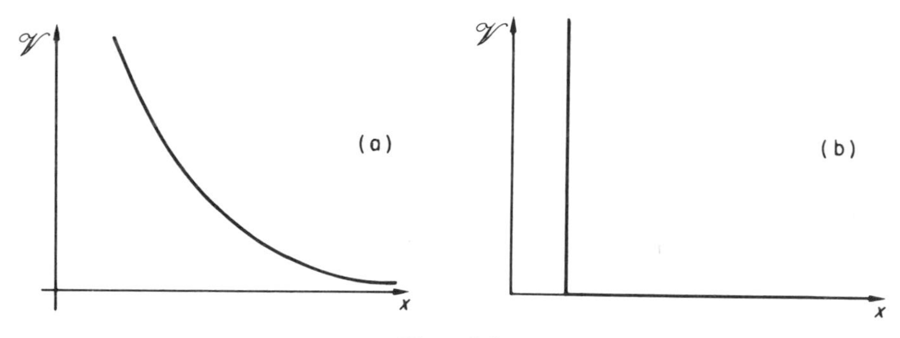

Figure 2.3

The essential features of the potential energy are that it should rise sufficiently high for the particles to repel each other away from some position of nearest approach and that it should tend to some limiting value (zero in the cases illustrated) as the particles recede towards an infinite separation. If this is so then the two conservation laws (2.33) and (2.36) become

$$P_{1(1)} + P_{2(1)} = P_{1(2)} + P_{2(2)} \tag{2.37}$$

and

$$\mathscr{T}_{1(1)} + \mathscr{T}_{2(1)} = \mathscr{T}_{1(2)} + \mathscr{T}_{2(2)}, \tag{2.38}$$

where (1) and (2) refer to conditions before and after the collision, when the separation is very large. In terms of the masses m_1, m_2 and the velocities V_1, V_2 of the two particles, these may be rewritten as

$$m_1(V_{1(1)} - V_{1(2)}) = m_2(V_{2(2)} - V_{2(1)}) \tag{2.39}$$

and

$$\tfrac{1}{2}m_1(V^2_{1(1)} - V^2_{1(2)}) = \tfrac{1}{2}m_2(V^2_{2(2)} - V^2_{2(1)}). \tag{2.40}$$

Dividing the second by the first,

$$V_{1(1)} + V_{1(2)} = V_{2(1)} + V_{2(2)}. \tag{2.41}$$

Now $V_{1(1)} - V_{2(1)}$ and $V_{1(2)} - V_{2(2)}$ are the velocities of particle 1 relative to particle 2 before and after the collision. The result (2.41) therefore shows that the relative velocity is simply reversed as a consequence of an elastic collision. (See Section 2.9.10, page 48.)

Equations (2.39) and (2.41) may now be solved for the final velocities:

$$V_{1(2)} = \frac{m_1 - m_2}{m_1 + m_2} V_{1(1)} + \frac{2m_2}{m_1 + m_2} V_{2(1)}$$

and

$$V_{2(2)} = \frac{2m_1}{m_1 + m_2} V_{1(1)} + \frac{m_2 - m_1}{m_1 + m_2} V_{2(1)}.$$

2.6.4. *Inelastic collisions*

Many collisions do not conserve kinetic energy, as described by equation (2.38), with the consequent reversal of velocity. Two lumps of putty will stick together on collision and therefore have zero final relative velocity. It is easy to devise models that will give a diminished speed. For example if the two particles act on each other through the medium of a spring which is compressed as they approach but which is prevented from fully expanding afterwards, their final relative speed will be less than the initial speed. In both cases kinetic energy will be lost, in the first case by increasing the heat content of the combined lump and in the second by leaving stored potential energy in the spring.

Such collisions are called inelastic and are commonly described by a parameter called the *coefficient of restitution*, e. It measures the ratio of the final to the initial relative velocities:

$$e(V_{1(1)} - V_{2(1)}) = V_{1(2)} - V_{2(2)}. \tag{2.42}$$

This, together with equation (2.39) (which is derived from the conservation of momentum (2.37)—a principle which still holds even when the collision is inelastic), are sufficient to determine the final velocities. We can see that $e = 1$ corresponds to the elastic case; $e = 0$ implies that the particles stick together and the collision is then said to be *totally inelastic*.

2.6.5. *The centre-of-mass*

Collisions, particularly inelastic collisions, are more easily described by introducing the idea of the *centre-of-mass* (see Figure 2.4). For two particles with masses m_1, m_2 and coordinates X_1, X_2 this is the point X defined by

$$(m_1 + m_2)X = m_1 X_1 + m_2 X_2. \tag{2.43}$$

By differentiating this we obtain

$$(m_1 + m_2)\frac{\mathrm{d}X}{\mathrm{d}t} = m_1 \frac{\mathrm{d}X_1}{\mathrm{d}t} + m_2 \frac{\mathrm{d}X_2}{\mathrm{d}t}. \tag{2.44}$$

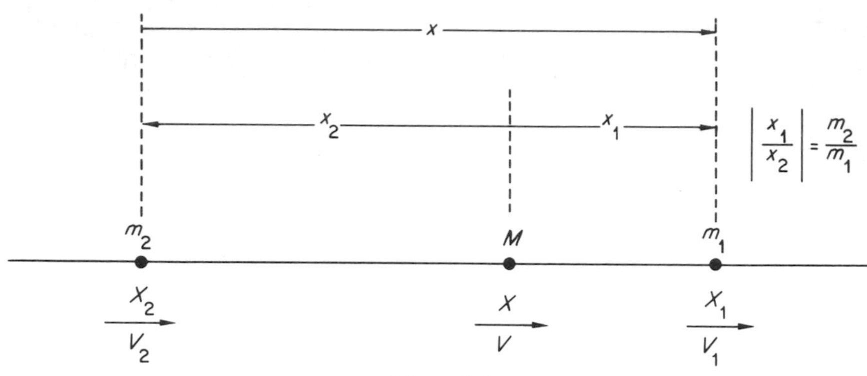

Figure 2.4

The right-hand side of this equation is

$$m_1 V_1 + m_2 V_2 = P_1 + P_2, \tag{2.45}$$

the sum of the two momenta. Thus equations (2.44) and (2.45) show that the total momentum of the two-particle system is the same as that of a single particle whose mass, M, is the sum of those of the two particles,

$$M = m_1 + m_2,$$

moving with the velocity, V, of the centre-of-mass,

$$V = \frac{dX}{dt}.$$

We may express equations (2.44) and (2.45) as

$$MV = P_1 + P_2 = P. \tag{2.46}$$

Since M is unchanged throughout the collision, and we have seen from equation (2.33) that the same is true of the total momentum P, it follows that *the centre-of-mass moves at all times with a constant velocity.*

Relative to the centre-of-mass the coordinates x_1, x_2 of the two particles are given by

$$x_1 = X_1 - X, \qquad x_2 = X_2 - X. \tag{2.47}$$

By using equation (2.43) we can see that

$$m_1 x_1 + m_2 x_2 = 0, \tag{2.48}$$

which gives another interpretation to the centre-of-mass: it is that point for which the relative coordinates satisfy equation (2.48). Differentiating this,

$$m_1\frac{dx_1}{dt} + m_2\frac{dx_2}{dt} = m_1v_1 + m_2v_2$$

$$= p_1 + p_2 = 0, \tag{2.49}$$

where v_1, v_2 are the velocities and p_1, p_2 are the momenta of the two particles relative to the centre-of-mass. Thus another property of the centre-of-mass is that it is a (moving) point relative to which the total momentum is zero.

The total kinetic energy is

$$\begin{aligned}
\mathscr{T} &= \mathscr{T}_1 + \mathscr{T}_2 \\
&= \frac{1}{2}m_1\left(\frac{dX_1}{dt}\right)^2 + \frac{1}{2}m_2\left(\frac{dX_2}{dt}\right)^2 \\
&= \frac{1}{2}m_1\left(\frac{dX}{dt} + \frac{dx_1}{dt}\right)^2 + \frac{1}{2}m_2\left(\frac{dX}{dt} + \frac{dx_2}{dt}\right)^2 \\
&= \frac{1}{2}(m_1 + m_2)\left(\frac{dX}{dt}\right)^2 + \frac{1}{2}m_1\left(\frac{dx_1}{dt}\right)^2 + \frac{1}{2}m_2\left(\frac{dx_2}{dt}\right)^2 \\
&\quad + \left(m_1\frac{dx_1}{dt} + m_2\frac{dx_2}{dt}\right)\frac{dX}{dt}.
\end{aligned}$$

The last term vanishes according to equation (2.49), so

$$\mathscr{T} = \tfrac{1}{2}MV^2 + \tfrac{1}{2}m_1v_1^2 + \tfrac{1}{2}m_2v_2^2$$

$$= \mathscr{T}_{\text{cm}} + \mathscr{T}_{\text{rel}}, \tag{2.50}$$

where $\mathscr{T}_{\text{cm}}$ is the kinetic energy of a single particle whose mass is the sum of those of the two particles, moving with the velocity of the centre-of-mass, and where $\mathscr{T}_{\text{rel}}$ is the total kinetic energy of the two particles relative to the centre-of-mass (i.e. calculated in a frame of reference moving with the velocity of the centre-of-mass).

Since M and V are unchanged throughout the collision $\mathscr{T}_{\text{cm}}$ will be constant, and any changes in the total kinetic energy $\mathscr{T}$ can arise only from changes in the relative kinetic energy $\mathscr{T}_{\text{rel}}$, from before to after the collision. For an elastic collision $\mathscr{T}_{\text{rel}}$ also will have an unchanged final value. When the particles stick together thay must do so at the centre-of-mass, and therefore have no velocity relative to it. Hence in a totally inelastic collision $\mathscr{T}_{\text{rel}}$ disappears.

The position, x, of particle 1 relative to particle 2 is given by

$$x = X_1 - X_2 = (X + x_1) - (X + x_2) = x_1 - x_2.$$

Using equation (2.48) we obtain

$$x_1 = \frac{m_2}{m_1 + m_2}x, \qquad x_2 = -\frac{m_1}{m_1 + m_2}x. \tag{2.51}$$

Hence

$$v_1 = \frac{m_2}{m_1 + m_2}v, \qquad v_2 = -\frac{m_1}{m_1 + m_2}v, \tag{2.52}$$

where v is the velocity of particle 2 relative to particle 1. We may therefore express the kinetic energy relative to the centre-of-mass as

$$\begin{aligned}\mathscr{T}_{\text{rel}} &= \tfrac{1}{2}m_1v_1^2 + \tfrac{1}{2}m_2v_2^2 \\ &= \frac{1}{2}\frac{m_1m_2}{m_1 + m_2}v^2.\end{aligned} \tag{2.53}$$

If, as a result of the collision, v changes from $v_{(1)}$ to $v_{(2)}$ according to equation (2.42),

$$v_{(2)} = -ev_{(1)}.$$

Then $\mathscr{T}_{\text{rel}}$ changes from

$$\mathscr{T}_{\text{rel}(1)} = \frac{1}{2}\frac{m_1m_2}{m_1 + m_2}v_{(1)}^2$$

to

$$\mathscr{T}_{\text{rel}(2)} = \frac{1}{2}\frac{m_1m_2}{m_1 + m_2}e^2v_{(1)}^2 = e^2\mathscr{T}_{\text{rel}(1)}. \tag{2.54}$$

Thus, instead of considering the coefficient of restitution e primarily in terms of the relative velocity we can, according to equation (2.54), use e^2 as a measure of the change in $\mathscr{T}_{\text{rel}}$.

2.7. Non-conservative forces

In Section 2.5 we saw that a force, $F(x)$, depending only upon the position of its point of application, could be derived from a potential energy, and that in this case the total energy consisted only of this and the kinetic energy, and was conserved.

However, examples were given in Section 2.3 of electrical forces which could not be expressed simply as $F(x)$. When time and velocity are involved it is clear that a potential energy cannot be calculated as in equation (2.18) since the expression

$$\int_x^{x_0} F\left(x', \frac{dx'}{dt}, t\right)dx'$$

cannot be evaluated until a relationship between x' and t is known or assumed. Physically this means that we cannot identify a single store of static potential energy which can be exchanged with the kinetic energy of the moving particle.

To illustrate this let us suppose that a particle of mass m and position x is attached to one end of a spring. The other end is not fixed, however, but is being moved in a predetermined way, its position being $X(t)$, as shown in Figure 2.5.

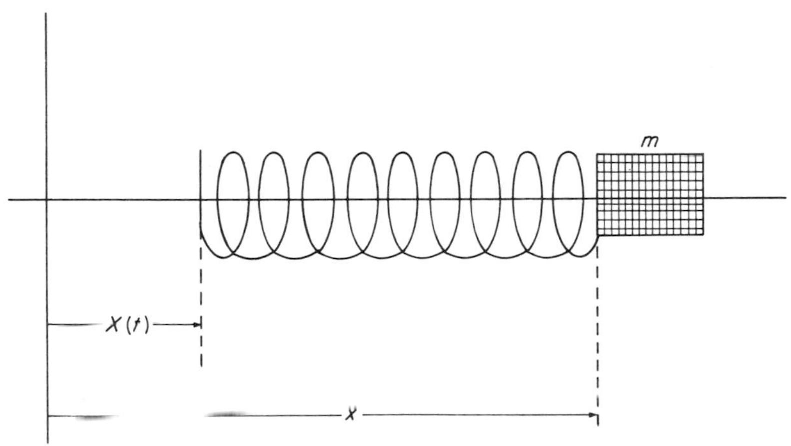

Figure 2.5

The length of the spring is then $x - X$, so if l is its natural length and K its strength constant, the force on the particle will depend upon both x and t:

$$F(x, t) = -K[(x - X) - l],$$

the explicit time-dependence arising from the fact that X is a known function of t, independent of the position of the particle x.

At any stage the potential energy stored in the spring is

$$\mathscr{V} = \tfrac{1}{2}K[(x - X) - l]^2.$$

If after a short interval of time the ends of the spring have moved to $x + \delta x$ and $X + \delta X$, the increase in the potential energy of the spring will be

$$\begin{aligned} \delta\mathscr{V} &= \tfrac{1}{2}K[(x + \delta x - X - \delta X - l)^2 - (x - X - l)^2] \\ &\approx K(x - X - l)(\delta x - \delta X) \\ &= -F\,\delta x + F\,\delta X. \end{aligned} \tag{2.55}$$

Now $F\,\delta x$ is the work done by the spring on the particle and will give the particle, as we know from Section 2.3, an increase, $\delta\mathscr{T}$, in kinetic energy. The force exerted by the spring at the end X is $-F$, and therefore the force exerted by whatever it is that is causing this end of the spring to move is F. Hence $F\,\delta X$ is the work, $\delta\mathscr{W}$, done by *this* force. Thus equation (2.55) is equivalent to

$$\delta\mathscr{T} + \delta\mathscr{V} = \delta\mathscr{W} \tag{2.56}$$

and not

$$\delta\mathscr{T} + \delta\mathscr{V} = 0, \tag{2.57}$$

which we would expect if just the moving particle and the strained spring were conserving total energy. The essential point is that there is another source of energy associated with the force moving the end X of the spring, and the force experienced by the particle itself is not a conservative force since the source of this force (the spring) is no longer capable by itself of accounting for the changes in kinetic energy of the particle.

2.7.1. *Frictional forces*

Even when a force does not appear at first sight to depend upon the time or the velocity of the particle on which it is acting, a physically meaningful potential energy may not exist.

Suppose, for example, that the particle is moving on a rough surface and that the frictional force is independent of its speed. It will also be independent of time since the surface properties remain constant. Thus as the particle moves from x to X_0 (Figure 2.6) we could express the force as

$$F(x) = -A, \qquad A > 0,$$

and say that the work done by this force during the motion is

$$V = -\int_x^X A\,\mathrm{d}x = -A(X_0 - x).$$

The integral will have a definite value that apparently depends only upon x and X_0, and we therefore appear to have defined a potential energy just as in Section 2.4 with its corollary of conservation of total energy. However, we know that the frictional force opposes the motion and therefore always extracts energy; it cannot give back energy as a compressed spring can to *increase* the kinetic energy of the particle.

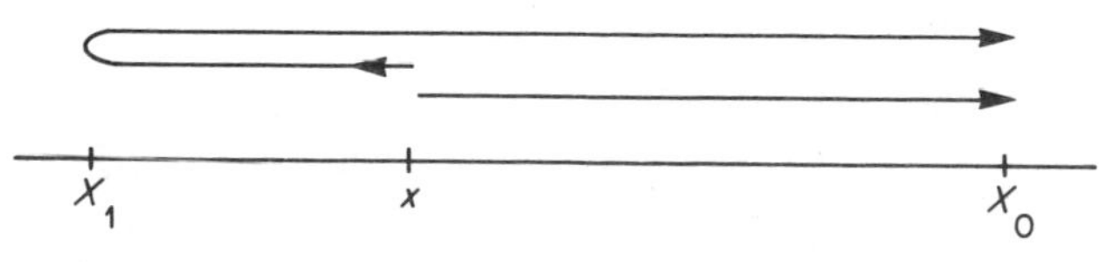

Figure 2.6

To make watertight our definition of a conservative force we must stipulate that the work done is not only independent of time but of *how* the particle moves from x to X_0. In the case under consideration the work done is independent of how the particle moves in the sense that its speed does not matter. However, if the particle moves backwards to X_1, say, before moving on to X_0 it will lose energy to the frictional force during the motion from x to X_1 *and* from X_1 back to x, as well as that which it loses in going directly from x to X_0 (see Figure 2.6).

Thus in a wider sense the work done *does* depend upon how the particle moves from x to X_0. If it were a spring force acting on the particle, the work done against the force in moving from x to X_1 would be exactly returned to the particle as it returned from X_1 to x. The integral then really would be independent of how the particle moves. We shall see in Chapter 4 how this concept of a conservative force applies in three-dimensional motion.

It may be observed that, strictly speaking, the frictional force just considered is *not* independent of *velocity* v, since the direction of the force is opposite to the direction of the velocity. To give the force a mathematical description that really accords with the physics we should write it as

$$F = -A\frac{v}{|v|},$$

and it then falls within the category of non-conservative forces discussed in Section 2.7.

2.8. Simulated one-dimensional motion

One-dimensional motion considered so far has meant linear motion, since the coordinate x represents distance along a straight line. However, when the motion is really two- or even three-dimensional, one measurement may be sufficient to specify the position of a particle. For example a bead sliding on a smooth wire in the form of a vertical circle, of radius l, will have its position determined by the distance s along the wire from the lowest point (Figure 2.7).

If it is then possible to find an equation of motion that involves only that one measurement we can use the results of this chapter to analyse the motion. In this example the bead will be acted on by the force mg vertically downwards, and the reaction R of the wire normal to the circumference. Only the former has a component tangential to the wire,

$$F_s = -mg\sin\theta.$$

Since the acceleration in this direction is $\mathrm{d}^2s/\mathrm{d}t^2$, the equation of motion will therefore be

$$m\frac{\mathrm{d}^2s}{\mathrm{d}t^2} = -mg\sin\theta = -mg\sin\left(\frac{s}{l}\right). \tag{2.58}$$

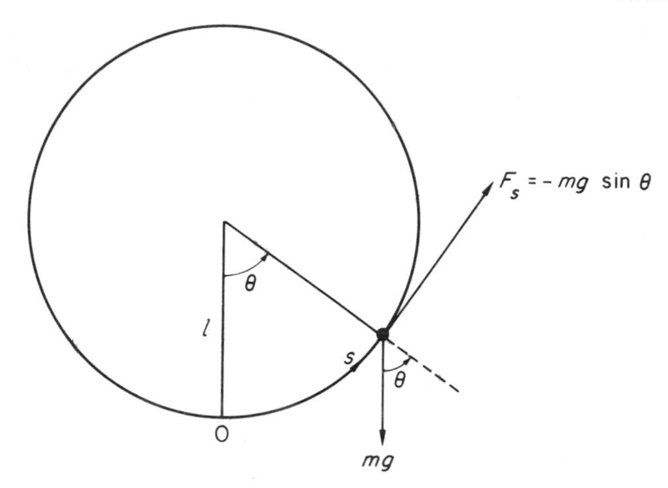

Figure 2.7

Thus although the motion of the bead is not linear its equation of motion (2.58) is one-dimensional and the results obtained in this chapter will all apply. (See Section 2.9.11, page 48.)

2.9. Comments and worked examples

2.9.1. See Section 2

It may be noted that since external forces are only considered as means by which the rest of the solar system affects the small part of it under investigation, and not vice versa, the restrictions implied by Newton's third law need not apply to them. Internal forces, by contrast, we shall generally take to satisfy that law.

2.9.2. See Section 2.2

It is remarkable that gravity should depend upon *inertial* mass. Just as the electric force on a body is proportional to its electric charge, so the gravitational force is proportional to a gravitational quantity possessed by the body, which we might call gravitational 'charge', and there is no more reason to expect this gravitational charge to be related to inertial mass than electric charge is.

Galileo's experiments, showing that bodies of different masses fell together from the leaning tower of Pisa, established the connexion, and later experiments have shown to very great accuracy that the same mass can be used to describe either inertial or gravitational phenomena. This is called the *principle of equivalence*.

The principle is of great practical importance, since the direct setting-up of an inertial mass scale based upon the standard kilogramme (as outlined in

Sections 1.4.2 and 1.11.7), which is normally necessary for analysing dynamical problems, is experimentally difficult to achieve with any great accuracy. It is much easier and more accurate to *weigh* objects, which is equivalent to comparing the gravitational force on any object with that on the standard kilogramme. What the principle of equivalence tells us is that an *inertial mass scale* and a *weight scale*, based upon the same standard, are exactly the same for all objects.

2.9.3. See Section 2.4

Example 2

A straight wire of mass m is initially at rest in the gap of a permanent magnet producing a uniform induction B over a length l. The separation of the poles is small enough for the fringing field at the ends of the gap to be neglected. A varying current flows through the wire for a certain time during which the motion of the wire does not take it beyond the gap. If q is the total charge

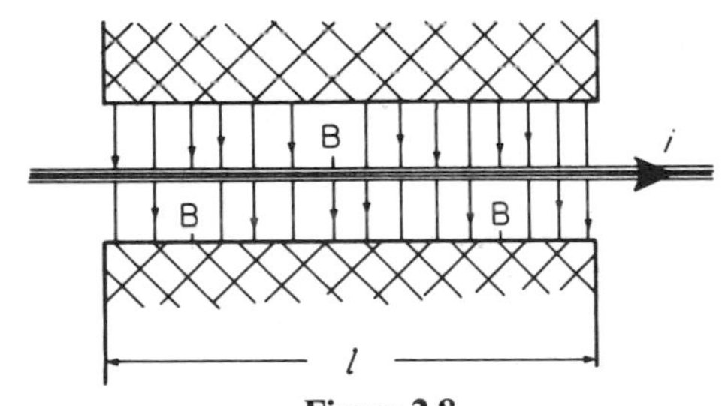

Figure 2.8

passing through the wire in this time and v is the velocity of the wire at the end of it, show that

$$v = Blq/m.$$

The force on the wire when the current through it is i is

$$F = Bli,$$

directed perpendicularly into the plane of the paper. Since the wire does not move out of the gap while the current flows, B and l will remain constant. However, as i varies with time so does F. Thus the impulse delivered to the wire between t_1 and t_2 will be

$$I_{12} = \int_{(1)}^{(2)} F(t)\,\mathrm{d}t = Bl\int_{(1)}^{(2)} i(t)\,\mathrm{d}t.$$

The time integral of current is the charge,

$$\int_{(1)}^{(2)} i(t)\,\mathrm{d}t = \int_{(1)}^{(2)} \frac{\mathrm{d}q}{\mathrm{d}t}\,\mathrm{d}t = \int_{(1)}^{(2)} \mathrm{d}q = q,$$

and the impulse is the increase in momentum,

$$I_{12} = p_2 - p_1 = mv,$$

since the initial velocity is zero. Hence

$$mv = Blq \quad \text{or} \quad v = Blq/m.$$

2.9.4. See Section 2.4

Example 3

Air of density $1{\cdot}3 \text{ kg m}^{-3}$ flows through a sheet of fine-mesh netting at right angles to the air-flow. The effect of the netting is to halve the velocity of the air. Show that when the velocity of the air (before reaching the netting) is 5 m s^{-1} then the pressure on the netting is $16{\cdot}25 \text{ N m}^{-2}$.

Consider a small volume of air, $\delta\sigma$, with mass $\rho\,\delta\sigma$, ρ being the density. This suffers a sudden change of velocity from v to $\frac{1}{2}v$, and consequently a sudden change in momentum from $\rho\,\delta\sigma v$ to $\frac{1}{2}\rho\,\delta\sigma v$, in passing through the netting. Hence the netting must deliver an impulse δI to this volume of air, where

$$\delta I = \tfrac{1}{2}\rho\,\delta\sigma v - \rho\,\delta\sigma v = -\tfrac{1}{2}\rho\,\delta\sigma v.$$

In a time t the volume of air passing through an area, A, of the netting will be Avt, as shown in Figure 2.9. The number of volume elements $\delta\sigma$ in this is

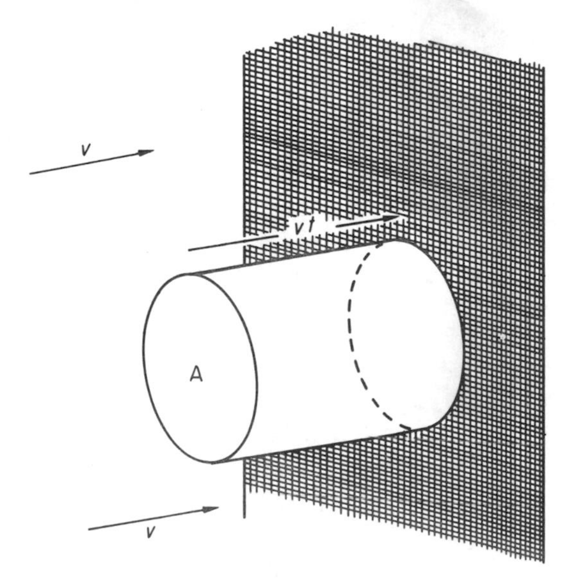

Figure 2.9

$Avt/\delta\sigma$, and since each of these gains the same impulse δI in passing through the netting, the total impulse, I, delivered by the area A to the air in time t is

$$I = \frac{Avt}{\delta\sigma}\,\delta I = -\tfrac{1}{2}\rho\, Av^2 t.$$

Hence the average force exerted on the air by the area A is

$$\bar{F} = \frac{I}{t} = -\tfrac{1}{2}\rho\, Av^2,$$

and by Newton's third law the force exerted on the netting will be the negative of this. Since this is spread over an area A the pressure P will be

$$P = \frac{-\bar{F}}{A} = \tfrac{1}{2}\rho\, v^2.$$

Substituting the values given,

$$P = \tfrac{1}{2} \times 1{\cdot}3 \times 5^2 = 16{\cdot}25 \text{ N m}^{-2}.$$

2.9.5. See Section 2.5

Example 4

A narrow beam of length l and specific density $\frac{1}{2}$ is held vertically with its lower end just touching the surface of some deep water. It is then allowed to fall freely from rest into the water while remaining upright. Show that if motion of the water and any viscous forces it might exert on the beam can be neglected, then the speed of the beam when it is half submerged is $\sqrt{gl/2}$.

Since water motion and viscous forces may be neglected, the only effect of the water is the upthrust of the displaced liquid. If m is the mass of the beam, then when a length, x, is submerged the mass of displaced water is $2mx/l$, since the water has twice the density of the beam. The upthrust exerted on the beam is therefore $2mgx/l$ while the downward force is the constant gravitational force mg. Thus the resultant force in the positive x direction (downwards) is $mg(1 - 2x/l)$, which depends only upon x. The increase in kinetic energy in moving from x_1 to x_2 will therefore be the work done,

$$\mathscr{W}_{12} = \int_{x_1}^{x_2} F(x)\,dx = mg\int_{x_1}^{x_2}\left(1 - \frac{2x}{l}\right)dx$$

$$= mg\left[x - \frac{x^2}{l}\right]_{x_1}^{x_2}.$$

If the beam moves from just touching the water to half-submersion,

$$x_1 = 0 \quad \text{and} \quad x_2 = \tfrac{1}{2},$$

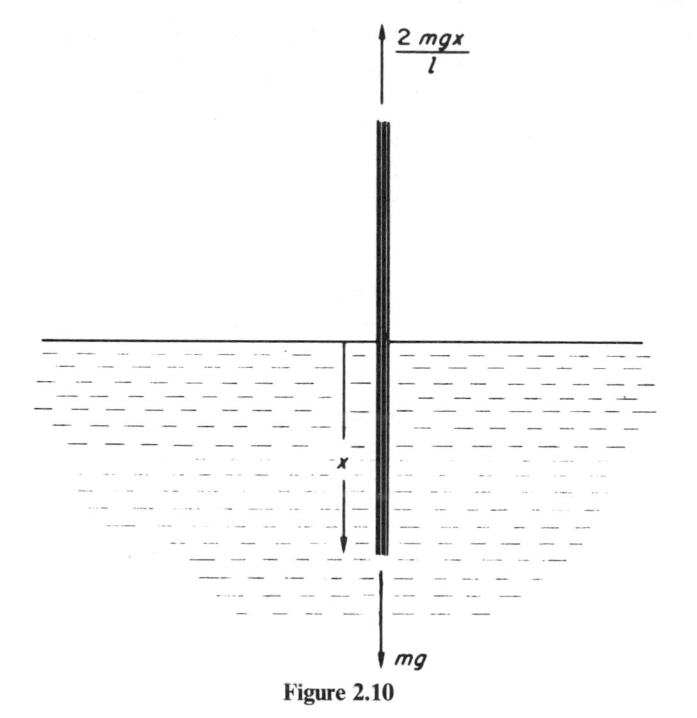

Figure 2.10

and the work done is then

$$\mathscr{W}_{12} = \frac{mgl}{4}.$$

The initial speed is zero, and we shall denote the speed at half-submersion by v. The increase in kinetic energy is therefore

$$\mathscr{T}_2 - \mathscr{T}_1 = \frac{mv^2}{2} = \mathscr{W}_{12} = \frac{mgl}{4},$$

and from this

$$v = \sqrt{\frac{gl}{2}}.$$

2.9.6. See Section 2.5

It is important, for some readers at least, to note that when we are discussing the potential energy at some point x we *must not* use the same symbol when integrating between x and some other point. This accounts for the use of x' in equation (2.18).

The distinction is not a point of mathematical pedantry. If we were to write

$$\mathscr{V}(x, X_0) = \int_x^{X_0} F(x)\,\mathrm{d}x,$$

we would be using x to denote (as the lower limit of the integral) a particular position for which we are evaluating the potential energy and at the same time (as the variable in the integrand $F(x)$) a whole range of other positions which are involved in that calculation. It is just as illiterate to use the same symbol to denote two different things within the same sentence when it involves mathematical ideas as it would be to use the same word to mean two different things within the same prose sentence.

One of the greatest blocks to the successful tackling of physical problems is the belief that the transformation of a physical statement into mathematical symbols and equations enables one to dispense with precision and clarity in the original thought. On the contrary, it is only when that thought has been sharpened into a clear and unambiguous statement that it is ready to be represented mathematically.

2.9.7. See Section 2.5.2

This furnishes an example of one of the limitations of classical mechanics. The particles that make up an atomic nucleus may be considered as bound within a potential hollow. However, in the case of radioactive nuclei they can sometimes escape, apparently 'burrowing through' the potential barrier, in contradiction to the classical prediction. Quantum mechanics is necessary for a description of this phenomenon.

2.9.8. See Section 2.5.3

The energy content of a field may still seem a nebulous, even 'unreal', concept by contrast with the energy stored in a spring. However, we must take care not to equate 'reality' with what is directly apparent to the senses. If we believe that physics would go on whether or not human beings were present we must test the reality of concepts by the importance they have in the general structure and description of physical phenomena.

By this test fields are as necessary and real as particles in many branches of physics; indeed, in some cases they may be considered the more fundamental concept.

2.9.9. See Section 2.6.2

Kinetic energy and momentum were introduced in Section 2.3 on level terms, one expressing the 'distance' effect of a force and the other the 'time' effect. It may therefore seem strange that the conservation law for the latter (equation 2.33) concerns only momentum, while that for the former (equation 2.36) requires in addition to kinetic energy the concept of potential energy.

This difference between the two laws arises from the limitations inherent in Newton's third law, which, as we pointed out in Section 1.11.12, is restricted to low velocity phenomena. At low velocities the capacity of mechanical transmitters of forces such as springs, and of gravitational and electric fields, to carry momentum can be neglected. Momentum can therefore be thought of as being directly exchangeable between particles, whereas this is usually not true of energy. In consequence, conservation of energy is, in practice, often difficult to formulate exactly, while conservation of momentum is simple to express; both conservation laws are, however, of equal and fundamental importance.

2.9.10. See Section 2.6.3

The derivation of the result (2.41) rests upon the validity of dividing by equation (2.39), which we may do only if both sides are not zero. If they are, then we have

$$V_{1(1)} - V_{1(2)} = 0 = V_{2(2)} - V_{2(1)},$$

instead of (2.41). This alternative result implies that the final velocities of both particles are unaltered by the collision and therefore that they interpenetrate each other. This would be possible, for example, if one of the 'particles' were in fact a ring, with the motion of itself and of the other particle directed along the axis of the ring. A repulsive force could exist between these two which would diminish their relative velocity as they approached each other but was insufficient to reverse the motion.

2.9.11. See Section 2.8

Example 5

Does equation (2.58) represent motion under a conservative force and, if so, determine the corresponding potential energy? What is the minimum speed that the bead must have at the bottom of the wire to enable it to traverse the whole circle?

Since the circumferential force acting on the particle is given by

$$F_s = -mg \sin\left(\frac{s}{l}\right),$$

there is no dependence on time or velocity. The force is therefore a conservative one and the potential energy, taking $s = 0$ as the reference point, will be

$$\begin{aligned} \mathscr{V}(s) &= \int_s^0 F(s')\,\mathrm{d}s' \\ &= -mg \int_s^0 \sin\frac{s'}{l}\,\mathrm{d}s' = mgl\left[1 - \cos\left(\frac{s}{l}\right)\right]. \end{aligned}$$

This has the form shown in Figure 2.11.

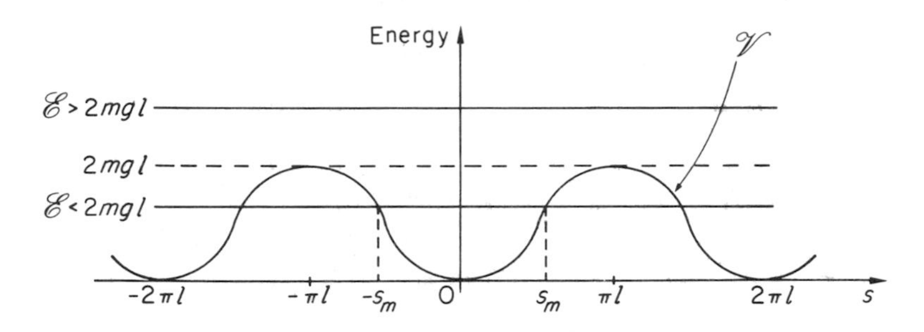

Figure 2.11

The motion of the bead will therefore be bound when the total energy $\mathscr{E}$ satisfies

$$0 < \mathscr{E} < 2mgl,$$

and it will oscillate between

$$s = \pm s_m, \quad \text{where} \quad s_m < \pi l.$$

The motion is unbound when

$$\mathscr{E} > 2mgl,$$

and the bead will then continue to encircle the wire in the positive or negative direction indefinitely. At the dividing value,

$$\mathscr{E} = 2mgl, \tag{2.59}$$

the bead will attain the values

$$s = \pm \pi l,$$

which implies that it will just reach the top of the circle.

If v_m is the corresponding velocity at the bottom of the wire, the total energy there will be

$$\mathscr{E} = \tfrac{1}{2}mv_m^2 + V(0) = \tfrac{1}{2}mv_m^2. \tag{2.60}$$

From equations (2.59) and (2.60)

$$\tfrac{1}{2}mv_m^2 = 2mgl,$$

and the least speed that ensures traversal of the whole circle is therefore

$$|v_m| = 2(gl)^{1/2}.$$

2.10. Problems

2.1. Sketch each of the following potential energy functions and determine the corresponding force:

(a) $\mathscr{V}(x) = ax + b; a > 0, b > 0$ (b) $\mathscr{V}(x) = x^2 - 4x - 21$

(c) $\mathscr{V}(x) = |x|$ (d) $\mathscr{V}(x) = \begin{cases} 5x + 6 \ ; x \geqslant 2 \\ -3x + 22; x \leqslant 2 \end{cases}$

(e) $\mathscr{V}(x) = -a/x + b; a > 0, b > 0.$

Where appropriate, choose new axes whose origin coincides with the potential energy minimum, and find the expression for $\mathscr{V}$ referred to the new axes.

2.2. What new zeros for x and $\mathscr{V}$ would you use to give the simplest form for the potential energy:

$$\mathscr{V}(x) = \frac{2}{x - 4} + 10\,?$$

Sketch the form of $\mathscr{V}$ for the new axes. A particle with this potential energy moves from infinity towards the origin of the potential, first with an initial velocity of 2 units and then with a velocity of 4 units. If it travels a distance 3 units more in the second case, show that the mass of the particle is $\frac{1}{4}$ unit.

2.3. From what potential energy functions may the following forces be derived:

(a) $F(x) = ax + b$ (b) $F(x) = F_0 \exp\left(-\dfrac{x}{a}\right)$

(c) $F(x) = F_0 \exp\left(-\left|\dfrac{x}{a}\right|\right)$ (d) $F(x) = F_0 \sin kx$

(e) $F(x) = \begin{cases} +1; \sin \pi x > 0 \\ -1; \sin \pi x < 0\,? \end{cases}$

2.4. What are the forces corresponding to:

(a) $\mathscr{V}(x) = -\left|\dfrac{A}{x}\right|$ (b) $\mathscr{V}(x) = -\left|\dfrac{A}{x}\right| \exp\left(-\left|\dfrac{x}{a}\right|\right)$?

Show that for $|x| \ll |a|$ the forces are approximately the same, but that for $|x| \gg |a|$ the second is negligible compared with the first. (These are the one-dimensional versions of the 'inverse square' and 'screened inverse square' forces.)

2.5. A particle of unit mass moves along the x-axis with a potential energy:

$$\mathscr{V}(x) = x^3 - 5x^2 + 2x + 12.$$

(a) Where would the particle experience its maximum (positive) force?

(b) What types of motion are possible if the total energy of the particle is 4 units?

(c) Show that, with a suitable choice of axes, small oscillations of the particle about its position of stable equilibrium are governed by the energy equation

$$\mathscr{E} = \tfrac{1}{2}\dot{x}^2 + \sqrt{19}x^2.$$

What is its period of oscillation?

2.6. (a) For which values of x would the beam in Example 4 be (i) above the water, (ii) partially submerged, (iii) totally submerged?

(b) What is the force in each of these three regions?

(c) Show that a potential energy giving these forces is

$$\mathscr{V}(x)\begin{cases} = -mgx; & x \leqslant 0 \\ = -mg(x - x^2/l); & 0 \leqslant x \leqslant l \\ = \quad mg(x - l); & l \leqslant x. \end{cases}$$

(d) Hence show that if the beam is released from rest when its lower end is a height h above the surface it will come to rest again when its upper end is a distance h below the surface.

(e) Suppose that in (d) the beam is given a velocity v instead of being released at rest. Find the maximum depth reached by the beam and explain why this is independent of whether the initial velocity is directed upwards or downwards.

2.7. A charge q in an electric field E experiences the force qE. In one dimension the *electrostatic potential*, Φ, is defined by

$$E = -\frac{d\Phi}{dx}.$$

(In S.I. units Φ is measured in volts and E in volts metre^{-1}).

(a) What is the *potential energy* of an electron, which has a negative charge $-e$, in the field E?

(b) An electron is emitted with zero speed from a cathode C and moves under the influence of the electric field arising from potentials Φ_C, Φ_{A1}, Φ_{A2}, at the cathode C, and two parallel grids, A1, A2, as shown in Figure 2.12. If it is to reach a speed v_A at A1 and then to travel with constant speed between A1 and A2, what must the potentials Φ_{A1}, Φ_{A2} be relative to Φ_C? Do these values depend upon the spacings of the electrodes?

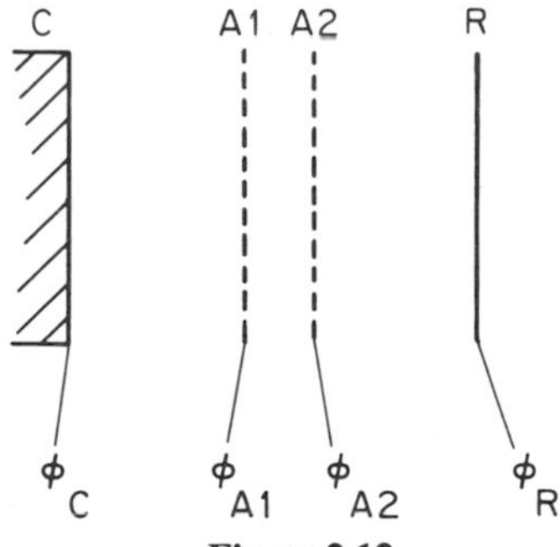

Figure 2.12

(c) If the electrode R is used as a reflector, what must its potential Φ_R be, in terms of Φ_C and Φ_{A2}, if the electron is to be turned back (i) half-way, (ii) two-thirds of the way, between A1 and R?

(d) For the case

$$\Phi_C = -300 \text{ V}, \qquad \Phi_{A1} = \Phi_{A2} = 0 \text{ V}, \qquad \Phi_R = -500 \text{ V}, \qquad l = 2 \text{ mm},$$

determine the time the electron spends in the reflexion region (the transit time) and the distance it penetrates. ($\sqrt{2e/m} = 5{\cdot}93 \times 10^5$ m s^{-1} V$^{-1/2}$, for an electron.)

2.8. Figure 2.12 represents in outline the electrode structure of a microwave reflexion oscillator (or reflex klystron). A high frequency oscillatory field exists across the small gap between the grids A1 and A2. This does not affect the steady electric potentials of the electrodes, but accelerates or decelerates an electron beam passing through the gap so that an electron traversing A2 at time t will have, in addition to the velocity v_A it acquires from the difference in potential $\Phi_{A2} - \Phi_C$, a small sinusoidally varying velocity,

$$v(t) = v_0 \sin 2\pi f t,$$

where f is the frequency of the oscillatory field.

(a) Show that if two electrons pass A2 at a small time interval δt, this changes to

$$\delta t \left[1 + \frac{2\pi l m f v_0 \cos 2\pi f t}{e(\phi_{A2} - \phi_R)} \right]$$

on their return.

(b) Arising from (a), an electron beam which initially has a uniform density will have a variable density on its return through A2, these parts with greater density being known as 'bunches'. What phase conditions (values of $2\pi f t$) must be satisfied by those electrons forming bunches, and which electrons will be at their centres? You may assume

$$\frac{2\pi m l f v_0}{e(\phi_{A2} - \phi_R)} < 1$$

—why?

(c) As a bunch returns through the gap the electrons comprising it will again be accelerated or decelerated. If the bunch is to give energy to the oscillating field and thereby maintain it in oscillation, show that the mode of operation must satisfy

$$f\tau_c = n + \tfrac{3}{4}; \qquad n = 0, 1, 2, \ldots,$$

where τ_c is the transit time in the reflecting region for the central electron of the bunch.

(d) Show that the reflector potential necessary to maintain oscillations is one of the values

$$\phi_R = \phi_{A2} - \frac{4fl}{n + \frac{3}{4}} \sqrt{\frac{m}{2e}} \sqrt{\phi_{A2} - \phi_c}.$$

(e) For the case

$$f = 10^{10}\ \text{Herz}, \qquad l = 1\ \text{mm}, \qquad \phi_{A2} = 0\ \text{V}, \qquad \phi_c = -300\ \text{V},$$

determine how many modes of operation are possible and calculate the corresponding reflector potentials.

2.9. A particle of mass m and charge q is ejected vertically upwards from the earth's surface with velocity v towards a charge Q of opposite sign fixed at a height h; h is small enough for the gravitational field to be assumed constant. Show that the potential energy determining the particle's motion may be written in the form

$$\mathscr{V}(z) = mgz - \frac{\beta}{h - z}$$

where z is the distance above the earth's surface and β is a positive constant. Hence show that if h is less than some minimum value, $h_{\min}$, the charges will always collide, but that if it exceeds this they will only do so if v exceeds a certain minimum value,

$$v_{\min} = \sqrt{2gh}\left(1 - \frac{h_{\min}}{h}\right).$$

2.10. A pile is driven into the ground by repeatedly allowing a mass m to fall a distance h on to the end of the pile. If τ is the time for which the mass is in contact with the pile during each collision, show that the peak force experienced by the pile will exceed $\sqrt{2ghm/\tau}$. If the pile when fully set is to support a mass of 10^5 kg, how high must a mass of 10^3 kg be raised to drive it in, assuming that the collision time is 10^{-2} s?

2.11. In problem 2.10 the following mass will have a certain velocity relative to the earth just before it hits the stationary pile. Although the mass will bounce, possibly several times, the collisions are not perfectly elastic and eventually both the mass and the pile will be stationary. How do you account for the apparent loss of momentum in this process?

2.12. A particle of mass m travels with velocity v between two parallel walls a distance l apart, rebounding from them back and forth along the same straight line.

(a) Show that if the force F experienced by either wall is plotted against time it will consist of a series of peaks spaced a time $2l/v$ apart.

(b) How does the shape of the peaks change with the rigidity of the wall? Show that if the collision is perfectly elastic the area under each peak is independent of the nature of the wall, and has the value $2mv$.

(c) Hence show that the mean force on the wall is $2T/l$, where T is the kinetic energy of the particle.

(d) Suppose the walls are of finite area A and there are many similar particles travelling between them on parallel paths spaced uniformly apart. Show that the pressure on each wall is $2\sigma T$, where σ is the spatial density of the particles (number per unit volume).

(e) Now suppose that the particles have a spread of velocities. Use the arguments above to show that the pressure is $2\sigma\overline{T}$, where $\overline{T}$ is the mean kinetic energy of the particles. (This is a one-dimensional model for a gas.)

(f) So far the particles have been travelling independently of each other. Elastic collisions between them may be simulated by supposing that sudden changes occur in the velocities of pairs of particles such that their joint momentum and kinetic energy are unchanged. Would the above derivation of the result (e) remain valid if such collisions were (i) rare, (ii) frequent, compared with collisions with the walls?

2.13. A group of particles, of masses $m_1, m_2, \ldots, m_n$, move along the same straight line, with positions $x_1, x_2, \ldots, x_n$. Their centre-of-mass, X_{cm}, is defined by

$$MX_{\text{cm}} = \sum_{i=1}^{n} m_i x_i \quad \text{where} \quad M = \sum_{i=1}^{n} m_i.$$

Show that if p is the total momentum and $\mathscr{T}$ the total kinetic energy of the group, then

$$p = p_{\text{cm}}, \qquad p_{\text{rel}} = 0, \quad \text{and} \quad \mathscr{T} = \mathscr{T}_{\text{cm}} + \mathscr{T}_{\text{rel}},$$

where p_{cm} is the momentum and $\mathscr{T}_{\text{cm}}$ the kinetic energy of a single particle of mass M moving with the velocity of the centre-of-mass, and where p_{rel} and $\mathscr{T}_{\text{rel}}$ are the total momentum and total kinetic energy of the n particles relative to the centre-of-mass.

Hence show that, if the particles are viewed from such a great distance that the group appears as a single point, their observed response to a given total external force is independent of how this is apportioned among the individual particles and of how they interact with, and move relative to, each other. Is the motion of this point an adequate guide to the energy content of the group?

Chapter 3

Oscillatory Motion in One Dimension

In the previous chapter we have established the concepts of momentum and energy, and the corresponding conservation laws. We saw that these are powerful aids to understanding the motion of simple systems in one dimension. When more complicated systems in two or three dimensions are to be discussed an extension of these ideas gives a deep insight into the qualitative nature of their motion, and these concepts often provide a good starting point for detailed quantitative calculations. Chapters 5 and 6 will illustrate this.

For the present, however, we shall concentrate on bound motion in one dimension, which is necessarily oscillatory, as was shown in Section 2.5.2. In many cases it is possible to analyse this precisely and to extend the analysis to phenomena which are widespread and of considerable practical importance in physics and engineering. Most of the theory will be developed in terms of a variable x describing motion in a straight line. However, the discussion of Section 2.8 can be applied here, too, and shows that the analysis is equally valid for any motion that can be described by a single variable.

3.1. Simple harmonic motion

The force exerted by a spring is proportional to the extension or compression, $\pm x$, relative to its natural length l, provided this is small, and is always directed towards the equilibrium or unstrained position. Thus a particle attached to the moving end which is free to move only along the spring axis will have for its equation of motion

$$m\frac{\mathrm{d}^2x}{\mathrm{d}t^2} = -Sx.$$

The strength constant S is characteristic of the spring. Since it is a positive constant this equation may be rewritten as

$$\frac{\mathrm{d}^2x}{\mathrm{d}t^2} + \omega_0^2 x = 0, \qquad (3.1)$$

where

$$\omega_0 = \left(\frac{S}{m}\right)^{1/2} \qquad (3.2)$$

is a real number with dimensions

$$\left[\frac{S}{m}\right]^{1/2} = \left(\frac{\mathrm{MLT}^{-2}}{\mathrm{ML}}\right)^{1/2} = \mathrm{T}^{-1}.$$

Equation (3.1) can describe many situations in addition to the motion arising from a spring. Consider any bound motion under the action of a conservative force. As we saw in Section 2.5.2 the particle will oscillate about the position of potential energy minimum. Let us take this as zero for the x coordinate. Then, provided x is sufficiently small, we may expand the potential energy as a power series in x,

$$\mathscr{V}(x) = \mathscr{V}(0) + \left(\frac{\mathrm{d}\mathscr{V}}{\mathrm{d}x}\right)_0 x + \frac{1}{2}\left(\frac{\mathrm{d}^2\mathscr{V}}{\mathrm{d}x^2}\right)_0 x^2 + \dots. \tag{3.3}$$

(In one dimension it is usually clear that x is the only variable and the distinction between $\partial V/\partial x$, as used in equation (2.22), and $\mathrm{d}\mathscr{V}/\mathrm{d}x$ is unnecessary.)

Now we are free to choose the potential energy to be zero at $x = 0$ by taking

$$\mathscr{V}(0) = 0.$$

Also, since $\mathscr{V}(x)$ has a *minimum* at $x = 0$, we must have

$$\left(\frac{\mathrm{d}\mathscr{V}}{\mathrm{d}x}\right)_0 = 0 \quad \text{and} \quad \left(\frac{\mathrm{d}^2\mathscr{V}}{\mathrm{d}x^2}\right)_0 > 0.$$

Thus when x is small enough for x^3 and higher powers to be neglected, the potential energy may be written as

$$\mathscr{V}(x) = \frac{1}{2}\left(\frac{\mathrm{d}^2\mathscr{V}}{\mathrm{d}x^2}\right)_0 x^2 = \frac{1}{2}Kx^2, \tag{3.4}$$

where K is a positive constant. The corresponding force is

$$F(x) = -\frac{\mathrm{d}\mathscr{V}}{\mathrm{d}x} = -Kx,$$

giving the equation of motion

$$m\frac{\mathrm{d}^2x}{\mathrm{d}t^2} = -Kx. \tag{3.5}$$

This, too, can be written as equation (3.1),

$$\frac{\mathrm{d}^2x}{\mathrm{d}t^2} + \omega_0^2 x,$$

where, in this case,

$$\omega_0 = \left(\frac{K}{m}\right)^{1/2}. \tag{3.6}$$

(See Section 3.5.1, page 68.)

Thus within any potential energy hollow that is expanded in the form (3.3) bound motion of sufficiently small extent will be described by the equation of motion (3.1) (see Section 3.5.2, page 70).

The general solution to this equation of motion may be expressed in any of the forms:

$$x(t) = a\cos\omega_0 t + b\sin\omega_0 t \tag{3.7}$$

$$= A\cos(\omega_0 t + \alpha) \tag{3.8}$$

$$= B\sin(\omega_0 t + \beta). \tag{3.9}$$

Motion described by any of these alternative expressions is called *simple harmonic motion.* Each description contains two adjustable real constants: a and b, A and α, B and β. This is a further instance of the observation made in Section 2.2. Two conditions (the initial position and velocity, for example) are required to determine any such pair of constants.

3.1.1. Complex representation

A very convenient method of dealing with equation (3.1) is to note that if we consider z as a *complex* quantity,

$$z = x + jy,$$

satisfying that equation, it will automatically be true that both the real and imaginary parts will separately satisfy it. We may then take either part (usually the real part) to represent the physical measurement. If we attempt a complex solution in the form

$$z = Ce^{pt}$$

for the equation

$$\frac{d^2z}{dt^2} + \omega_0^2 z = 0, \tag{3.10}$$

we must have

$$(p^2 + \omega_0^2)z = 0.$$

For there to be any motion at all z must not be zero. Hence

$$p^2 + \omega_0^2 = 0, \qquad p = \pm j\omega_0.$$

A solution, then, is

$$z = C\,e^{j\omega_0 t}, \tag{3.11}$$

and the physical motion is given by the real part of this,

$$x = \mathrm{Re}(C\,e^{j\omega_0 t}). \tag{3.12}$$

Although this omits the $\exp(-j\omega_0 t)$ solution, nothing of physical significance has been lost since the real and imaginary parts of complex C provide the two adjustable constants required to fit two chosen physical conditions. (See Sections 3.5.3 and 3.5.4, pages 70 and 71.)

Since a complex solution, z, is so often used for equations of oscillatory motion, it would become rather wearying to qualify it each time with the statement (3.12). Whenever complex z occurs subsequently it will be implied that the real part, x, is the description of the physical motion.

3.1.2. *Angular frequency and amplitude*

The fundamental parameter in any simple harmonic motion is the *angular frequency*, ω_0, present in the equation of motion (3.1) and therefore necessarily in any of the forms of solutions (3.7), (3.8), (3.9) or (3.11). As we can see from any of these equations, it measures the constant rate at which the angle $\omega_0 t$ is varying (in radians s^{-1}). When $\omega_0 t$ increases by 2π the solutions repeat, so that the *period* τ of the motion (the time taken for the motion to go through a complete oscillation and commence the next) is given by

$$\tau = \frac{2\pi}{\omega_0}. \tag{3.13}$$

The number of complete oscillations in unit time,

$$f = \frac{1}{\tau} = \frac{\omega_0}{2\pi}, \tag{3.14}$$

is called the *frequency*.

The other chief characteristic of simple harmonic motion is the *amplitude*, or maximum displacement of the particle from the equilibrium position. For the solutions (3.8), (3.9) and (3.11) the amplitudes are simply A, B and $|C|$, remembering that in the last case $|C|$ is the length or magnitude of a complex number. For solution (3.7) we may write

$$x(t) = (a^2 + b^2)^{1/2}\left[\frac{a\cos\omega_0 t}{(a^2 + b^2)^{1/2}} + \frac{b\sin\omega_0 t}{(a^2 + b^2)^{1/2}}\right]$$

$$= (a^2 + b^2)^{1/2}\sin(\omega_0 t + \phi),$$

where

$$\tan\phi = \frac{a}{b},$$

from which we can see that the amplitude is $(a^2 + b^2)^{1/2}$.

The angles α, β and ϕ are usually called *phase angles.*

3.1.3. *Kinetic and potential energy*

Whenever a one-dimensional force can be written as $-Kx$ there will be a potential energy

$$\mathscr{V}(x, X_0) = -K\int_x^{X_0} x'\,\mathrm{d}x' = \tfrac{1}{2}K(x^2 - X_0^2).$$

By choosing X_0 at the origin we obtain the particular potential energy, given in equation (3.4),

$$\mathscr{V} = \tfrac{1}{2}Kx^2.$$

This process is the converse of that given in Section 3.1, where it was shown that at any minimum the potential will be of the form (3.4) and is equivalent to a force $-Kx$.

The kinetic energy is

$$\mathscr{T} = \frac{1}{2}m\left(\frac{\mathrm{d}x}{\mathrm{d}t}\right)^2,$$

and we may therefore express conservation of energy for a simple harmonic oscillator as

$$\mathscr{T} + \mathscr{V} = \frac{1}{2}m\left(\frac{\mathrm{d}x}{\mathrm{d}t}\right)^2 + \frac{1}{2}Kx^2 = \mathscr{E} = \text{constant}. \tag{3.15}$$

By differentiating this with respect to time we obtain the equation of motion (3.5). We can therefore regard either the force equation (3.5) or the energy equation (3.15) as equally valid starting points for describing simple harmonic motion. In the first, K defines the strength of the force and m the inertia of the response to it. In the second, K measures the capacity for storing potential energy and m the capacity for storing kinetic energy. Whichever pair of roles is chosen for these constants it is their ratio which determines the angular frequency ω_0, as given in equation (3.6),

$$\omega_0 = \left(\frac{K}{m}\right)^{1/2}.$$

According to the energy equation (3.15) we can think of the motion as characterized by a cyclic flow of energy from the kinetic to the potential form and then back again. When the particle reaches its maximum distance, x_0, from the equilibrium point, it is momentarily at rest and the energy is entirely potential in form:

$$\mathscr{E} = \tfrac{1}{2}Kx_0^2.$$

The total energy of the oscillator is therefore proportional to the *square of its amplitude.* (See Section 3.5.5, page 73.)

3.2. Forced oscillations; resonance

In many practical uses of springs one end is *not* held fixed while a body moves with the other end. Instead the end is moved about by some external agency and the spring links this motion to the body at its other end. Consider the simple case shown in Figure 3.1, where this motion is simple harmonic, of amplitude X_0 and angular frequency ω,

$$X = X_0 \cos \omega t.$$

The extension of the spring is now not the displacement, x, of the particle at the other end from its equilibrium position, but $x - X_0 \cos \omega t$. Then the equation of motion is

$$m\frac{d^2x}{dt^2} = -S(x - X_0 \cos \omega t)$$

or

$$\frac{d^2x}{dt^2} + \omega_0^2 x = \frac{SX_0}{m} \cos \omega t.$$

This is an example of *forced oscillations*, which we can write generally (in complex form) as

$$\frac{d^2z}{dt^2} + \omega_0^2 z = B\,e^{j\omega t} = \frac{F}{m}, \tag{3.16}$$

where F is that part of the applied force which has the given, or *forcing*, angular frequency ω.

We might expect that the particle could oscillate at the forcing frequency. If such motion is denoted by z', it will be of the form

$$z' = D\,e^{j\omega t}, \tag{3.17}$$

and since it must satisfy the equation of motion (3.16), then

$$\frac{d^2z'}{dt^2} + \omega_0^2 z' = B\,e^{j\omega t}. \tag{3.18}$$

Equations (3.17) and (3.18) show that

$$(\omega_0^2 - \omega^2)D\,e^{j\omega t} = B\,e^{j\omega t}$$

or

$$D = \frac{B}{\omega_0^2 - \omega^2}. \tag{3.19}$$

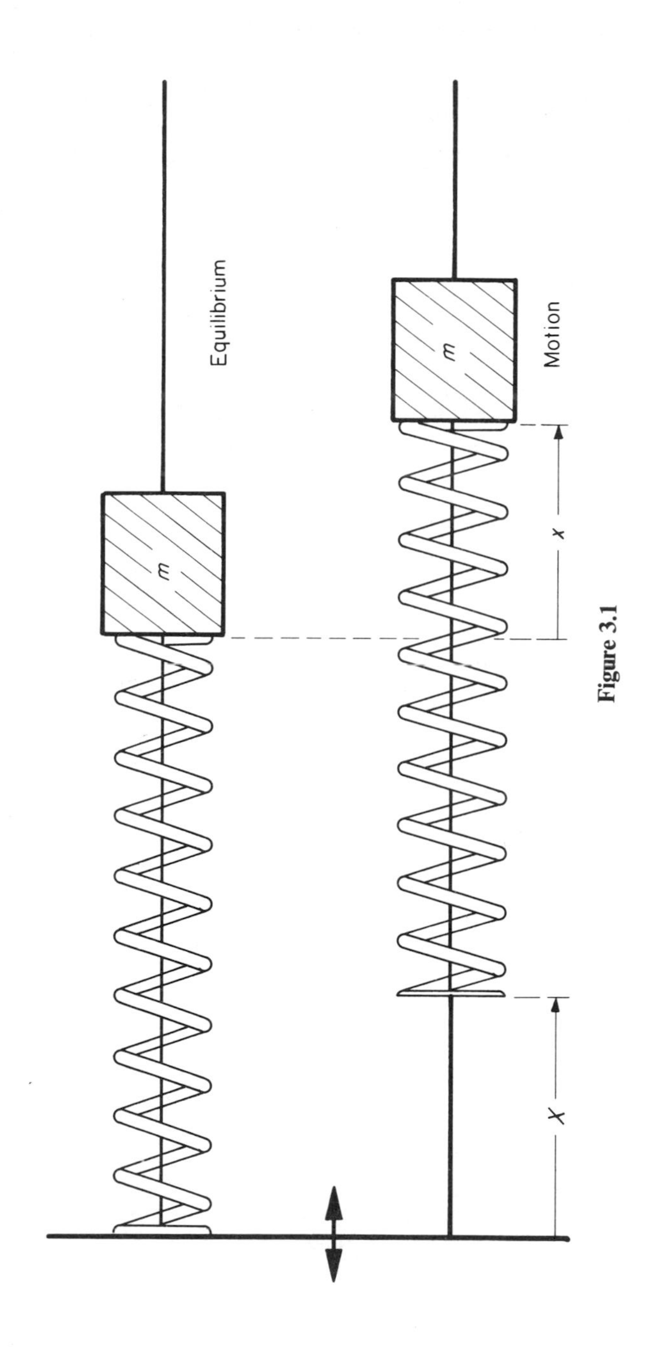

Figure 3.1

This gives z' as a completely defined solution, provided ω and ω_0 are not equal, with no adjustable constants. However, this cannot be a full description of all possible solutions of (3.16), since it is obvious that the form of motion must depend upon how it is started. Suppose, then, that z describes a more general solution. By subtracting equation (3.18) from equation (3.16) we obtain

$$\frac{d^2(z - z')}{dt^2} + \omega_0^2(z - z') = 0,$$

which is equation (3.10) again, with a solution

$$z - z' = C\,e^{j\omega_0 t}.$$

This contains two adjustable constants, as we saw in Section 3.1.1. Thus the general solution to equation (3.16) capable of describing any appropriate pair of physical conditions is

$$z = C\,e^{j\omega_0 t} + z'.$$

Using equations (3.17) and (3.19) this is

$$z = C\,e^{j\omega_0 t} + \frac{B}{\omega_0^2 - \omega^2}e^{j\omega t}. \tag{3.20}$$

The second term is called a *particular* solution or integral of equation (3.16). To obtain the general solution we must always add to this the solution to the free oscillations equation (3.10). This free oscillation has the natural frequency, ω_0, of the oscillator and its amplitude and phase are determined by the initial conditions. The particular solution represents *forced* oscillations, which have the frequency of that part of the force described by the term $B\,e^{j\omega t}$. Their amplitude $B/(\omega_0^2 - \omega^2)$ is strongly dependent upon the difference between the natural frequency of the oscillator and that of the applied force. However weak the force (i.e. however small the magnitude of B), the amplitude of the forced oscillations may be made arbitrarily large by choosing ω sufficiently near ω_0.

When the two frequencies are actually equal the solution of equation (3.16) is

$$z = \left(C - \frac{jB}{2\omega_0}t\right)e^{j\omega_0 t}, \tag{3.21}$$

as the reader may readily check (see problem 3.9). We can see that the amplitude of the oscillations will increase indefinitely with t.

The ability to produce a large oscillation with a weak force is a familiar phenomenon. To set a child swinging high we need push him only gently, provided we do so every time he passes a certain point, thus exerting a force whose frequency is approximately that of the swing. Similarly, if a car is stuck in a deep rut, we rock it backwards and forwards, giving it regular pushes and pulls to produce a motion sufficiently large for it to surmount the rut.

These are examples of *resonance* oscillations. In fact, neither of them is adequately described by equations (3.20) or (3.21). Firstly, the potential energy does not increase indefinitely with displacement as is true of the expression (3.4). When the amplitude of the swing is 180° it will reach the top of its circular path and the maximum of its gravitational potential energy. When the car surmounts the rut, it will be on the horizontal road surface in a region of constant potential energy. Any increase of total energy above these values will therefore give rise to *unbound* motion; the swing will turn over and over and the car will continue to move away from the rut.

Although these considerations set a finite limit to the amplitude of the oscillatory motion that is possible in such cases, it should still be possible, according to the preceding analysis, to reach these limits however small the resonant driving force. This prediction indicates a second failing of the analysis, since we often find that while the motion does build up at resonance, the amplitude will only reach a steady value that is below such a limit; one man may be enough to set a large car in oscillation, but if the rut is deep the oscillations may never be great enough for the car to be freed. To see how this second limitation occurs, let us take another look at free oscillations.

3.3. Damped oscillations

In Section 3.1.3 it was shown that the total energy of simple harmonic motion is constant and proportional to the square of the amplitude. Any diminution in amplitude must therefore represent a loss in energy. Now, in practice, apparently free oscillations do always decrease in amplitude with time, and to represent such motion correctly it is necessary to modify the equation of motion (3.1) in a way that will describe a continuing drain of energy from the system.

This requires the addition of a *damping force*, acting in such a way as always to slow down the particle and therefore diminish its kinetic energy. This is quite different from the spring force which we used to typify the simple harmonic motion of Section 3.1. That force sometimes slows down the particle, thereby extracting energy from it, but this is returned later in the cycle when the force speeds up the particle.

A simple description of a force which always acts against the velocity, v, is $-\alpha v$, where α is a positive constant. (See Section 3.5.6, page 74.) When this is added to the simple harmonic force, the equation of motion becomes

$$m\frac{\mathrm{d}^2x}{\mathrm{d}t^2} + \alpha\frac{\mathrm{d}x}{\mathrm{d}t} + Kx = 0.$$

The complex form of this, corresponding to equation (3.10), is then

$$\frac{\mathrm{d}^2z}{\mathrm{d}t^2} + 2\gamma\frac{\mathrm{d}z}{\mathrm{d}t} + \omega_0^2 z = 0, \tag{3.22}$$

where

$$\omega_0 = \left(\frac{K}{m}\right)^{1/2}$$

is the angular frequency of the oscillation in the absence of any damping force and

$$\gamma = \frac{\alpha}{2m}$$

is a positive constant of dimension T^{-1} introduced by that damping force.

Substituting the trial solution,

$$z = C\,e^{pt}, \tag{3.23}$$

we have

$$(p^2 + 2\gamma p + \omega_0^2)z = 0.$$

Thus there will be a physically significant solution ($z \neq 0$) only if

$$p^2 + 2\gamma p + \omega_0^2 = 0;$$

that is, if

$$p = -\gamma \pm (\gamma^2 - \omega_0^2)^{1/2}.$$

For a small damping force,

$$\gamma < \omega_0,$$

we can define an angular frequency

$$\omega_1^2 = \omega_0^2 - \gamma^2,$$

and then the solution (3.23) becomes

$$z = C\,e^{-\gamma t}\,e^{j\omega_1 t}, \tag{3.24}$$

the real part of which represents the oscillator motion. It is reasonable to speak of an oscillator still because of the factor $\exp(j\omega_1 t)$, which has a frequency ω_1 somewhat less than that of the undamped oscillations, ω_0. The important difference from the undamped case is the amplitude $C\exp(-\gamma t)$ which decays with time as shown in Figure 3.2. $1/\gamma$ has the dimension T and is known as the decay time; over any interval of time $1/\gamma$ the amplitude diminishes by the factor $1/e$.

As the damping increases, γ increases, and the oscillations die away more rapidly until when

$$\gamma = \omega_0, \qquad \omega_1 = 0,$$

the condition of *critical damping* is reached at which no oscillations occur. It is inadequate simply to substitute

$$\omega_1 = 0$$

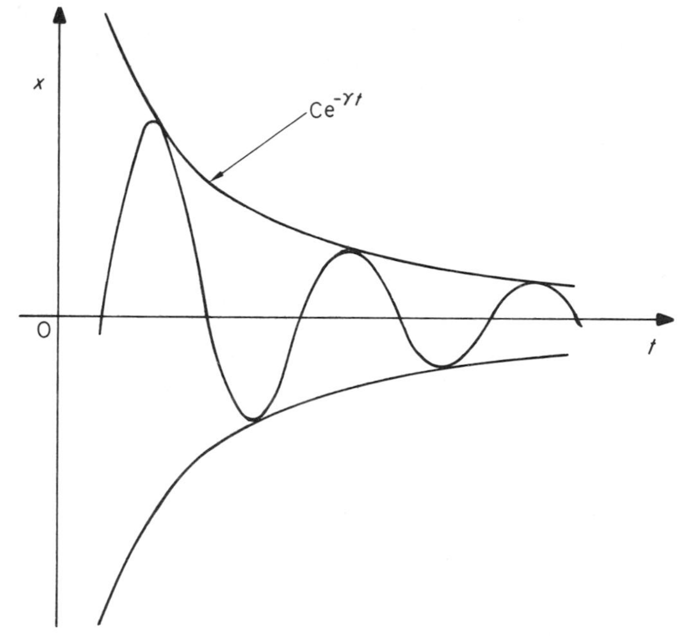

Figure 3.2

in the solution (3.24), for it would then contain only one adjustable constant, Re (C). Physically two initial conditions can be specified and this requires two constants. In this case, as can be shown by substitution, the general solution is

$$z = (C + Dt)\,e^{-\gamma t}. \tag{3.25}$$

A critically damped 'oscillator' allowed to move from a displaced position at rest back towards equilibrium will have the motion shown in Figure 3.3. (See Section 3.5.7, page 74.) When

$$\gamma > \omega_0$$

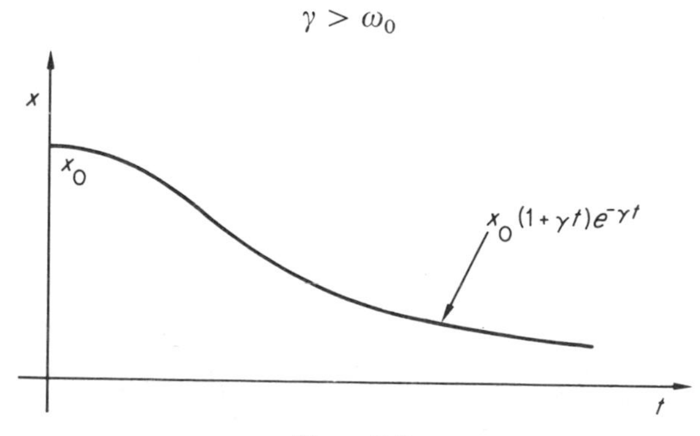

Figure 3.3

there is *over-damping* and the general solution becomes

$$z = A\,\mathrm{e}^{-(\gamma-\beta)t} + B\,\mathrm{e}^{-(\gamma+\beta)t},$$

where

$$\beta = +(\gamma^2 - \omega_0^2)^{1/2}.$$

The first term now has a time constant larger than $1/\gamma$. Thus, increasing the damping above critical increases the time the system takes to return to equilibrium.

3.4. Forced damped oscillations

We saw that the description of forced oscillations given in Section 3.2 was unrealistic when the forcing angular frequency, ω, approached, or became equal to, the natural angular frequency, ω_0. It can be made more realistic by adding a damping force term to equation (3.16). The equation of motion then becomes

$$\frac{\mathrm{d}^2 z}{\mathrm{d}t^2} + 2\gamma\frac{\mathrm{d}z}{\mathrm{d}t} + \omega_0^2 z = B\,\mathrm{e}^{j\omega t}. \tag{3.26}$$

If this is satisfied by oscillations of the forcing frequency,

$$z' = D\,\mathrm{e}^{j\omega t},$$

we find, on substitution, that their amplitude will be

$$D = \frac{B}{\omega_0^2 - \omega^2 + 2j\gamma\omega}.$$

An argument similar to that of Section 3.3 shows that free damped oscillations must be added to give the general solution

$$z = C\,\mathrm{e}^{-\gamma t}\,\mathrm{e}^{j\omega_1 t} + \frac{B}{\omega_0^2 - \omega^2 + 2j\gamma\omega}\,\mathrm{e}^{j\omega t}, \tag{3.27}$$

provided the damping is small enough to satisfy

$$\gamma < \omega_0.$$

When, in Section 3.3, damping was neglected, both parts of the motion could be considered to have equal importance. Now, however, although the first must be included to allow for the variety of ways in which the oscillations may be started, this motion is only *transient*, for its amplitude decays with the time constant $1/\gamma$. For times much greater than this the transient motion becomes negligible, and the motion settles down, however it is started, to the *steady state*:

$$z = \frac{B}{\omega_0^2 - \omega^2 + 2j\gamma\omega}\,\mathrm{e}^{j\omega t} = \frac{F}{m(\omega_0^2 - \omega^2 + 2j\gamma\omega)}. \tag{3.28}$$

The corresponding term in the undamped case is

$$z = \frac{B}{\omega_0^2 - \omega^2} e^{j\omega t} = \frac{F}{m(\omega_0^2 - \omega^2)}.$$

Thus z is always linked to F in the latter case by the *real* factor $1/[m(\omega_0^2 - \omega^2)]$. For small values of ω, $\omega < \omega_0$, this is positive and the oscillations are in phase with the force; for large values, $\omega > \omega_0$, the factor is negative and the oscillations are in antiphase to the force. Between these, at $\omega = \omega_0$, occurs the result (3.21)—unrealistic oscillations of infinitely increasing amplitude.

In our present case the linking factor is the *complex* quantity $1/[m(\omega_0^2 - \omega^2 + 2j\gamma\omega)]$. For ω very small or very large compared with ω_0 the situation is practically the same as before, but the phase now changes continuously in the intermediate region. If we write

$$\frac{1}{m(\omega_0^2 - \omega^2 + 2j\gamma\omega)} = G\,e^{j\phi}; \qquad G \text{ real,} \tag{3.29}$$

then

$$G = \frac{1}{m[(\omega_0^2 - \omega^2)^2 + 4\gamma^2\omega^2]^{1/2}} \tag{3.30}$$

and

$$\tan\phi = -\frac{2\gamma\omega}{\omega_0^2 - \omega^2}. \tag{3.31}$$

(See Section 3.5.8, page 76.)

The form of G, which gives the amplitude of the oscillation, and of ϕ, which gives its phase, are shown in Figure 3.4. The phenomenon of resonance is still evident, but now, more realistically, with finite amplitude:

$$G_{\text{max}} = \frac{1}{2m\gamma(\omega_0^2 - \gamma^2)^{1/2}}, \tag{3.32}$$

which occurs when

$$\omega^2 = \omega_0^2 - 2\gamma^2 \tag{3.33}$$

and

$$\tan\phi = -\frac{(\omega_0^2 - 2\gamma^2)^{1/2}}{\gamma}. \tag{3.34}$$

The smaller γ is compared with ω_0, the sharper the resonance curve, the greater its peak value, and the narrower the region over which ϕ changes

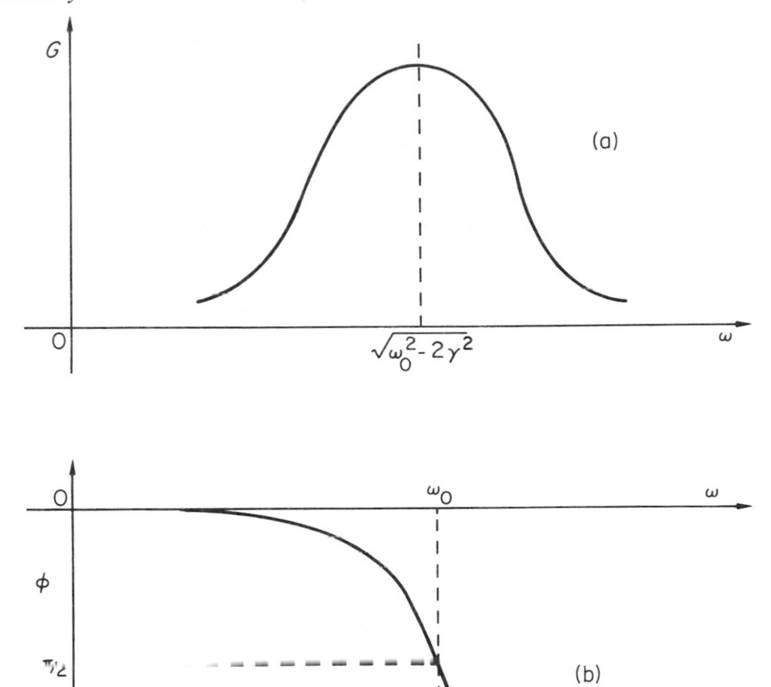

Figure 3.4

from 0 to $-\pi$. For very small damping resonance occurs at the approximate values

$$\omega \simeq \omega_0 \quad \text{and} \quad \phi \simeq -\frac{\pi}{2}.$$

We saw in Section 3.1.3 that the energy stored in an oscillator is proportional to the square of the amplitude. A convenient way of describing the sharpness of the resonance peak is to measure its width at the *half-energy points* (i.e. where its amplitude is $1/\sqrt{2}$ that of the amplitude at resonance). For small values of γ this occurs at

$$\omega = \omega_0 \pm \gamma,$$

which gives 2γ as the width of the peak.

Conditions of critical damping and overdamping, similar to those of Section 3.3, can occur here too, for the values

$$\gamma = \omega_0 \quad \text{and} \quad \gamma > \omega_0,$$

and these will change the form of the first term in equation (3.27). To fit given initial conditions the appropriate form must be used. However, in all cases such initially determined motion is only transient, and the steady-state motion described by equation (3.28) is still the ultimate result. (See Section 3.5.9, page 79.)

3.5. Comments and worked examples

3.5.1. See Section 3.1

Example 6

A spring of natural length l and strength constant S is fixed at its upper end. Attached to the lower end is a mass m, acted upon by the earth's gravity. Show that when the (stretched) length of the spring is X, the total potential energy may be written as

$$\mathscr{V} = \tfrac{1}{2}S(X - l)^2 - mgX.$$

Use this result to determine the position of equilibrium of the mass and the period of oscillation about this. Show that the same results are obtained by starting with the (force) equation of motion.

(a) The spring force $-S(X - l)$ (in the X-direction, downwards) is conservative, derivable from the potential energy stored in the spring:

$$\mathscr{V}_s = -S\int_X^{X_0} (X' - l)\,dX' = \tfrac{1}{2}S(X - l)^2 + \text{constant}. \tag{3.35}$$

The gravitational force mg is also conservative, the corresponding potential energy of the gravitational field being

$$\mathscr{V}_g = mg\int_X^{X_0} dX' = -mgX + \text{constant}. \tag{3.36}$$

Note that this expression for $\mathscr{V}_g$ is only true when a particle keeps sufficiently near to a particular point of the earth's surface for the gravitational force to remain *unchanged in direction* and to have the *constant magnitude mg*. Adding equations (3.35) and (3.36), and choosing a suitable value of X_0 to give zero for the constant term,

$$\mathscr{V} = \mathscr{V}_s + \mathscr{V}_g = \tfrac{1}{2}S(X - l)^2 - mgX,$$

as shown in Figure 3.5.

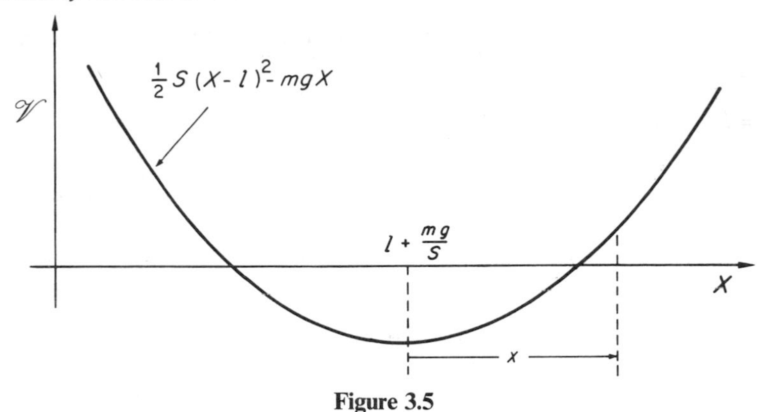

Figure 3.5

Now

$$\frac{d\mathscr{V}}{dX} = S(X - l) - mg, \qquad \frac{d^2\mathscr{V}}{dX^2} = S.$$

Hence, the equilibrium position X_m is given by

$$\left(\frac{d\mathscr{V}}{dX}\right)_{X_m} = S(X_m - l) - mg = 0, \qquad X_m = l + \frac{mg}{S}. \tag{3.37}$$

The potential energy for small displacements,

$$x = X - X_m,$$

about this will be

$$\mathscr{V} = \frac{1}{2}\left(\frac{d^2\mathscr{V}}{dX^2}\right)_{X_m} x^2 + \text{constant} = \tfrac{1}{2}Sx^2 + \text{constant}.$$

Conservation of energy may therefore be written as

$$\frac{1}{2}m\left(\frac{dx}{dt}\right)^2 + \frac{1}{2}Sx^2 = \text{constant}. \tag{3.38}$$

This shows that the motion is simple harmonic, with angular frequency ω_0 and period τ given by

$$\omega_0 = \left(\frac{S}{m}\right)^{1/2}, \qquad \tau = \frac{2\pi}{\omega_0} = 2\pi\left(\frac{m}{S}\right)^{1/2}.$$

(b) The total force in the X-direction (downwards) is

$$F(X) = -S(X - l) + mg \tag{3.39}$$

and the equation of motion is therefore

$$m\frac{d^2X}{dt^2} = -S(X - l) + mg. \tag{3.40}$$

Equilibrium will be at the position X_m at which the force $F(X)$ vanishes. From equation (3.39) this is given by

$$X_m = l + \frac{mg}{K},$$

which agrees with the result (3.37). By substituting the displacement x relative to this,

$$x = X - X_m,$$

into equation (3.39), the equation of motion becomes

$$m\frac{d^2x}{dt^2} = -Sx. \tag{3.41}$$

Comparing equations (3.38) and (3.41) we see that they both describe simple harmonic motion with the same period.

3.5.2. See Section 3.1

It would in principle be possible to have a potential energy of the form

$$\mathscr{V} = \alpha x^n; \qquad \alpha > 0, \ n = 4, 6, 8, \ldots,$$

which has a (physical) minimum at the origin. Motion about this would be bound and oscillatory, but not simple harmonic. In practice, such cases are unusual.

3.5.3. See Section 3.1.1

Example 7

Show that the inclusion of both solutions,

$$z_1 = C\,e^{j\omega_0 t}, \qquad z_2 = D\,e^{-j\omega_0 t},$$

of the equation of simple harmonic motion,

$$\frac{d^2z}{dt^2} + \omega_0^2 z = 0, \tag{3.42}$$

gives a physical result no more general than that of either result on its own.

The most general *mathematical* solution of equation (3.42) is

$$z = C\,e^{j\omega_0 t} + D\,e^{-j\omega_0 t} \tag{3.43}$$

where C and D are arbitrary *complex* constants. We note, firstly, that if any complex quantity, w, is written as the sum of its real and imaginary parts,

$$w = u + jv; \qquad u, v \text{ real},$$

then these may be found from

$$\left.\begin{aligned} w + w^* &= (u + jv) + (u - jv) = 2u \\ w - w^* &= (u + jv) - (u - jv) = 2jv \end{aligned}\right\} \qquad (3.44)$$

and

where w^* is the complex conjugate of w.

Thus if x is the real part of z and is used to describe the physical motion, equations (3.44) enable us to express it as

$$\begin{aligned} x &= \tfrac{1}{2}(z + z^*) \\ &= \tfrac{1}{2}[(C + D^*)\,e^{j\omega_0 t} + (C^* + D)\,e^{-j\omega_0 t}] \\ &= \tfrac{1}{2}[(C + D^*)(\cos\omega_0 t + j\sin\omega_0 t) + (C^* + D)(\cos\omega_0 t - j\sin\omega_0 t)] \\ &= \tfrac{1}{2}(C + C^* + D + D^*)\cos\omega_0 t + \tfrac{1}{2}j(C - C^* + D - D^*)\sin\omega_0 t. \qquad (3.45) \end{aligned}$$

Now C and D are arbitrary complex constants. Equations (3.44) show that $C + C^*$ and $D + D^*$ will therefore be (arbitrary) real constants, while $C - C^*$ and $D - D^*$ are (arbitrary) imaginary constants. Thus if we define a', b' by

$$a' = \tfrac{1}{2}(C + C^* + D + D^*), \qquad b' = \tfrac{1}{2}j(C - C^* + D - D^*),$$

these will be two arbitrary real constants which will allow us to fit the solution

$$x = a'\cos\omega_0 t + b'\sin\omega_0 t$$

to two physical conditions.

If, however, we take x to be the real part of only the solution z_1, then

$$\begin{aligned} x &= \tfrac{1}{2}(z_1 + z_1^*) \\ &= \tfrac{1}{2}\{C\,e^{j\omega_0 t} + C^*\,e^{-j\omega_0 t}\} \\ &= \tfrac{1}{2}(C + C^*)\cos\omega_0 t + \tfrac{1}{2}j(C - C^*)\sin\omega_0 t. \qquad (3.46) \end{aligned}$$

Although the expression (3.46) is less complicated than (3.45) it still contains two arbitrary real constants,

$$a = \tfrac{1}{2}(C + C^*), \qquad b = \tfrac{1}{2}j(C - C^*),$$

and the corresponding solution,

$$x = a\cos\omega_0 t + b\sin\omega_0 t,$$

can equally well be fitted to two physical conditions. A similar result can be obtained by starting with the solution z_2.

3.5.4. See Section 3.1.1

Example 8

A particle at rest at a position of stable equilibrium is suddenly given a velocity V at time zero and thereafter oscillates with simple harmonic motion of angular frequency ω_0. Determine the constants in each of the two descriptions

$$x = a \cos \omega_0 t + b \sin \omega_0 t \tag{3.47}$$

and

$$z = C\, e^{j\omega_0 t}, \tag{3.48}$$

and show that they represent the same motion.

The corresponding velocities are

$$\frac{dx}{dt} = -a\omega_0 \sin \omega_0 t + b\omega_0 \cos \omega_0 t \tag{3.49}$$

and

$$\frac{dz}{dt} = j\omega_0 C\, e^{j\omega_0 t}. \tag{3.50}$$

We are given *two* physical conditions,

$$x = 0 \quad \text{and} \quad \frac{dx}{dt} = V \quad \text{at} \quad t = 0. \tag{3.51}$$

(a) Using the solutions (3.47) and (3.49),

$$a = 0, \qquad V = b\omega_0,$$

thus fixing the solution as

$$x = \frac{V}{\omega_0} \sin \omega_0 t. \tag{3.52}$$

(b) Remembering that the real parts of z and of dz/dt are physically significant, the two given conditions at $t = 0$ imply, for the solutions (3.48) and (3.50),

$$\tfrac{1}{2}(z + z^*)_{t=0} = \tfrac{1}{2}(C + C^*) = 0$$

$$\frac{1}{2}\left(\frac{dz}{dt} + \frac{dz^*}{dt}\right)_{t=0} = \tfrac{1}{2}(j\omega_0 C - j\omega_0 C^*) = V$$

or

$$\tfrac{1}{2}(C - C^*) = -\frac{jV}{\omega_0}.$$

Hence

$$C = -\frac{jV}{\omega_0},$$

thus fixing the complex form of the solution as

$$z = -\frac{jV}{\omega_0} e^{j\omega_0 t} = -\frac{jV}{\omega_0}(\cos \omega_0 t + j \sin \omega_0 t). \tag{3.53}$$

The real part of this is $(V/\omega_0) \sin \omega_0 t$, in agreement with equation (3.52).

3.5.5. See Section 3.1.3

Example 9

Verify by direct calculation that the oscillatory motion described (a) by equation (3.8) and (b) by equations (3.11) and (3.12) conserves total energy.

(a) If

$$x = A \cos (\omega_0 t + \alpha)$$

then

$$\frac{dx}{dt} = -\omega_0 A \sin (\omega_0 t + \alpha).$$

Hence the total energy is

$$\mathscr{E} = \mathscr{T} + \mathscr{V} = \frac{1}{2} m \left(\frac{dx}{dt}\right)^2 + \frac{1}{2} K x^2$$

$$= \tfrac{1}{2} A^2 [m\omega_0^2 \sin^2 (\omega_0 t + \alpha) + K \cos^2 (\omega_0 t + \alpha)].$$

From equation (3.6)

$$m\omega_0^2 = K. \tag{3.54}$$

Hence

$$\mathscr{E} = \tfrac{1}{2} K A^2 = \tfrac{1}{2} m\omega_0^2 A^2 = \text{constant}.$$

(b) If

$$z = C\, e^{j\omega_0 t}$$

then

$$x = \text{Re}\,(z) = \tfrac{1}{2}(z + z^*) = \tfrac{1}{2}(C\, e^{j\omega_0 t} + C^*\, e^{-j\omega_0 t})$$

$$\frac{dx}{dt} = \text{Re}\left(\frac{dz}{dt}\right) = \frac{1}{2}\left(\frac{dz}{dt} + \frac{dz^*}{dt}\right) = \tfrac{1}{2}(j\omega_0 C\, e^{j\omega_0 t} - j\omega_0 C^*\, e^{-j\omega_0 t}).$$

Hence the potential and kinetic energies are

$$\mathscr{V} = \tfrac{1}{2} K x^2 = \tfrac{1}{8} K (C^2\, e^{2j\omega_0 t} + C^{*2}\, e^{-2j\omega_0 t} + 2CC^*)$$

$$\mathscr{T} = \frac{1}{2} m \left(\frac{dx}{dt}\right)^2 = -\tfrac{1}{8} m\omega_0^2 (C^2\, e^{2j\omega_0 t} + C^{*2}\, e^{-2j\omega_0 t} - 2CC^*).$$

Adding these and using the result (3.54),

$$\mathscr{E} = \mathscr{T} + \mathscr{V} = \tfrac{1}{2}KCC^* = \tfrac{1}{2}K|C|^2 = \text{constant}.$$

3.5.6. See Section 3.3

There are many other types of force which would continuously drain energy from the system. The particular force, $F = -\alpha v$, has the property that when acting alone upon the particle it extracts energy at a rate proportional to the kinetic energy. For

$$\frac{d\mathscr{T}}{dt} = \frac{d}{dt}\left[\frac{1}{2}m\left(\frac{dx}{dt}\right)^2\right] = m\frac{d^2x}{dt^2}\frac{dx}{dt} = F\frac{dx}{dt}$$

$$= -\alpha\left(\frac{dx}{dt}\right)^2 = -\frac{2\alpha}{m}\mathscr{T}.$$

Conversely, since

$$\mathscr{T} = \frac{1}{2}m\left(\frac{dx}{dt}\right)^2 \quad \text{and} \quad \frac{d\mathscr{T}}{dt} = F\frac{dx}{dt},$$

then any process by which $d\mathscr{T}/dt$ is maintained negative and proportional to $\mathscr{T}$,

$$\frac{d\mathscr{T}}{dt} = -4\gamma\mathscr{T}, \tag{3.55}$$

say, is equivalent to a force

$$F = -2m\gamma\frac{dx}{dt} = -\alpha v.$$

3.5.7. See Section 3.3

Example 10

Attached to a linear oscillator of mass m and natural frequency f is a single metal loop of resistance R and negligible mass. A length l of this loop is within, and perpendicular to, a constant field of induction B between the poles of a magnet, and is also perpendicular to the oscillator motion, x, as shown in Figure 3.6.

Show that this *electromagnetic damping* is equivalent to a damping force of the form $-\alpha(dx/dt)$ and that the resistance necessary to achieve critical damping is

$$R = \frac{l^2B^2}{4\pi mf}.$$

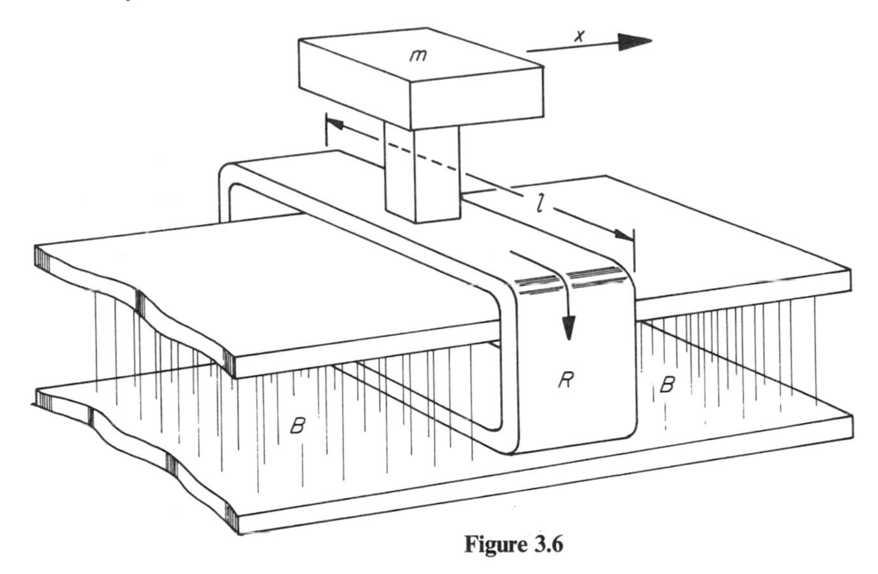

Figure 3.6

Calculate this resistance for the values $l = 1$ cm $= 0{\cdot}01$ m, $B = 1000$ Gauss $= 0{\cdot}1$ Tesla (Weber m^{-2}), $m = 100$ g $= 0{\cdot}1$ kg, and $f = 4\,\mathrm{s}^{-1}$, and comment on the order of magnitude of the result.

The rate at which lines of induction are cut is $Bl(\mathrm{d}x/\mathrm{d}t)$, and this will be the E.M.F. generated in the loop. The current in the loop will therefore be $Bl(\mathrm{d}x/\mathrm{d}t)/R$, and since this passes along a length l perpendicular to the field B, the force exerted on the loop will be

$$F = -\frac{B^2l^2(\mathrm{d}x/\mathrm{d}t)}{R} = -\alpha\frac{\mathrm{d}x}{\mathrm{d}t},$$

where

$$\alpha = \frac{B^2l^2}{R}.$$

The corresponding damping constant is

$$\gamma = \frac{\alpha}{2m} = \frac{B^2l^2}{2mR}. \tag{3.56}$$

Alternatively, we may say that the E.M.F. $Bl(\mathrm{d}x/\mathrm{d}t)$ will give a power dissipation (in resistance heating) of amount $B^2l^2(\mathrm{d}x/\mathrm{d}t)^2/R$, and since this loss of energy can come only from the kinetic energy, $\mathscr{T}$, of the oscillator,

$$\frac{\mathrm{d}\mathscr{T}}{\mathrm{d}t} = -\frac{B^2l^2(\mathrm{d}x/\mathrm{d}t)^2}{R}. \tag{3.57}$$

The kinetic energy itself is

$$\mathscr{T} = \frac{1}{2}m\left(\frac{\mathrm{d}x}{\mathrm{d}t}\right)^2. \qquad (3.58)$$

Thus kinetic energy and its rate of loss are proportional to each other. In conformity with equation (3.55),

$$\frac{\mathrm{d}\mathscr{T}}{\mathrm{d}t} = -4\gamma\mathscr{T},$$

where, from equations (3.57) and (3.58),

$$\gamma = \frac{B^2l^2}{2mR}, \qquad (3.59)$$

in agreement with the previous derivation. Critical damping occurs when

$$\gamma = \omega_0.$$

Using equation (3.59), the corresponding resistance is

$$R = \frac{B^2l^2}{2m\omega_0} = \frac{B^2l^2}{4\pi mf}.$$

For the given values,

$$R = \frac{10^{-5}}{16\pi} \simeq 2 \times 10^{-7}\,\Omega. \qquad (3.60)$$

Such a very low value demands in practice a complete 'short'. In fact, if the length of the whole loop were 4 cm, and if it were made of copper with resistivity $1{\cdot}7 \times 10^{-8}\,\Omega$m, the cross-section of the wire would need to have the impracticably large area of over 30 cm^2 to achieve the value (3.60).

3.5.8. See Section 3.4

Example 11

Discuss the physical significance of the steady-state motion (a) at very low driving frequencies, (b) at very high driving frequencies, of a system similar to that of Figure 3.1 but with a small damping force added.

When the damping term is added the equation of motion is

$$\frac{\mathrm{d}^2x}{\mathrm{d}t^2} + 2\gamma\frac{\mathrm{d}x}{\mathrm{d}t} + \omega_0^2x = \frac{SX_0}{m}\cos\omega t,$$

which is the real part of

$$\frac{\mathrm{d}^2z}{\mathrm{d}t^2} + 2\gamma\frac{\mathrm{d}z}{\mathrm{d}t} + \omega_0^2z = B\,\mathrm{e}^{j\omega t}; \qquad B = \frac{SX_0}{m}.$$

After the transient motion has died away the steady-state motion (equation 3.28) remains:

$$z = \frac{B}{\omega_0^2 - \omega^2 + 2j\gamma\omega} e^{j\omega t}.$$

(a) When the driving frequency ω is very small the predominant term in the denominator (remembering that the damping, represented by γ, is also small) is ω_0^2. Thus

$$z \to \frac{B}{\omega_0^2} e^{j\omega t} = \frac{SX_0}{m\omega_0^2} e^{j\omega t} = X_0 e^{j\omega t},$$

since

$$\omega_0^2 = \frac{S}{m}.$$

Hence the motion, x, of the particle is given by

$$x = \mathrm{Re}\,(z) \to X_0 \cos \omega t \quad \text{as} \quad \omega \to 0,$$

which is the same as that of the driven end of the spring.

(b) When ω is very large the predominant term in the denominator is $-\omega^2$. Thus

$$z \to -\frac{B}{\omega^2} e^{j\omega t} = -\frac{\omega_0^2}{\omega^2} X_0 e^{j\omega t},$$

and the motion of the particle is given by

$$x = \mathrm{Re}\,(z) \to -\frac{\omega_0^2}{\omega^2} X_0 \cos \omega t \to 0 \quad \text{as} \quad \omega \to \infty.$$

So the particle remains stationary in the limit.

At very low frequencies there is time for the force to be transmitted through the spring, and it behaves effectively as a rigid rod, storing no energy. The particle moves in phase with, and directly controlled by, the external force which oscillates the end of the spring.

At very high frequencies the oscillatory force acting on the particle changes direction so rapidly that very little change in its momentum and therefore in its velocity or position is possible. Only the spring, and not the particle, shows response to the external force, alternately storing and releasing energy. The motion of the particle is vanishingly small and in antiphase to the external force (see Figure 3.7).

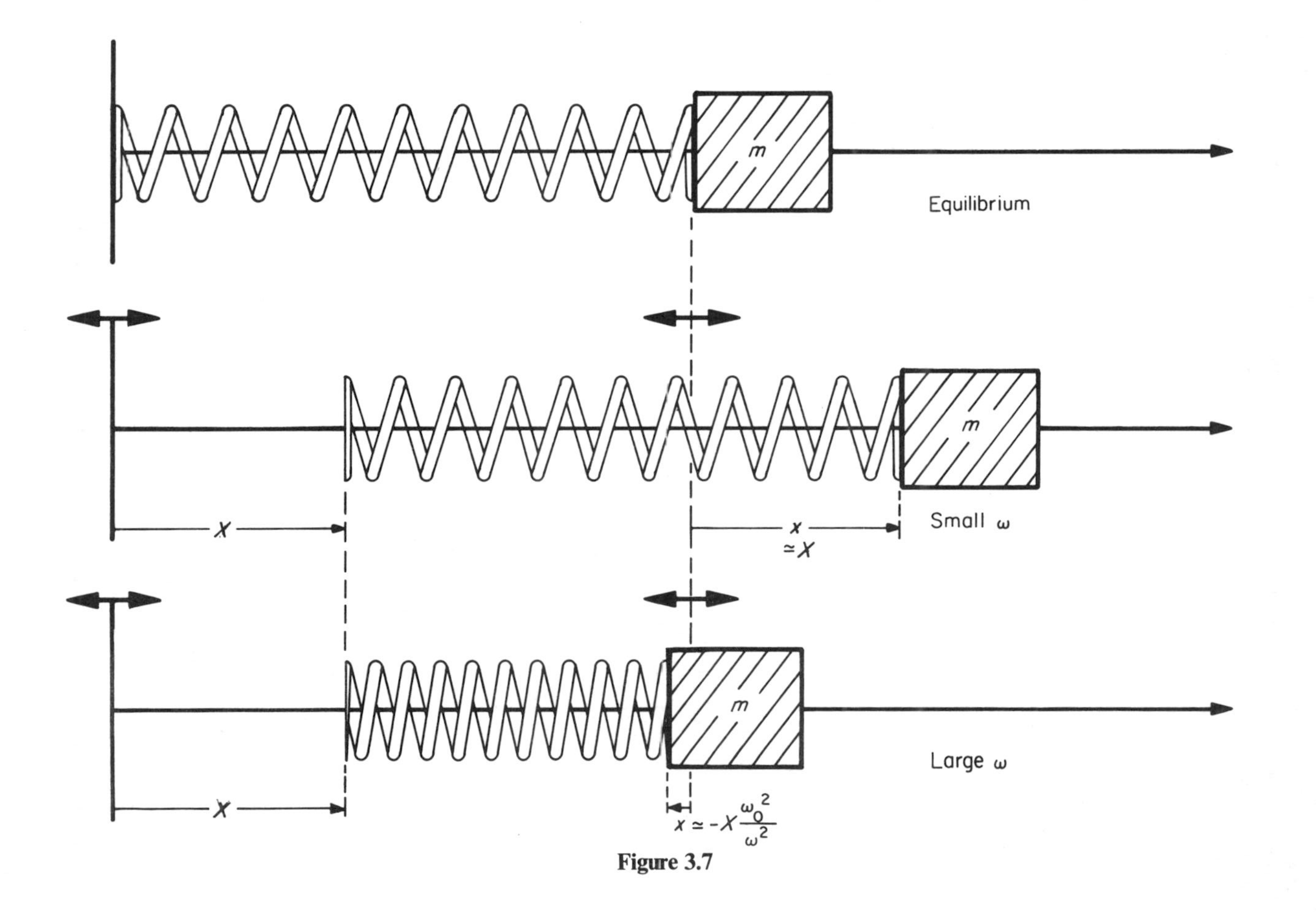

m
Equilibrium
m
Small ω
X
$x \simeq X$
m
Large ω
X
$x \simeq -X\frac{\omega_0^2}{\omega^2}$

Figure 3.7

3.5.9. See Section 3.4

Example 12

Discuss the steady-state motion when damping is large.

Whatever the value of the damping coefficient γ, the steady-state motion is that given by equations (3.28) to (3.31):

$$z = FG\,e^{j\phi},$$

where F is the driving force, of angular frequency ω, and G and ϕ determine the amplitude and phase of the motion according to

$$G = \frac{1}{m[(\omega_0^2 - \omega^2)^2 + 4\gamma^2\omega^2]^{1/2}}; \qquad \tan\phi = -\frac{2\gamma\omega}{\omega_0^2 - \omega^2}.$$

At very low driving frequencies

$$G \to \frac{1}{m\omega_0^2}, \qquad \phi \to 0, \qquad \text{as } \omega \to 0,$$

and the motion is of finite amplitude in phase with the driving force.

At very high frequencies

$$G \to \frac{1}{m\omega^2} \to 0, \qquad \phi \to -\pi, \quad \text{as } \omega \to \infty,$$

and the motion, in antiphase to the driving force, tends towards zero.

These results are similar to those of Example 11. However, as we can see from equation (3.33), when

$$\gamma = \sqrt{2}\omega_0,$$

the maximum value of G, indicating maximum response to the force, occurs at zero driving frequency ω, and for

$$\gamma > \sqrt{2}\omega_0$$

there is no real non-zero value of ω giving a maximum value for G. Thus for large damping, defined by

$$\gamma \geqslant \sqrt{2}\omega_0,$$

the response of the oscillator steadily diminishes as the forcing frequency increases, as shown in Figure 3.8a, and there is no resonance in the sense of that shown by Figure 3.4a when damping is small. The phase of the response, however, is similar in both cases, varying from 0 to $-\pi$ with the quadrature value, $-\pi/2$, occurring at the natural frequency, ω_0, as shown in Figure 3.8b.

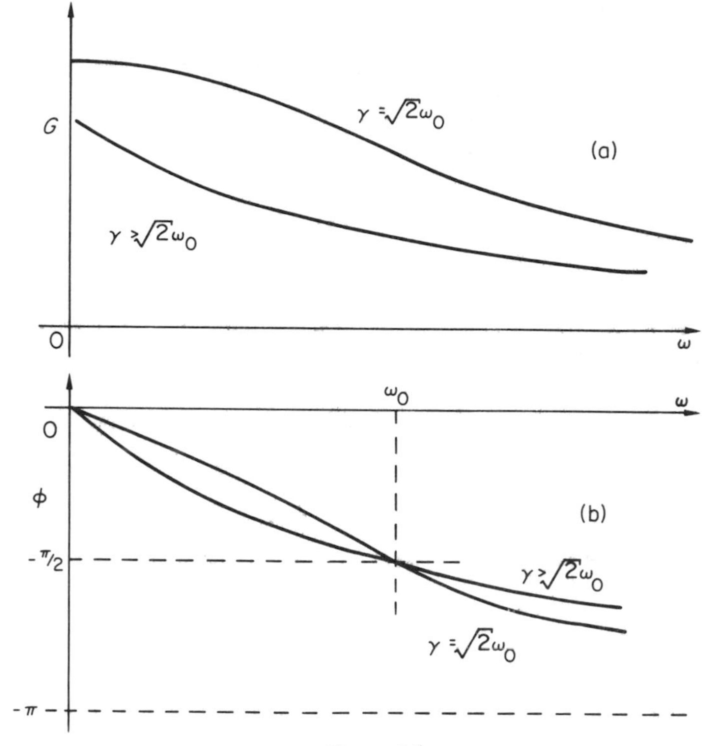

Figure 3.8

3.6. Problems

3.1. A spring of natural length b hangs vertically downwards from a fixed point. When a mass m is hung on it, it extends a further distance d in equilibrium. If the mass is now lifted a distance $d/2$ and released, calculate the period of the subsequent oscillations and the maximum velocity of the mass.

3.2. A particle of mass m oscillates sinusoidally about a fixed point O with angular frequency ω_0. Each time it passes O in the positive direction a force F is applied to the mass over a length Δx of its path. Show that the amplitude, x_0, increases according to the law

$$x_0^2 = \frac{F\,\Delta x}{\pi m \omega_0}(t - t_0).$$

3.3. Suppose that in the preceding problem the force is applied for a time Δt at each positive traversal of O. Show that x_0 now increases according to the law

$$x_0 = \frac{F\,\Delta t}{2\pi m}(t - t_0).$$

3.4. Show that if slight damping is added to the oscillator of problems 3.2 and 3.3, represented by the constant γ, then a steady amplitude, x_s, is reached in both cases, given in the first by

$$x_s^2 = \frac{F\,\Delta x}{2\pi m\omega_1\gamma},$$

and in the second by

$$x_s = \frac{F\,\Delta t}{2\pi\gamma m},$$

where $\omega_1^2 = \omega_0^2 - \gamma^2$.

3.5. Delicate apparatus is commonly protected from the vibration of the laboratory floor by mounting it on a heavy sprung base. If the floor starts oscillating sinusoidally with amplitude X and angular frequency ω at $t = 0$, show that the motion of the base about its static equilibrium position is

$$x = \frac{\omega_0 X}{\omega^2 - \omega_0^2}(\omega \sin \omega_0 t - \omega_0 \sin \omega t),$$

assuming that damping is negligible, where ω_0 is the natural angular frequency of the system.

Deduce from this that the greatest protection is afforded by choosing the smallest practicable value for ω_0, and that when $\omega_0 \ll \omega$ the maximum amplitude of vibration of the apparatus and its maximum acceleration will not exceed $\omega_0 X/\omega$ and $\omega_0^2 X$ respectively.

3.6. Suppose that in the preceding problem slight damping is introduced so that the free movement of the base would have as its equation of motion

$$\ddot{x} + 2\gamma\dot{x} + \omega_0^2 x = 0,$$

where

$$\gamma \ll \omega_0.$$

Show that the effect of this damping is noticeable only when the number of oscillations of the floor, n, becomes comparable with ω/γ, and determine the maximum amplitude of the vibration of the apparatus and its maximum acceleration when $n \gg \omega/\gamma$. In this case show that when $\omega_0 \ll \omega$ the maximum amplitude and acceleration are $\omega_0^2 X/\omega^2$ and $\omega_0^2 X$.

3.7. An additional force $F(t)$ is applied to a damped oscillator whose equation of motion for free oscillations is

$$m\ddot{x} + 2\gamma m\dot{x} + Kx = 0.$$

Show that the instantaneous power delivered to the oscillator by this force is

$$Q = \frac{d\mathscr{E}}{dt} + 4\gamma\mathscr{T},$$

where $\mathscr{T}$ is the kinetic energy of the mass m and $\mathscr{E}$ is the total energy of the oscillator. Hence prove that if the force is such as to maintain steady-state oscillations of constant amplitude, but not necessarily sinusoidal with time, then

$$\langle Q \rangle = 4\gamma \langle \mathscr{T} \rangle,$$

where $\langle\ \rangle$ denotes the time average over one cycle of oscillation.

3.8. A mass of 1 kg is attached to a spring on which, without damping, it would oscillate with a period of 1 s. A damping device is attached to the mass which absorbs its kinetic energy at a fractional rate $6\pi\ \mathrm{s}^{-1}$. The mass is initially at rest and an oscillatory force of magnitude $3\sqrt{2}\pi^2$ N and period 2 s is then applied to it, the phase of the force being such that at $t = 0$ the force is zero. Determine the subsequent motion of the mass.

3.9. An oscillator of natural angular velocity ω_0, at rest in its equilibrium position at time zero, is acted on by the force $F_0 \cos \omega t$. Show that equation (3.20) gives for the subsequent motion

$$x = \frac{F_0}{\omega^2 - \omega_0^2}(\cos \omega t - \cos \omega_0 t),$$

when $\omega \neq \omega_0$, and equation (3.21) gives

$$x = \frac{F_0}{2\omega_0} t \sin \omega_0 t,$$

when $\omega = \omega_0$. Describe the motion when ω and ω_0 are nearly equal and prove that in the limit $\omega \to \omega_0$ the two expressions for the motion agree.

Chapter 4

Three-dimensional Motion

By considering motion in one dimension we have gained some insight into methods by which Newton's laws can be used to analyse a problem. In particular, we have seen the importance of momentum, kinetic energy and potential energy. However, if our mechanics is to be more widely useful we must break away from the one-dimensional restriction and see how such concepts can be more generally applied and whether additional ones are called for.

4.1. Separable three-dimensional motion

The simplest and most common type of three-dimensional motion is that of a particle moving under gravity near the earth's surface. For example a ball of mass m may be thrown with a speed v_0 at an angle α to the horizontal. It will rise to some maximum height and then fall to earth, the trajectory being determined by the magnitude and direction of the initial velocity and by the gravitational force (we shall neglect any effects arising from air friction).

Let us take axes 1 and 2 in the earth's surface and axis 3 vertically upwards. The last choice ensures that the gravitational force has a non-zero component only in this direction:

$$F_1 = F_2 = 0, \qquad F_3 = -mg. \tag{4.1}$$

(See Section 4.10.1, page 104.) If we take the origin of the axes to coincide with the point from which the ball is projected, the initial position is given by

$$x_1 \equiv x_1(0) = 0, \qquad x_2 \equiv x_2(0) = 0, \qquad x_3 \equiv x_3(0) = 0. \tag{4.2}$$

Finally, if we choose the axis 2 so that the initial velocity $\mathbf{v}_0$ lies in the plane 2, 3, it will have only two non-zero components:

$$v_{01} \equiv \frac{\mathrm{d}x_1(0)}{\mathrm{d}t} = 0, v_{02} \equiv \frac{\mathrm{d}x_2(0)}{\mathrm{d}t} = v_0 \cos \alpha, v_{03} \equiv \frac{\mathrm{d}x_3(0)}{\mathrm{d}t} = v_0 \sin \alpha. \tag{4.3}$$

The problem is then that of a particle of mass m whose equations of motion are

$$\frac{d^2x_1}{dt^2} = 0, \qquad \frac{d^2x_2}{dt^2} = 0, \qquad \frac{d^2x_3}{dt^2} = -g, \tag{4.4}$$

and whose initial conditions are given by equations (4.2) and (4.3).

The fact that the equations for each coordinate do not involve the other coordinates enables us to solve each of them independently. The solutions are

$$x_1 = 0, \qquad x_2 = v_0 \cos \alpha\, t, \qquad x_3 = v_0 \sin \alpha\, t - \tfrac{1}{2}gt^2, \tag{4.5}$$

and from these any information about the behaviour of the particles can be calculated. (See Section 4.10.2, page 104.)

This is one example of those cases in which the components of force depend only on the corresponding coordinate,

$$m\frac{d^2x_1}{dt^2} = F_1(x_1), \qquad m\frac{d^2x_2}{dt^2} = F_2(x_2), \qquad m\frac{d^2x_3}{dt^2} = F_3(x_3). \tag{4.6}$$

For each coordinate there will be two initial conditions or their equivalents,

$$x_1(0) = x_{01}, \qquad x_2(0) = x_{02}, \qquad x_3(0) = x_{03} \tag{4.7}$$

and

$$\frac{dx_1(0)}{dt} = v_{01}, \qquad \frac{dx_2(0)}{dt} = v_{02}, \qquad \frac{dx_3(0)}{dt} = v_{03}. \tag{4.8}$$

We have, in effect, three separate one-dimensional problems which may be solved by the methods of the preceding chapters to give definite solutions for each coordinate:

$$x_1 \equiv x_1(t; x_{01}, v_{01}), \qquad x_2 \equiv x_2(t; x_{02}, v_{02}), \qquad x_3 \equiv x_3(t; x_{03}, v_{03}). \tag{4.9}$$

4.2. General three-dimensional motion

In many cases the motion is not separable. For example if a charge q is fixed at the origin of the frame of reference, a particle with charge q, whose position vector is $\mathbf{x}$ will experience a force

$$\mathbf{F} = \frac{1}{4\pi\varepsilon_0}\frac{q^2}{x^2}\hat{\mathbf{x}}, \tag{4.10}$$

where x is the magnitude of $\mathbf{x}$,

$$x = |\mathbf{x}|,$$

and $\hat{\mathbf{x}}$ is the unit vector in the direction of $\mathbf{x}$,

$$\hat{\mathbf{x}} = \frac{\mathbf{x}}{x}. \tag{4.11}$$

Using equation (4.11) to rewrite the force in the form

$$\mathbf{F} = \frac{q^2}{4\pi\varepsilon_0} \frac{\mathbf{x}}{x^3},$$

the equation of motion for the particle, with mass m, becomes

$$m\frac{\mathrm{d}^2\mathbf{x}}{\mathrm{d}t^2} = \mathbf{F} = \frac{q^2}{4\pi\varepsilon_0} \frac{\mathbf{x}}{x^3}. \tag{4.12}$$

In component form this is

$$m\frac{\mathrm{d}^2 x_1}{\mathrm{d}t^2} = F_1 = \frac{q^2}{4\pi\varepsilon_0} \frac{x_1}{(x_1^2 + x_2^2 + x_3^2)^{3/2}}$$

with two similar equations. More generally, when the force depends upon the (vector) position, $\mathbf{x}$, of the particle, the equation of motion,

$$m\frac{\mathrm{d}^2\mathbf{x}}{\mathrm{d}t^2} = \mathbf{F}(\mathbf{x}), \tag{4.13}$$

yields component equations of the form

$$m\frac{\mathrm{d}^2 x_1}{\mathrm{d}t^2} = F_1(x_1, x_2, x_3), \qquad m\frac{\mathrm{d}^2 x_2}{\mathrm{d}t^2} = F_2(x_1, x_2, x_3),$$

$$m\frac{\mathrm{d}^2 x_3}{\mathrm{d}t^2} = F_3(x_1, x_2, x_3). \tag{4.14}$$

Each of the force components F_1, F_2, F_3 may vary with all three of the coordinates x_1, x_2, x_3. Mathematically the problem is much more difficult than that expressed by equations (4.6), since the three differential equations are now linked with each other. There is usually no direct way of solving them. The complication is even greater when the force depends upon velocity and time.

However, the ideas introduced in Chapter 2 for one-dimensional motion have an extended scope in three dimensions which will often give descriptions and understanding that would be difficult or impossible to gain from an attempt to solve directly the equations of motion.

4.3. Impulse and momentum

Allowing the force to depend possibly upon the time t, and upon the position $\mathbf{x}$ and velocity $\mathbf{v}$ of the particle, as would be the case if the latter were charged and moving in a varying electromagnetic field, we may write the equation of motion as

$$\frac{\mathrm{d}\mathbf{p}}{\mathrm{d}t} = \mathbf{F}(\mathbf{x}, \mathbf{v}, t). \tag{4.15}$$

This may be integrated,

$$\int_{(1)}^{(2)} \mathrm{d}\mathbf{p} = \int_{(1)}^{(2)} \mathbf{F}(\mathbf{x}, \mathbf{v}, t)\, \mathrm{d}t$$

or

$$\mathbf{p}_{(2)} - \mathbf{p}_{(1)} = \mathbf{I}_{(1,2)}. \tag{4.16}$$

The left-hand side is the gain in the momentum vector of the particle, and $\mathbf{I}_{12}$ is the impulse vector,

$$\mathbf{I}_{(1,2)} = \int_{(1)}^{(2)} \mathbf{F}(\mathbf{x}, \mathbf{v}, t)\, \mathrm{d}t, \tag{4.17}$$

delivered by the force over the interval $t_{(1)}$ to $t_{(2)}$.

As in the one-dimensional case the impulse concept is particularly useful in collision processes. (See Section 4.10.3, page 105.)

4.4. Work and kinetic energy

In one dimension any small movement $\mathrm{d}x$ is necessarily either in the direction of the force F, in which case the force does a positive amount of work, $F\,\mathrm{d}x$, or it is in the opposite direction, in which case the work done is of the same magnitude but negative.

In three dimensions, however, both the movement $\mathbf{dx}$ and the force $\mathbf{F}$ are vectors and may have any orientation relative to each other. The definition of the work done in this more general case, expressing the 'distance' effect of the force introduced in Section 2.3 and according with the one-dimensional value, is the scalar product,

$$\mathrm{d}\mathscr{W} = \mathbf{F}\,.\,\mathbf{dx} = F_1\,\mathrm{d}x_1 + F_2\,\mathrm{d}x_2 + F_3\,\mathrm{d}x_3 = F\,\mathrm{d}x\cos\theta. \tag{4.18}$$

F and $\mathrm{d}x$ are the magnitudes of $\mathbf{F}$ and $\mathbf{dx}$ and θ is the angle between the two vectors. $\mathrm{d}\mathscr{W}$ may be thought of most simply as the product of the displacement, $\mathrm{d}x$, and of the component of the force, $F\cos\theta$, in the direction of the displacement. Using Newton's second law,

$$\begin{aligned}\frac{\mathrm{d}}{\mathrm{d}x_1}\left[\frac{m}{2}\left(\frac{\mathrm{d}x_1}{\mathrm{d}t}\right)^2\right] &= \frac{\mathrm{d}t}{\mathrm{d}x_1}\frac{\mathrm{d}}{\mathrm{d}t}\left[\frac{m}{2}\left(\frac{\mathrm{d}x_1}{\mathrm{d}t}\right)^2\right]\\ &= \frac{\mathrm{d}t}{\mathrm{d}x_1} m \frac{\mathrm{d}x_1}{\mathrm{d}t}\frac{\mathrm{d}^2x_1}{\mathrm{d}t^2}\\ &= m\frac{\mathrm{d}^2x_1}{\mathrm{d}t^2} = F_1.\end{aligned}$$

Hence

$$\mathrm{d}\left[\frac{m}{2}\left(\frac{\mathrm{d}x_1}{\mathrm{d}t}\right)^2\right] = F_1\,\mathrm{d}x_1.$$

Adding the similar expressions corresponding to the other components of the force we have

$$\mathrm{d}\left\{\frac{m}{2}\left[\left(\frac{\mathrm{d}x_1}{\mathrm{d}t}\right)^2 + \left(\frac{\mathrm{d}x_2}{\mathrm{d}t}\right)^2 + \left(\frac{\mathrm{d}x_3}{\mathrm{d}t}\right)^2\right]\right\} = F_1\,\mathrm{d}x_1 + F_2\,\mathrm{d}x_2 + F_3\,\mathrm{d}x_3. \quad (4.19)$$

If, therefore, we define the kinetic energy, $\mathscr{T}$, in three dimensions by

$$\begin{aligned}\mathscr{T} &= \frac{m}{2}\left[\left(\frac{\mathrm{d}x_1}{\mathrm{d}t}\right)^2 + \left(\frac{\mathrm{d}x_2}{\mathrm{d}t}\right)^2 + \left(\frac{\mathrm{d}x_3}{\mathrm{d}t}\right)^2\right] \\ &= \frac{m}{2}(v_1^2 + v_2^2 + v_3^2) \\ &= \frac{m}{2}\mathbf{v}\,.\,\mathbf{v} = \frac{m}{2}v^2, \end{aligned} \quad (4.20)$$

equations (4.18), (4.19) and (4.20) show that

$$\mathrm{d}\mathscr{T} = \mathbf{F}\,.\,\mathrm{d}\mathbf{x} = \mathrm{d}\mathscr{W}. \quad (4.21)$$

We saw in Sections 2.5 and 2.7.1 that in some circumstances it is possible to associate a potential energy with a one-dimensional force and to establish an energy conservation law. The change in kinetic energy as the particle moves from a point $x_{(1)}$ to another point $x_{(2)}$ is then uniquely determined.

In our present three-dimensional case the change in kinetic energy as the particle moves from a point $\mathbf{x}_{(1)}$ to another point $\mathbf{x}_{(2)}$ is obtained by integrating equation (4.21):

$$\mathscr{T}_{(2)} - \mathscr{T}_{(1)} = \int_{(1)}^{(2)} \mathrm{d}\mathscr{T} = \int_{\mathbf{x}(1)}^{\mathbf{x}(2)} \mathbf{F}\,.\,\mathrm{d}\mathbf{x}. \quad (4.22)$$

The integrand on the right-hand side of this equation is the amount of work done by the force $\mathbf{F}$ as the particle moves $\mathrm{d}\mathbf{x}$ along its path, and the integral is the sum of such contributions over the complete path from $\mathbf{x}_{(1)}$ to $\mathbf{x}_{(2)}$. Such an integral is called a *line integral.*

4.5. Potential energy

We would expect the integral (4.22) to depend upon the end points, just as in the one-dimensional case. The crucial question is 'Does the integral depend *only* on the end points?'. We saw in the one-dimensional case that if the answer is to be 'Yes' the work (represented by the integral) must be independent not only of time and of the particle velocity but also of *how* the particle moves between the end points. We could express this by saying that the work done is *independent of the path between them.* In one dimension this is a rather artificial idea since the only way of obtaining a different path between x and X_0 is by going backwards and forwards over some part of the same straight line, as described in Section 2.7.1.

In three dimensions, however, we can choose quite different paths between points (1) and (2), as in Figure 4.1a. To ensure that the change in kinetic

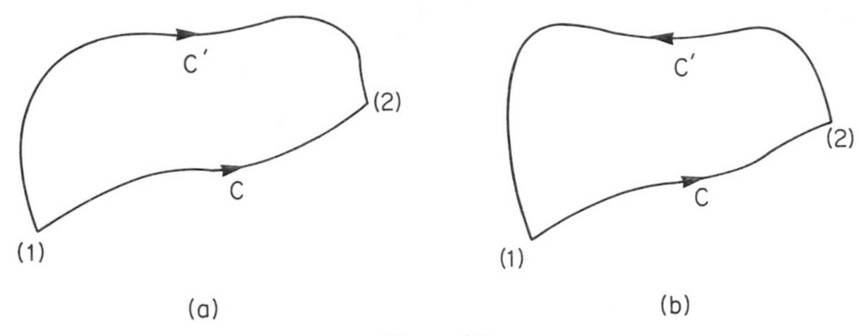

Figure 4.1

energy is always the same, the work done must therefore be independent of the path in the more general three-dimensional sense:

$$\int_{C\,(1)}^{(2)} \mathbf{F} \, . \, \mathbf{dx} = \int_{C'\,(1)}^{(2)} \mathbf{F} \, . \, \mathbf{dx}$$

for all paths C, C′ connecting (1) to (2). Then if C′ is traversed in the opposite direction,

$$\int_{C\,(1)}^{(2)} \mathbf{F} \, . \, \mathbf{dx} = -\int_{C'\,(2)}^{(1)} \mathbf{F} \, . \, \mathbf{dx}$$

or

$$\int_{C\,(1)}^{(2)} \mathbf{F} \, . \, \mathbf{dx} + \int_{C'\,(2)}^{(1)} \mathbf{F} \, . \, \mathbf{dx} = 0.$$

But, as we can see from Figure 4.1b, this expression represents a line integral carried round the closed path, C plus C′, from position (1) back to itself. We may write this condition as

$$\oint \mathbf{F} \, . \, \mathbf{dx} \equiv \oint (F_1 \, dx_1 + F_2 \, dx_2 + F_3 \, dx_3) = 0. \tag{4.23}$$

Since there was no restriction on C to begin with, the integral denoted by equation (4.23) refers to any closed path through (1) and (2). Finally, if we wish all positions of the particle to correspond to a definite kinetic energy we can interpret equation (4.23) as a line integral around *any closed path.* It is this property which is necessary in three dimensions to define a *conservative force* for which a potential energy can be found. This potential energy at a point $\mathbf{x}$ or (x_1, x_2, x_3) is then the work done by the force as the particle moves from $\mathbf{x}$ to some arbitrary fixed point $\mathbf{X}_0$ or (X_{01}, X_{02}, X_{03}) *along any path*:

$$\mathscr{V}(\mathbf{x}; \mathbf{X}_0) = \int_{\mathbf{x}}^{\mathbf{X}_0} \mathbf{F} \, . \, \mathrm{d}\mathbf{x}' = -\int_{\mathbf{X}_0}^{\mathbf{x}} \mathbf{F} \, . \, \mathrm{d}\mathbf{x}'. \tag{4.24}$$

As in one dimension, $\mathbf{X}_0$ may be chosen to give a convenient zero for $\mathscr{V}$, but its choice will not affect the difference in potential energy at two points $\mathbf{x}_{(1)}$ and $\mathbf{x}_{(2)}$:

$$\begin{aligned} \mathscr{V}_{(1)} - \mathscr{V}_{(2)} &= \mathscr{V}(\mathbf{x}_{(1)}; \mathbf{X}) - \mathscr{V}(\mathbf{x}_{(2)}; \mathbf{X}) \\ &= \int_{\mathbf{x}(1)}^{\mathbf{X}} \mathbf{F} \, . \, \mathrm{d}\mathbf{x} - \int_{\mathbf{x}(2)}^{\mathbf{X}} \mathbf{F} \, . \, \mathrm{d}\mathbf{x} \\ &= \int_{\mathbf{x}(1)}^{\mathbf{x}(2)} \mathbf{F} \, . \, \mathrm{d}\mathbf{x}. \end{aligned} \tag{4.25}$$

From equations (4.22) and (4.25)

$$\mathscr{T}_{(2)} + \mathscr{V}_{(2)} = \mathscr{T}_{(1)} + \mathscr{V}_{(1)} = \mathscr{E} = \text{constant}, \tag{4.26}$$

where $\mathscr{E}$ is the *total energy*. (See Section 4.10.4, page 106.)

To find the force from the potential energy let us take (1) as the point (x_1, x_2, x_3) and (2) as $(x_1 + \mathrm{d}x_1, x_2, x_3)$. Then

$$\begin{aligned} \mathscr{V}_{(1)} - \mathscr{V}_{(2)} &\equiv \mathscr{V}(x_1, x_2, x_3; \mathbf{X}_0) - \mathscr{V}(x_1 + \mathrm{d}x_1, x_2, x_3; \mathbf{X}_0) \\ &\simeq -\frac{\partial \mathscr{V}}{\partial x_1} \mathrm{d}x_1. \end{aligned} \tag{4.27}$$

If we choose as the path of the line integral between the two points the short straight line of length $\mathrm{d}x_1$ connecting them, then

$$\mathscr{V}_{(1)} - \mathscr{V}_{(2)} = \int_{\mathbf{x}(1)}^{\mathbf{x}(2)} \mathbf{F} \, . \, \mathrm{d}\mathbf{x} = F_1 \, \mathrm{d}x_1. \tag{4.28}$$

Comparing equations (4.27) and (4.28) in the limit $\mathrm{d}x_1 \to 0$ we have

$$F_1 = -\frac{\partial \mathscr{V}}{\partial x_1}.$$

In a similar way

$$F_2 = -\frac{\partial \mathscr{V}}{\partial x_2}, \qquad F_3 = -\frac{\partial \mathscr{V}}{\partial x_3}. \tag{4.29}$$

These three equations are summarized in the expression

$$\mathbf{F} = -\mathbf{grad}\, \mathscr{V}. \tag{4.30}$$

An alternative notation comes from regarding the three partial differentiations $\partial/\partial x_1, \partial/\partial x_2, \partial/\partial x_3$ as components of a vector operator $\mathbf{\nabla}$. In terms of this,

$$\mathbf{F} = -\mathbf{\nabla}\mathscr{V}. \tag{4.31}$$

(See Section 4.10.5, page 107.)

The surfaces in three-dimensional space,

$$\mathscr{V}(x_1, x_2, x_3) = \text{constant},$$

are called *equipotentials*. The direction of greatest variation in $\mathscr{V}$, and the direction therefore of the force at any point, is normal to the equipotential at that point.

A convenient way of describing the property (4.23) is to consider an infinitesimally small rectangle, parallel with the (2, 3) plane, of sides dx_2, dx_3 (Figure 4.2). If the centre of the rectangle is at (x_1, x_2, x_3) the work done by the force around the rectangle, traversed counterclockwise, is

$$\begin{aligned} d\mathscr{W}_1 = {} & F_2(x_1, x_2, x_3 - \tfrac{1}{2}dx_3)\,dx_2 + F_3(x_1, x_2 + \tfrac{1}{2}dx_2, dx_3) \\ & - F_2(x_1, x_2, x_3 + \tfrac{1}{2}dx_3)\,dx_2 - F_3(x_1, x_2 - \tfrac{1}{2}dx_2, dx_3). \end{aligned} \tag{4.32}$$

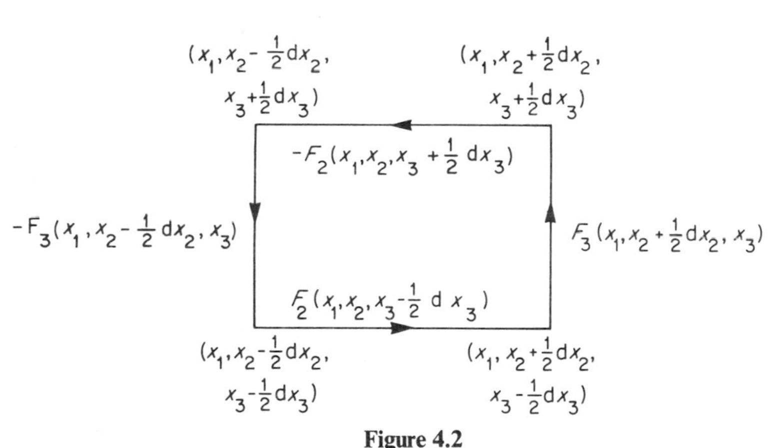

Figure 4.2

(Since the rectangle is infinitesimally small, we may assume that the component of force along each side is constant and equal to its value at the centre of that side.) Now

$$F_2(x_1, x_2, x_3 - \tfrac{1}{2}dx_3) = F_2(x_1, x_2, x_3) - \frac{1}{2}dx_3\frac{\partial F_2}{\partial x_3} + \ldots$$

$$F_2(x_1, x_2, x_3 + \tfrac{1}{2}dx_3) = F_2(x_1, x_2, x_3) + \frac{1}{2}dx_3\frac{\partial F_2}{\partial x_3} + \ldots$$

Substituting these and similar expressions for F_3 into equation (4.32) we find

$$\mathrm{d}\mathscr{W}_1 = \left(\frac{\partial F_3}{\partial x_2} - \frac{\partial F_2}{\partial x_3}\right) \mathrm{d}x_2\, \mathrm{d}x_3 + \ldots .$$

Now $\mathrm{d}x_2\, \mathrm{d}x_3$ is the area, $\mathrm{d}\sigma_1$, of the rectangle. In the limit as $\mathrm{d}\sigma_1$ approaches zero

$$\frac{\mathrm{d}\mathscr{W}_1}{\mathrm{d}\sigma_1} = \frac{\partial F_3}{\partial x_2} - \frac{\partial F_2}{\partial x_3}. \tag{4.33}$$

By considering similar small rectangles of areas $\mathrm{d}\sigma_2, \mathrm{d}\sigma_3$ parallel with the other coordinate planes and calculating the line integrals $\mathrm{d}\mathscr{W}_2, \mathrm{d}\mathscr{W}_3$, of the force **F** around them, we may show that

$$\frac{\mathrm{d}\mathscr{W}_2}{\mathrm{d}\sigma_2} = \frac{\partial F_1}{\partial x_3} - \frac{\partial F_3}{\partial x_1}, \qquad \frac{\mathrm{d}\mathscr{W}_3}{\mathrm{d}\sigma_3} = \frac{\partial F_2}{\partial x_1} - \frac{\partial F_1}{\partial x_2}. \tag{4.34}$$

The three quantities of equations (4.33) and (4.34) are the components of a vector written as

$$\mathbf{curl\ F} = \left(\frac{\partial F_3}{\partial x_2} - \frac{\partial F_2}{\partial x_3}, \frac{\partial F_1}{\partial x_3} - \frac{\partial F_3}{\partial x_1}, \frac{\partial F_2}{\partial x_1} - \frac{\partial F_1}{\partial x_2}\right), \tag{4.35}$$

and therefore in terms of the vector operator $\boldsymbol{\nabla}$,

$$\mathbf{curl\ F} \equiv \boldsymbol{\nabla} \wedge \mathbf{F}. \tag{4.36}$$

Now if the force is to be conservative, its line integral round any rectangle (which is the work done in traversing a closed path) must be zero. Hence a necessary condition for the force to be conservative is

$$\mathbf{curl\ F} \equiv \boldsymbol{\nabla} \wedge \mathbf{F} = 0. \tag{4.37}$$

(See Section 4.10.6, page 108.)

Conversely, when this condition is satisfied the line integral round *any* closed path is zero. Such a path, C, can be divided into a network of small rectangular loops D, E, F, . . . , as in Figure 4.3. Now

$$\oint_{\mathrm{C}} \mathbf{F}\,.\,\mathrm{d}\mathbf{x} = \int_{\mathrm{D}} \mathbf{F}\,.\,\mathrm{d}\mathbf{x} + \int_{\mathrm{E}} \mathbf{F}\,.\,\mathrm{d}\mathbf{x} + \int_{\mathrm{F}} \mathbf{F}\,.\,\mathrm{d}\mathbf{x} + \ldots,$$

since in the summation the integrations along the common sides of neighbouring loops (such as D and E) are in *opposite* directions and therefore cancel each other. The only non-cancelling contributions are those along the boundary of C.

Thus a necessary and sufficient condition for **F** to be a conservative force can be expressed either in the *integral* form (equation 4.23),

$$\oint_{\mathrm{C}} \mathbf{F}\,.\,\mathrm{d}\mathbf{x} = 0,$$

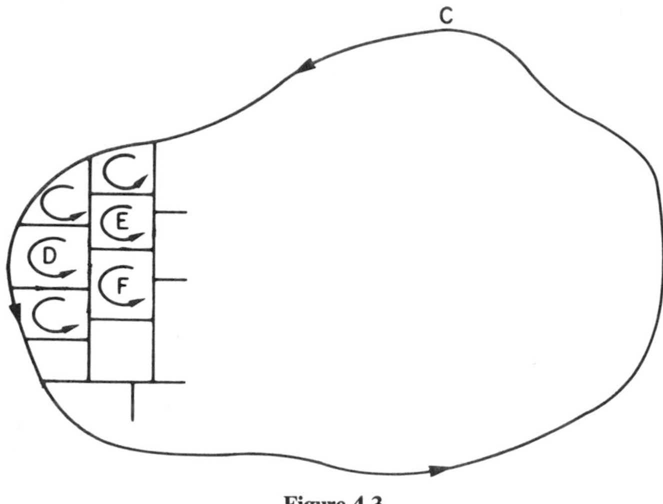

Figure 4.3

for all closed paths C, or in the *differential* form (equation 4.37),

$$\mathbf{curl\,F} \equiv \nabla \wedge \mathbf{F} = 0,$$

throughout the region under consideration.

4.6. Constant force

As we saw in the case of the gravitational force near the earth's surface, Cartesian axes can be chosen so that a constant force, **F**, has components

$$F_1 = F_2 = 0, \qquad F_3 = F = \text{constant}.$$

The conditions (4.35) and (4.37) are clearly satisfied; **F** is therefore conservative. In using

$$\mathscr{V}(\mathbf{x}, \mathbf{X}_0) = \int_{\mathbf{x}}^{\mathbf{X}_0} \mathbf{F}\,.\,\mathrm{d}\mathbf{x}'$$

to calculate the potential energy we may choose any path. A convenient one is shown in Figure 4.4. This has three straight parts along each of which only one coordinate varies:

$$\begin{aligned}\int_{\mathbf{x}}^{\mathbf{X}_0} \mathbf{F}\,.\,\mathrm{d}\mathbf{x}' &= \int_{(x_1,x_2,x_3)}^{(X_{01},x_2,x_3)} F_1\,\mathrm{d}x_1' + \int_{(X_{01},x_2,x_3)}^{(X_{01},X_{02},x_3)} F_2\,\mathrm{d}x_2' + \int_{(X_{01},X_{02},x_3)}^{(X_{01},X_{02},X_{03})} F_3\,\mathrm{d}x_3' \\ &= 0 + 0 + F\int_{(X_{01},X_{02},x_3)}^{(X_{01},X_{02},X_{03})} \mathrm{d}x_3' \\ &= F(X_{03} - x_3).\end{aligned}$$

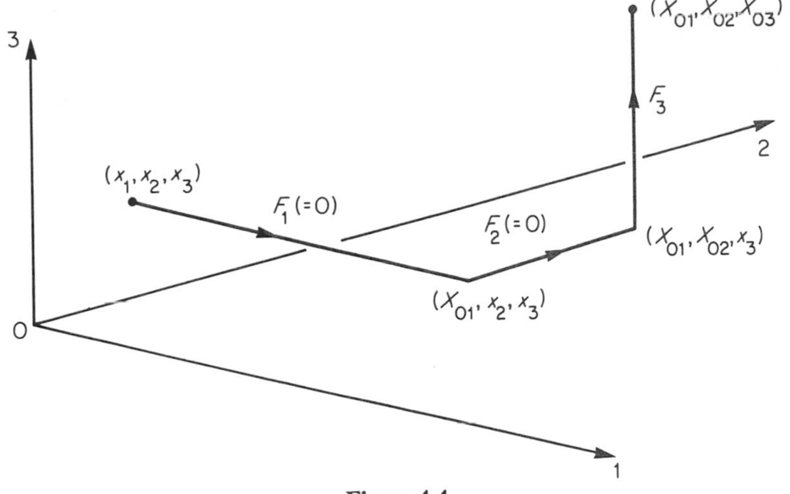

Figure 4.4

Thus

$$\mathscr{V}(x) \equiv \mathscr{V}(x; X_0) \equiv \mathscr{V}(x_1, x_2, x_3; X_{01}, X_{02}, X_{03})$$
$$= F(X_{03} - x_3).$$

We can check at once that

$$F_1 = -\frac{\partial \mathscr{V}}{\partial x_1} = 0, \qquad F_2 = -\frac{\partial \mathscr{V}}{\partial x_2} = 0, \qquad F_3 = -\frac{\partial \mathscr{V}}{\partial x_3} = F.$$

The most convenient choice for $\mathbf{X}_0$ is usually any point on the (1, 2) plane. Then

$$X_{03} = 0$$

and

$$\mathscr{V} = -Fx_3.$$

4.7. Radial force

When a particle is attached to an elastic string whose other end is fixed, or if it is electrically charged and is attracted or repelled by another charge which is fixed, the force it experiences is always directed towards or away from the fixed point, and its magnitude depends *only* upon its distance from that point. Such forces, which we shall call *radial*, are of the form

$$\mathbf{F}(\mathbf{x}) = f(r)\hat{\mathbf{r}},$$

or, written in component form, using a spherical polar frame,

$$F_r = f(r), \qquad F_\theta = F_\phi = 0.$$

$\hat{\mathbf{r}}$ is a unit vector directed away from the fixed point towards the particle and gives the direction of the force; $f(r)$ gives the magnitude as a function of the radial distance, r.

The two angles θ and ϕ are necessary, as well as r, to define position in spherical polar coordinates. We can obtain the three closed paths shown in Figure 4.5 by varying two of these while keeping the third fixed. Now

$$F_\theta = F_\phi = 0$$

so that

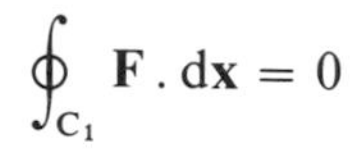

$$\oint_{C_1} \mathbf{F} \, . \, d\mathbf{x} = 0$$

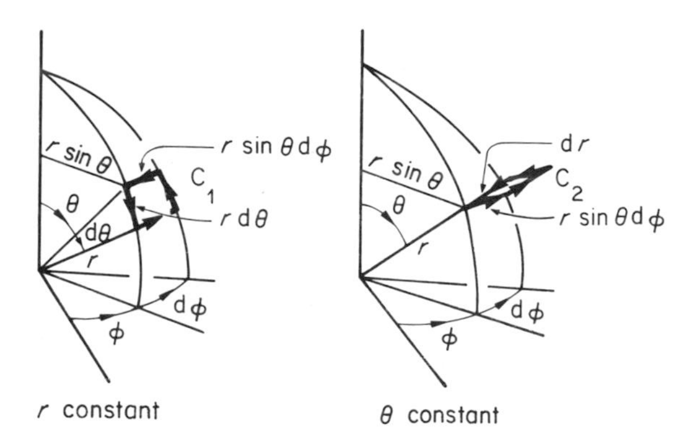

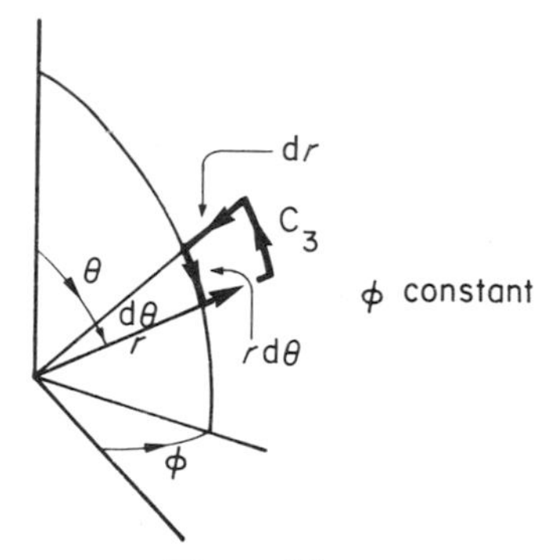

Figure 4.5

immediately. In the case of C_2 and C_3 the contributions to the line integral are non-zero only along the radial parts of the paths, and along these the contributions are equal and opposite and therefore cancel. Thus

$$\oint_{C_2} \mathbf{F} \, . \, d\mathbf{x} = \oint_{C_3} \mathbf{F} \, . \, d\mathbf{x} = 0.$$

Hence the force is conservative. Alternatively, we can use the results of Sections 4.5 and 4.10.6 to show that

$$\mathbf{curl\ F} = 0$$

and thus prove the same result.

Here, too, the potential energy,

$$\mathscr{V}(\mathbf{x}; \mathbf{X}_0) = \int_{\mathbf{x}}^{\mathbf{X}_0} \mathbf{F} \cdot \mathrm{d}\mathbf{x}',$$

can be found most simply by choosing segments of the path from $\mathbf{x}$ to $\mathbf{X}_0$ over which only one coordinate varies, as shown in Figure 4.6.

$$\begin{aligned}\int_{\mathbf{x}}^{\mathbf{X}_0} \mathbf{F} \,.\, \mathrm{d}\mathbf{x}' &= \int_{r,\theta,\phi}^{R,\theta,\phi} F_{r'}\, \mathrm{d}r' + \int_{R,\theta,\phi}^{R,\Theta,\phi} F_{\theta'}\, R\, \mathrm{d}\theta' + \int_{R,\Theta,\phi}^{R,\Theta,\Phi} F_{\phi'}\, R \sin\Theta\, \mathrm{d}\phi' \\ &= \int_r^R f(r')\, \mathrm{d}r'.\end{aligned}$$

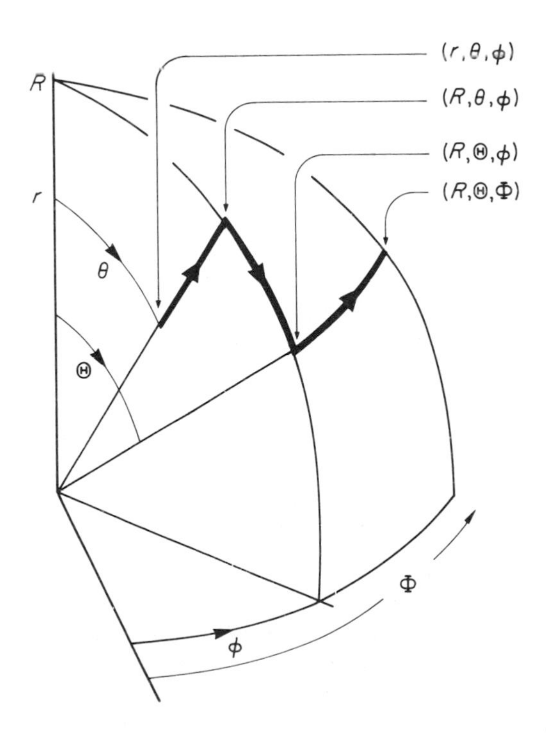

Figure 4.6

Here it is often convenient to use 'infinity' for the reference point $\mathbf{X}_0$. This is not in fact *a* point—any point on the 'infinite sphere' ($R \to \infty$) will give the same result. Sometimes the origin is a better choice.

4.8. Inverse square force

By far the most important radial forces observed in nature are the gravitational and electrostatic forces. A point particle of mass m_1, whose position vector relative to another point particle of mass m_2 is $\mathbf{x}$, experiences a gravitational force,

$$\mathbf{F}_g = -Gm_1m_2\frac{\hat{\mathbf{x}}}{x^2}, \tag{4.38}$$

where

$$G = 6{\cdot}67 \times 10^{-11}\,\mathrm{kg^{-1}\,m^3\,s^{-2}} \tag{4.39}$$

is the universal gravitational constant.

A point particle with electric charge q_1, whose position vector relative to another point particle with charge q_2 is $\mathbf{x}$, experiences an electrostatic force in a vacuum,

$$\mathbf{F}_e = \frac{q_1q_2}{4\pi\varepsilon_0}\frac{\hat{\mathbf{x}}}{x^2}, \tag{4.40}$$

where, if q_1, q_2 are measured in Coulombs and x in metres,

$$\varepsilon_0 = 4\pi \times 10^{-9}\,\mathrm{kg^{-1}\,m^{-3}\,s^4\,A^2} \tag{4.41}$$

is the permittivity or electric constant of free space.

These are both *inverse square law* forces of the type

$$\mathbf{F} = k\frac{\hat{\mathbf{x}}}{x^2} \equiv k\frac{\mathbf{x}}{x^3} \tag{4.42}$$

or

$$F_r = \frac{k}{r^2}, \qquad F_\theta = F_\phi = 0, \tag{4.43}$$

where k is a constant. x and r are interchangeable since they both represent distance from the origin of the force which is taken to be also the origin of the coordinate system. k is always negative in the gravitational case (representing an attraction), but can be positive (representing a repulsion) or negative in the electrostatic case, depending upon whether the charges are of the same or opposite sign.

From the previous section the potential energy is

$$\begin{aligned}\mathscr{V}(\mathbf{x}\,;\mathbf{X}_0) &\equiv \mathscr{V}(r, \theta, \phi\,; R, \Theta, \Phi)\\ &= k\int_r^R \frac{\mathrm{d}r'}{r'^2}\\ &= k\left(\frac{1}{r} - \frac{1}{R}\right).\end{aligned} \tag{4.44}$$

If the reference point, or zero of potential energy, is taken at infinity, we may then write simply

$$\mathscr{V}(\mathbf{x}) \equiv \mathscr{V}(\mathbf{x}, \infty) = \frac{k}{r}. \tag{4.45}$$

Thus the gravitational potential energy is

$$\mathscr{V}_{\mathrm{g}} = -\frac{Gm_1m_2}{r} \tag{4.46}$$

and the electrostatic potential energy is

$$\mathscr{V}_{\mathrm{e}} = \frac{q_1q_2}{4\pi\varepsilon_0 r} \tag{4.47}$$

Both these quantities are energies and therefore have dimensions $\mathrm{ML^2T^{-2}}$. If we write them as

$$\mathscr{V}_{\mathrm{g}} = m_1\left(-\frac{Gm_2}{r}\right) = m_1\phi_{\mathrm{g}},$$

$$\mathscr{V}_{\mathrm{e}} = q_1\left(\frac{q_2}{4\pi\varepsilon_0 r}\right) = q_1\phi_{\mathrm{e}},$$

we see that these energies are products of a quantity possessed by the particle 1 (its mass or its charge) and a quantity possessed by the point in space occupied by this particle, arising from the presence of particle 2. We say that particle 2 creates a *field* around itself which possesses a *potential*. In the one case this is the *gravitational potential*

$$\phi_{\mathrm{g}} = -\frac{Gm_2}{r}, \tag{4.48}$$

with dimensions $\mathrm{L^2T^{-2}}$, and in the other it is the *electrostatic potential*

$$\phi_{\mathrm{e}} = \frac{q_2}{4\pi\varepsilon_0 r}, \tag{4.49}$$

with dimensions $\mathrm{ML^2T^{-3}A^{-1}}$, commonly measured in Volts.

Experiments show that as well as following the inverse square law, the gravitational and electrostatic forces are *additive*. This means that if we have masses m_2 and m_3 (or charges q_2 and q_3) the force experienced by the particle 1 is the vector sum of those arising from the particles 2 and 3 *considered separately*. This is equivalent to saying that the potential energy is the (scalar) sum of the separate potential energies. We can add further particles with a similar result.

Thus if we have particles of mass $m_1, m_2, \ldots, m_N$ at positions $x_1, x_2, \ldots, x_N$, the gravitational potential energy of a particle of mass m at position r is

$$\begin{aligned}\mathscr{V}_g &= -\frac{Gmm_1}{|\mathbf{x}-\mathbf{x}_1|} - \frac{Gmm_2}{|\mathbf{x}-\mathbf{x}_2|} - \cdots - \frac{Gmm_N}{|\mathbf{x}-\mathbf{x}_N|} \\ &= -mG\sum_{i=1}^{N}\frac{m_i}{l_i},\end{aligned}$$

where

$$l_i = |\mathbf{x}-\mathbf{x}_i|.$$

The gravitational potential is then

$$\phi_g = -G\sum\frac{m_i}{l_i}. \tag{4.50}$$

The electrostatic potential is, similarly,

$$\phi_e = \frac{1}{4\pi\varepsilon_0}\sum\frac{q_i}{l_i}. \tag{4.51}$$

When we are dealing with continuously distributed mass or charge these potentials take the corresponding integral forms,

$$\phi_g = -G\int\frac{\mathrm{d}m}{l}, \tag{4.52}$$

$$\phi_e = \frac{1}{4\pi\varepsilon_0}\int\frac{\mathrm{d}q}{l}. \tag{4.53}$$

In problems involving many particles, or bodies of finite shape, it is usually much easier to find their influence upon another particle or body by finding the total potential energy first and then, if it is necessary, deriving from it the force. This involves only a single (scalar) summation or integration. If the individual forces are found first, not only are these usually more complicated, but their vector sum, involving three summations or integrations, is required in order to calculate the total effect.

4.8.1. Spherical shell of uniform density

To illustrate the preceding results let us consider the gravitational effect of a shell of material of constant density ρ lying between radii r' and $r' + dr'$. A small element of this, lying between θ and $\theta + d\theta$ and between ϕ and $\phi + d\phi$, will have a volume

$$d\tau = dr' r' \, d\theta r' \sin\theta \, d\phi,$$

as can be seen from Figure 4.7, and a mass

$$dm = \rho \, d\tau = \rho r'^2 \sin\theta \, dr' \, d\theta \, d\phi.$$

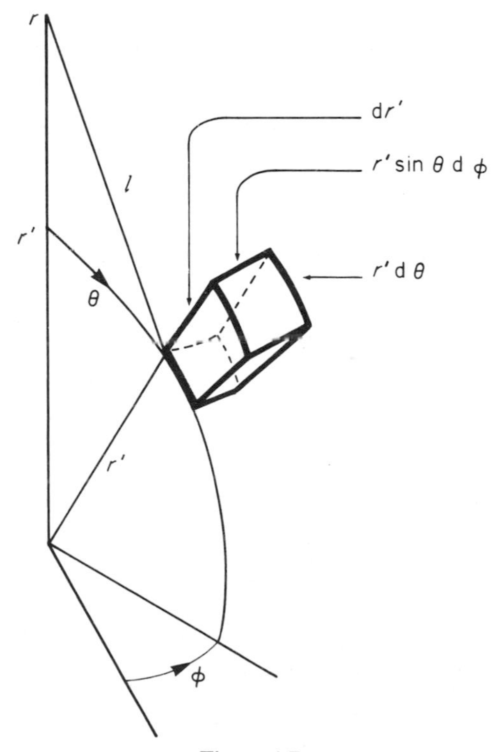

Figure 4.7

The distance of this element from a point $(r, 0, 0)$ will be

$$l = (r^2 + r'^2 - 2rr' \cos\theta)^{1/2},$$

where the *positive* square root is always implied. Using the expression (4.52), the gravitational potential at $(r, 0, 0)$ will therefore be

$$\phi_g = -G \int_{\phi=0}^{2\pi} \int_{\theta=0}^{\pi} \frac{\rho r'^2 \sin\theta \, dr' \, d\theta \, d\phi}{(r^2 + r'^2 - 2rr' \cos\theta)^{1/2}}.$$

The integrand is independent of ϕ. Hence

$$\phi_g = -2\pi G\rho r'^2 \,\mathrm{d}r' \int_0^{\pi} \frac{\sin\theta \,\mathrm{d}\theta}{(r^2 + r'^2 - 2rr'\cos\theta)^{1/2}}$$

$$= -\frac{2\pi G\rho r' \,\mathrm{d}r'}{r}\left[(r^2 + r'^2 - 2rr'\cos\theta)^{1/2}\right]_0^{\pi}.$$

The value of the integral depends upon whether the point $(r, 0, 0)$ is outside or inside the shell. In either case the upper limit is the same,

$$(r^2 + r'^2 - 2rr'\cos\theta)^{1/2}_{\theta=\pi} = (r^2 + r'^2 + 2rr')^{1/2} = r + r',$$

but if

$$r > r'$$

the physically meaningful (positive square root) of the lower limit is

$$(r^2 + r'^2 - 2rr'\cos\theta)^{1/2}_{\theta=0} = (r^2 + r'^2 - 2rr')^{1/2} = r - r',$$

while if

$$r < r'$$

it is

$$(r^2 + r'^2 - 2rr'\cos\theta)^{1/2}_{\theta=0} = r' - r.$$

Hence

$$\phi_g = -\frac{4\pi G\rho r'^2 \,\mathrm{d}r'}{r} = -\frac{Gm}{r}, \qquad r > r' \tag{4.54}$$

$$\phi_g = -4\pi G\rho r' \,\mathrm{d}r' = -\frac{Gm}{r'}, \qquad r < r', \tag{4.55}$$

where

$$m = 4\pi\rho r'^2 \,\mathrm{d}r'$$

is the total mass of the shell. Now although we chose to evaluate the potential at a point on the spherical polar axis

$$\theta = 0,$$

it is clear from the spherical symmetry of the shell that physically this has no special properties and that the results (4.54) and (4.55) will be true for any point at a distance r from the origin. Comparing them with the expression (4.48) we see that for points *outside* the shell, the gravitational potential is the same as if the whole mass were concentrated at its centre and force follows an inverse square law, while *within* the shell the potential is constant and the force is therefore zero.

4.8.2. Sphere of uniform density

We divide the sphere, of radius a, into a series of concentric shells, typically between radii r' and $r' + dr'$. A point outside or on the sphere,

$$r \geqslant a,$$

will be outside all the shells, each of which will behave gravitationally as if all its mass were concentrated at the centre. Adding all the contributions the potential will simply be

$$\phi_g = -\frac{GM}{r}, \qquad r \geqslant a, \tag{4.56}$$

where M is the total mass of the sphere.

For a point within or on the sphere,

$$r \leqslant a,$$

the contributions from shells of radius smaller than r will be that of their total mass M' concentrated at the centre,

$$\phi'_g = -\frac{GM'}{r}.$$

The contribution from shells of larger radius is, from the result (4.55),

$$\begin{aligned}\phi''_g &= -4\pi G\rho \int_r^a r'\,dr' \\ &= -2\pi G\rho(a^2 - r^2).\end{aligned}$$

Now the mass of the whole sphere is

$$M = \frac{4\pi}{3}\rho a^3$$

and the mass of the inner part is

$$M' = M\frac{r^3}{a^3}.$$

Hence the total gravitational potential is

$$\phi_g = \phi'_g + \phi''_g = -GM\left(\frac{3}{2a} - \frac{r^2}{2a^3}\right), \qquad r \leqslant a. \tag{4.57}$$

Hence the gravitational force on a mass m is

$$F_r = -\frac{GMm}{r^2}, \qquad F_\theta = F_\phi = 0, \qquad r \geqslant a, \tag{4.58}$$

$$F_r = -\frac{GMmr}{a^3}, \qquad F_\theta = F_\phi = 0, \qquad r \leqslant a. \tag{4.59}$$

4.8.3. Shell and sphere of uniform charge density

For these we substitute, according to expressions (4.52) and (4.53), charge for mass and $-1/(4\pi\varepsilon_0)$ for G.

Spherical shell of charge q' and radius r':

$$\phi_e = \frac{q'}{4\pi\varepsilon_0 r}, \qquad r \geqslant r' \tag{4.60}$$

$$\phi_e = \frac{q'}{4\pi\varepsilon_0 r'}, \qquad r \leqslant r'. \tag{4.61}$$

Hence the force on a charge q is

$$F_r = \frac{q'q}{4\pi\varepsilon_0 r^2}, \qquad F_\theta = F_\phi = 0, \qquad r \geqslant r', \tag{4.62}$$

$$F_r = F_\theta = F_\phi = 0, \qquad r \leqslant r'. \tag{4.63}$$

Sphere of charge Q and radius a:

$$\phi_e = \frac{Q}{4\pi\varepsilon_0 r}, \qquad r \geqslant a \tag{4.64}$$

$$\phi_e = \frac{Q}{4\pi\varepsilon_0}\left(\frac{3}{2a} - \frac{r^2}{2a^3}\right), \qquad r \leqslant a. \tag{4.65}$$

Hence the force on a charge q is

$$F_r = \frac{Qq}{4\pi\varepsilon_0 r^2}, \qquad F_\theta = F_\phi = 0, \qquad r \geqslant r', \tag{4.66}$$

$$F_r = \frac{Qqr}{4\pi\varepsilon_0 a^3}, \qquad F_\theta = F_\phi = 0, \qquad r \leqslant r'. \tag{4.67}$$

4.9. Constrained motion

It often happens that a particle is constrained to move upon a surface or upon a wire, while a field of force acting on it is not similarly restricted. For example a particle moving over hills and valleys under the earth's gravitational force will, provided its velocity is not too great, always slide upon the surface. However, gravity would act upon the particle even if it rose above or burrowed beneath the surface.

It is important to realize that in such cases, provided the constraining surface or wires are smooth, the concepts of kinetic and potential energy can be used in their full three-dimensional sense. Suppose a section through the surface is as shown in Figure 4.8. As well as the gravitational force, mg,

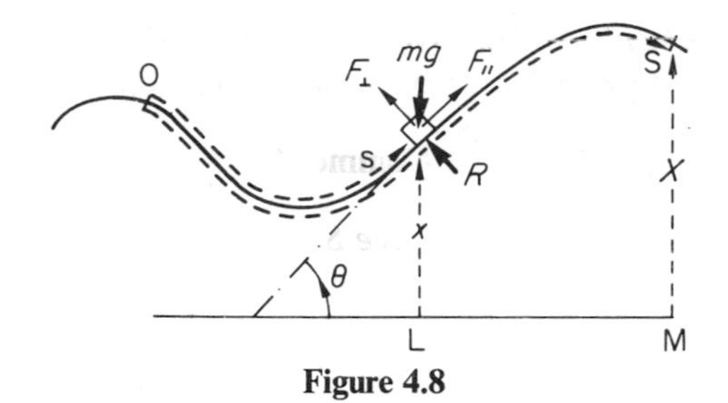

Figure 4.8

a reaction force, R, will be exerted on the particle by the surface. Since this is smooth the reaction will be perpendicular to the surface. Thus the total force on the particle has a component perpendicular to the surface, of magnitude

$$F_\perp = R - mg\cos\theta,$$

and one tangential to the motion, of magnitude

$$F_\parallel = -mg\sin\theta,$$

where θ gives the slope. Thus as the particle moves a short distance $\mathrm{d}\mathbf{s}'$ along the surface the work done on it is

$$\mathrm{d}\mathscr{W} = \mathbf{F}\,.\,\mathrm{d}\mathbf{s}' = F_\parallel\,\mathrm{d}s' = -mg\sin\theta\,\mathrm{d}s'.$$

The perpendicular component does no work and, as far as motion along the surface is concerned, it can be forgotten. We have a simulated one-dimensional motion defined by the single coordinate s, measured along the surface from some fixed point. Since θ is determined by s', the force, $-mg\sin\theta$, is of the form $F(s')$, independent of time and velocity. In principle, therefore, we could treat the motion as one-dimensional, calculate a potential energy relative to a fixed point S,

$$\mathscr{V}(s) = \int_s^S F_\parallel\,\mathrm{d}s',$$

and use this to analyse the motion. However, we already know that the gravitational force is conservative and that *any* path between s and S will give the same value. Consequently,

$$\mathscr{V}(s) = mg(x - X),$$

where x and X are the vertical coordinates of s and S.

Although this result is obtained most simply by considering the alternative path sLMS in Figure 4.8, the fact that the constrained particle cannot traverse this path does not invalidate it. All that matters is that the work done by the force as the particle moves along a particular path is the same whether or not it is constrained to do so. For this to be true the only requirement is that the constraint should be smooth. (See Section 4.10.7, page 110.)

4.10. Comments and worked examples

4.10.1. See Section 4.1

It must be emphasized that the description of the gravitational force given by equations (4.1), and the subsequent analysis of the motion derived from them, are true *only for restricted motion near the earth's surface.* The earth is (approximately) a sphere, and the gravitational force is directed towards its centre and depends upon the distance from that centre (as shown in Section 4.8.2). Consequently, as we move about the earth's surface the direction of the gravitational force changes, and if we move above or below the surface the magnitude changes.

When the effect of these changes becomes important (for example when the motion of rockets and satellites is under consideration), the methods of Chapter 6 must be used.

4.10.2. See Section 4.1

Example 13

Show that when a cricket ball is thrown with a given speed v_0, the maximum horizontal distance to its first bounce is achieved when the direction of the throw is at 45° to the horizontal. What is this maximum range?

Since a cricket ball is made of a dense material we shall neglect the effects of air resistance, and the magnitude of a normal throw is small enough in comparison with the earth's radius to assume the gravitational force on the ball to be constant throughout its motion. We may therefore use equations (4.4) and (4.5) as sufficiently accurate descriptions of the motion.

The ball returns to the ground when x_3 is zero, and therefore, from equation (4.5), at a time

$$t = 2v_0 \sin \alpha / g.$$

From equation (4.5), the horizontal distance will then be

$$x_2 = v_0 \cos \alpha\, t = 2v_0^2 \cos \alpha \sin \alpha\, t/g = v_0^2 \sin 2\alpha / g.$$

This will be a maximum when

$$\sin 2\alpha = 1 \quad \text{or} \quad \alpha = 45°,$$

and the maximum value will be

$$x_{2\,\max} = v_0^2/g.$$

4.10.3. See Section 4.3

Example 14

A nucleus of mass M and charge ze approaches at a high speed v a stationary electron of mass m and charge $-e$. If the electron remained stationary and the nucleus were undeflected, the distance of their nearest approach would be b.

Assuming that the deflexion and change of speed of the nucleus is negligible and that the displacement of the electron is negligible during the effective part of the interaction, show that the electron moves off at right angles to the path of the nucleus, and determine its kinetic energy.

Under the given assumptions the electron will be at a fixed position P and the nucleus at a point Q moving along a straight line. The force experienced by the electron will be of magnitude $ze^2/(4\pi\varepsilon_0 l^2)$ and in the direction $\overrightarrow{PQ}$ (see Figure 4.9). If perpendicular axes 1 and 3 are taken as shown in

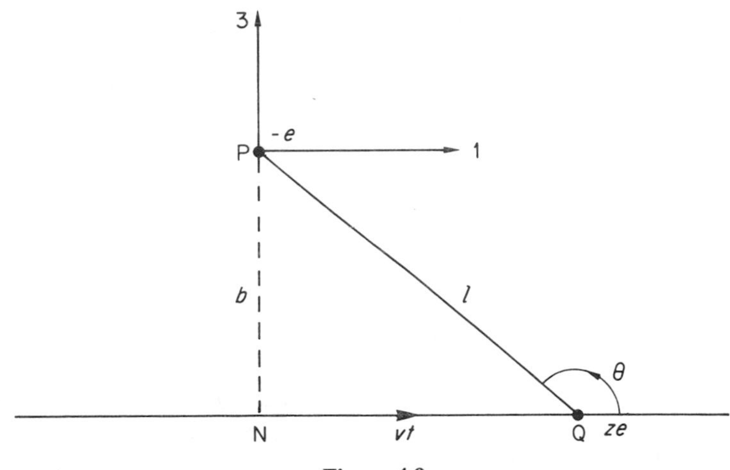

Figure 4.9

Figure 4.9, and axis 2 perpendicularly into the plane of the paper, the components of force will be

$$F_1 = -\frac{ze^2 \cos\theta}{4\pi\varepsilon_0 l^2}, \qquad F_2 = 0, \qquad F_3 = -\frac{ze^2 \sin\theta}{4\pi\varepsilon_0 l^2}. \tag{4.68}$$

The single variable θ is sufficient to describe the configuration since

$$l = b \operatorname{cosec} \theta, \tag{4.69}$$

and, if we take time zero to be when the nucleus passes N, the point of nearest approach,

$$vt = -b \cot\theta, \qquad v\,dt = b \operatorname{cosec}^2\theta\, d\theta. \tag{4.70}$$

Using equations (4.68), (4.69) and (4.70) the components of the impulse equation (4.17) may be written

$$
\begin{aligned}
I_{(1,2)1} &= \int_{(1)}^{(2)} F_1 \, \mathrm{d}t = -\frac{ze^2}{4\pi\varepsilon_0 bv} \int_0^\pi \cos\theta \, \mathrm{d}\theta = 0, \\
I_{(1,2)2} &= \int_{(1)}^{(2)} F_2 \, \mathrm{d}t \qquad\qquad\qquad\qquad\quad = 0, \\
I_{(1,2)3} &= \int_{(1)}^{(2)} F_3 \, \mathrm{d}t = -\frac{ze^2}{4\pi\varepsilon_0 bv} \int_0^\pi \sin\theta \, \mathrm{d}\theta = -\frac{ze^2}{2\pi\varepsilon_0 bv}.
\end{aligned} \tag{4.71}
$$

The conditions (1) and (2) here refer to the infinitely distant past and future, $\theta = 0, \pi$.

Since the electron is initially at rest we can see that its velocity component in direction 1 will remain zero since all the impulse contributions in this direction delivered by the nucleus when it is to the left of N are cancelled by equal and opposite contributions when it is in corresponding positions on the right. The velocity component in direction 2 remains zero because there is no force component at all in that direction. Only in direction 3 will the electron acquire a velocity, u, determined by equation (4.71):

$$mu = -\frac{ze^2}{2\pi\varepsilon_0 bv}.$$

The interaction therefore causes the electron to move in the direction $\overrightarrow{\mathrm{PN}}$, perpendicular to the path of the nucleus, and the kinetic energy it acquires is

$$\mathscr{T} = \tfrac{1}{2}mu^2 = \frac{z^2e^4}{4\pi^2\varepsilon_0^2 b^2 mv^2}.$$

This method of calculating the energy transferred to an electron is called the *impulse approximation* and is the basis upon which an approximate calculation of the ionization arising from the passage of a charged particle through a medium may be made.

4.10.4. See Section 4.5

It is most important to realize that energy, both kinetic and potential, is *a scalar quantity* with no implied direction or association with any coordinate axis.

When three-dimensional motion is separable and can be described by equations of the form (4.6), it is true that we may define

$$\mathscr{V}_1 = \int_{x_1}^{x_{01}} F_1(x_1') \, \mathrm{d}x_1', \qquad \mathscr{T}_1 = \frac{m}{2}\left(\frac{\mathrm{d}x_1}{\mathrm{d}t}\right)^2,$$

and show, as in Section 2.5, that

$$\mathcal{T}_1 + \mathcal{V}_1 = \mathcal{E}_1 = \text{constant},$$

with similar results for the other two coordinates.

It is unfortunate that such results happen to be true for this very special case, since they can lead to the use of phrases such as 'the energy in the direction 1', which seem to imply that energy has components. This quite incorrect attitude can lead to considerable confusion and should *always* be avoided.

4.10.5. See Section 4.5

It is often convenient to express results such as (4.30) or (4.31) in forms other than the Cartesian ones of equation (4.29).

In addition to the Cartesian frame there are two other three-dimensional frames in common use: the cylindrical polar and the spherical polar frames shown in Figures 4.10a and b. All three employ three coordinates, which

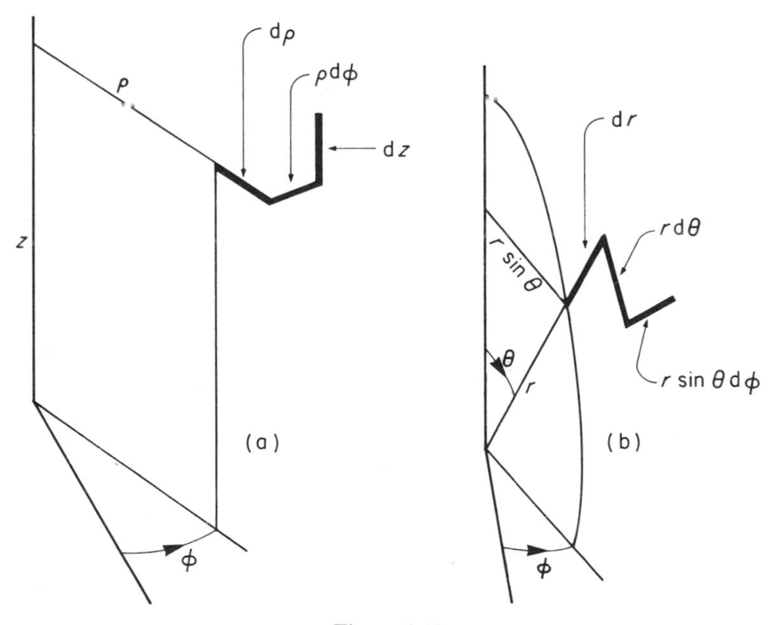

Figure 4.10

we shall call generally q_1, q_2, q_3. If each of these is varied slightly in turn while the others remain constant the point represented by them will be displaced in three mutually perpendicular directions. All three are therefore known as *orthogonal coordinate systems*.

In the Cartesian case the displacements ds_1, ds_2, ds_3 are strictly linear and equal to the changes in the corresponding coordinates dx_1, dx_2, dx_3. In the other two cases the displacements may be curved, but, provided the displacements are small enough, they may be taken as linear and will be proportional to the variations in the corresponding coordinates. Thus in all cases

$$ds_1 = h_1\,dq_1, \qquad ds_2 = h_2\,dq_2, \qquad ds_3 = h_3\,dq_3, \tag{4.72}$$

where, for the Cartesian system,

$$(q_1, q_2, q_3) \equiv (x_1, x_2, x_3); \qquad h_1 = h_2 = h_3 = 1, \tag{4.73}$$

for the cylindrical polar system,

$$(q_1, q_2, q_3) \equiv (\rho, \phi, z); \qquad h_1 = 1, \qquad h_2 = \rho, \qquad h_3 = 1, \tag{4.74}$$

and for the spherical polar system,

$$(q_1, q_2, q_3) \equiv (r, \theta, \phi); \qquad h_1 = 1, \qquad h_2 = r, \qquad h_3 = r\sin\theta \tag{4.75}$$

(see Figures 4.10a and b).

Now we know from equation (4.29) that the force components in three perpendicular directions are the negative gradients with respect to *distance* in each direction. For any orthogonal system, therefore, the components are

$$F_1 = -\frac{\partial V}{\partial s_1}, \qquad F_2 = -\frac{\partial V}{\partial s_2}, \qquad F_3 = -\frac{\partial V}{\partial s_3}. \tag{4.76}$$

Hence, from equations (4.72) and (4.74), for the cylindrical polar system,

$$\mathbf{F} = (F_\rho, F_\phi, F_z) = -\left(\frac{\partial V}{\partial \rho}, \frac{1}{\rho}\frac{\partial V}{\partial \phi}, \frac{\partial V}{\partial z}\right), \tag{4.77}$$

and, from equations (4.72) and (4.75), for the spherical polar system,

$$\mathbf{F} = (F_r, F_\theta, F_\phi) = -\left(\frac{\partial V}{\partial r}, \frac{1}{r}\frac{\partial V}{\partial \theta}, \frac{1}{r\sin\theta}\frac{\partial V}{\partial \phi}\right). \tag{4.78}$$

4.10.6. See Section 4.5

To derive an expression for **curl F** in a general orthogonal three-dimensional coordinate system, q_1, q_2, q_3, we consider the closed path shown in Figure 4.11 obtained by varying q_2 and q_3 while keeping q_1 constant. The work done by the force in traversing this counterclockwise is

$$\begin{aligned} d\mathscr{W}_1 = {} & F_2(q_1, q_2, q_3 - \tfrac{1}{2}dq_3)h_2(q_1, q_2, q_3 - \tfrac{1}{2}dq_3)\,dq_2 \\ & + F_3(q_1, q_2 + \tfrac{1}{2}dq_2, q_3)h_3(q_1, q_2 + \tfrac{1}{2}dq_2, q_3)\,dq_3 \\ & - F_2(q_1, q_2, q_3 + \tfrac{1}{2}dq_3)h_2(q_1, q_2, q_3 + \tfrac{1}{2}dq_3)\,dq_2 \\ & - F_3(q_1, q_2 - \tfrac{1}{2}dq_2, q_3)h_3(q_1, q_2 - \tfrac{1}{2}dq_2, q_3)\,dq_3. \end{aligned} \tag{4.79}$$

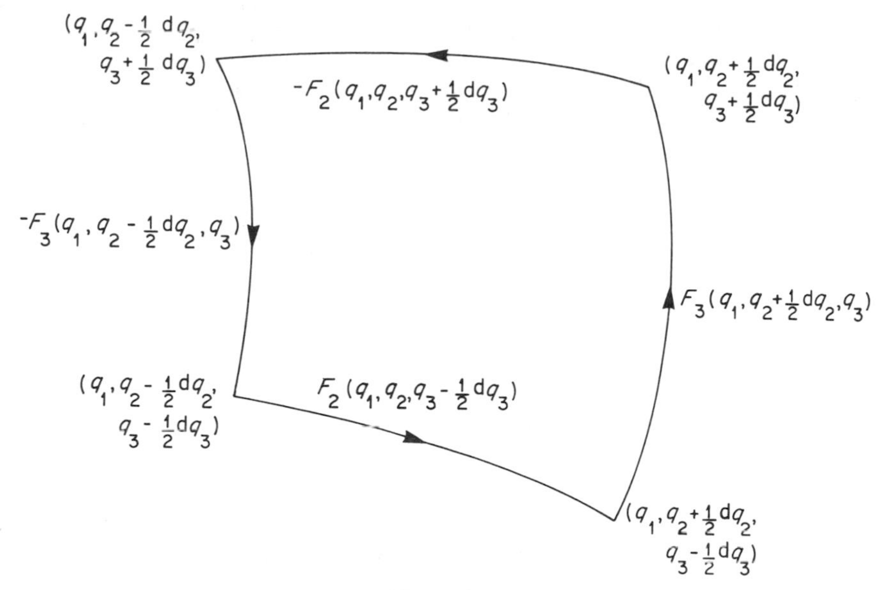

Figure 4.11

Note that h_2, h_3 as well as F_2, F_3 may vary with the q's. The expression (4.79) is similar to (4.32) with q_1, q_2, q_3 replacing x_1, x_2, x_3 and h_1F_1, h_2F_2, h_3F_3 replacing F_1, F_2, F_3. Proceeding as in Section 4.5 we find

$$d\mathscr{W}_1 = \left[\frac{\partial(h_3F_3)}{\partial q_2} - \frac{\partial(h_2F_2)}{\partial q_3}\right] dq_2\, dq_3 + \ldots.$$

The area, $d\sigma_1$, enclosed by the loop is approximately $h_2h_3\, dq_2\, dq_3$. Hence in the limit

$$\frac{d\mathscr{W}_1}{d\sigma_1} = \frac{1}{h_2h_3}\left[\frac{\partial(h_3F_3)}{\partial q_2} - \frac{\partial(h_2F_2)}{\partial q_3}\right].$$

Thus the component 1 of **curl F** in this general case is

$$(\mathbf{curl\ F})_1 = \frac{1}{h_2h_3}\left[\frac{\partial(h_3F_3)}{\partial q_2} - \frac{\partial(h_2F_2)}{\partial q_3}\right], \tag{4.80}$$

with similar expressions for the other components.

Substituting from equations (4.74) and (4.75) into (4.80) we find that, for cylindrical polar coordinates,

$$\mathbf{curl\ F} = \left\{\frac{1}{\rho}\frac{\partial F_z}{\partial \phi} - \frac{\partial F_\phi}{\partial z}, \frac{\partial F_\rho}{\partial z} - \frac{\partial F_z}{\partial \rho}, \frac{1}{\rho}\left[\frac{\partial(\rho F_\phi)}{\partial \rho} - \frac{\partial F_\rho}{\partial \phi}\right]\right\}, \tag{4.81}$$

and for spherical polar coordinates,

$$\mathbf{curl\,F} = \left\{\frac{1}{r\sin\theta}\left[\frac{\partial(\sin\theta F_\phi)}{\partial\theta} - \frac{\partial F_\theta}{\partial\phi}\right], \frac{1}{r\sin\theta}\left[\frac{\partial F_r}{\partial\phi} - \sin\theta\frac{\partial(rF_\phi)}{\partial r}\right], \frac{1}{r}\left[\frac{\partial(rF_\theta)}{\partial r} - \frac{\partial F_r}{\partial\theta}\right]\right\}. \tag{4.82}$$

4.10.7. See Section 4.9

Example 15

A charge q is fixed a distance h above a horizontal smooth rod. A particle of mass m and charge q, free to slide along the rod, is slightly displaced from its position of (unstable) equilibrium immediately below the fixed charge. What is its velocity when it has travelled a distance $\sqrt{3}h$?

If we were to think of this only in one-dimensional terms we would first find the force component along the rod at a distance x from the starting point (Figure 4.12). This force is

$$F_x = \frac{q^2\cos\theta}{4\pi\varepsilon_0(h^2 + x^2)} = \frac{q^2}{4\pi\varepsilon_0}\frac{x}{(h^2 + x^2)^{3/2}},$$

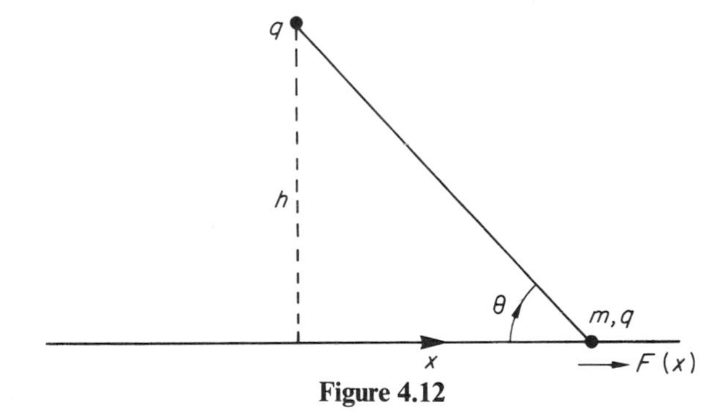

Figure 4.12

which depends only upon x and will therefore define a potential energy

$$\mathscr{V}(x) = \int_x^\infty F_{x'}\,\mathrm{d}x' = \frac{q^2}{4\pi\varepsilon_0}\left[\frac{-1}{(h^2 + x'^2)^{1/2}}\right]_x^\infty$$
$$= \frac{q^2}{4\pi\varepsilon_0}\frac{1}{(h^2 + x^2)^{1/2}}, \tag{4.83}$$

if we choose infinity to give the value zero. Since energy is conserved,

$$\mathscr{T}(0) + \mathscr{V}(0) = \mathscr{T}(\sqrt{3}h) + \mathscr{V}(\sqrt{3}h).$$

Using the result (4.83) and remembering that

$$\mathscr{T}(0) = 0$$

since the particle starts from rest,

$$\mathscr{T}(\sqrt{3}) = \mathscr{V}(0) - \mathscr{V}(\sqrt{3}h)$$

$$= \frac{q^2}{4\pi\varepsilon_0}\left(\frac{1}{h} - \frac{1}{2h}\right) = \frac{q^2}{8\pi\varepsilon_0 h}.$$

Thus if v is the required velocity,

$$\frac{m}{2}v^2 = \frac{q^2}{8\pi\varepsilon_0 h}$$

or

$$v = \frac{q}{(4\pi\varepsilon_0 mh)^{1/2}}.$$

Although in this case finding F_x and, from it, $\mathscr{V}(x)$ is not particularly difficult, it is nevertheless much simpler to consider the electrostatic field in three dimensions. The charges give rise to a potential energy $q^2/(4\pi\varepsilon_0 r)$ where r is the distance between them. Hence, using

$$\mathscr{Y} + \mathscr{V} = \text{constant},$$

$$\mathscr{T}(\sqrt{3}h) = \mathscr{T}(0) + \mathscr{V}(0) - \mathscr{V}(\sqrt{3}h)$$

$$= \frac{q^2}{4\pi\varepsilon_0}\left(\frac{1}{h} - \frac{1}{2h}\right) = \frac{q^2}{8\pi\varepsilon_0 h},$$

and therefore

$$v = \frac{q}{(4\pi\varepsilon_0 mh)^{1/2}}$$

as before.

4.11. Problems

4.1. Determine which of the following forces are conservative, and for those which are find a suitable potential energy:

(a) $F_1 = ax_2x_3 + bx_1^2,\ F_2 = ax_3x_1 + bx_2^2,\ F_3 = ax_1x_2 + bx_3^2$

(b) $F_1 = 3x_1^2 + 2x_2 - 2x_2x_3,\ F_2 = 4x_1^2 + x_3 + 2x_1x_2,\ F_3 = 2x_2(x_3 - x_1)$

(c) $F_r = -\dfrac{2k}{r^3}\cos\theta\cos\phi,\ F_\theta = \dfrac{k}{r^3}\sin\theta\sin\phi,\ F_\phi = \dfrac{k}{r^3}\cot\theta\sin\phi$

(d) $F_r = \dfrac{k}{r}\cos 2\theta,\ F_\theta = \dfrac{k}{r}\cos 2\theta\cos\phi,\ F_\phi = \dfrac{k}{r}\cos\theta\sin\phi$

(e) $F_\rho = \dfrac{k\sin\phi}{\rho},\ F_\phi = \dfrac{k\cos\phi}{\rho},\ F_z = 0.$

(a, b and k are constants.)

4.2. Show that if a conservative force shows no variation in one direction, then its component in that direction is the same everywhere. Hence show that with suitable choice of Cartesian axes the corresponding potential energy may be written in the form $\mathscr{V}(x_1, x_2) + Kx_3$, where K is a constant.

4.3. A mass m with electric charge $+e$ is free to move on a smooth vertical wire under the influence of a constant gravitational field and of another charge $+e$ held at a distance a from the wire. Show that oscillatory motion is possible only if

$$\frac{e^2}{6\sqrt{3}\pi\varepsilon_0 a^2} > mg.$$

4.4. A mass m, attached to two similar springs each of length b, may slide along a smooth horizontal wire. The other ends of the springs are fixed at a distance b from each other and from the wire. Both the springs and the wire lie in the same plane. The strength of each spring is such that it doubles in length when the mass hangs freely on it.

If initially the mass is held so that one spring is perpendicular to the wire, and it is then released, describe the subsequent motion and show that the maximum speed of the mass is $0{\cdot}38\sqrt{gb}$.

4.5. An electric charge q is at the position $(b, 0, 0)$ in a spherical polar coordinate frame. Show that the electric potential at any point (r, θ, ϕ) may be written in the form

$$\phi_e = \frac{q}{4\pi\varepsilon_0}\frac{1}{r}\left[1 + \cos\theta\frac{b}{r} + \frac{3\cos^2\theta - 1}{2}\left(\frac{b}{r}\right)^2 + \dots\right],$$

provided that

$$b < r.$$

A charge $-q$ is now placed at the position $(b, \pi, 0)$, forming with the positive charge an *electric dipole* of strength

$$\mu = 2qb.$$

Show that when

$$b \ll r,$$

the potential arising at any point from the dipole is

$$\phi_e = \frac{\mu\cos\theta}{4\pi\varepsilon_0 r^2}.$$

Hence show that if $\mathbf{x}$ is the displacement vector joining the origin to the point (r, θ, ϕ), and $\boldsymbol{\mu}$ is a vector representing the dipole in magnitude and direction (from negative to positive charge), then the potential may be written as

$$\phi_e = \frac{1}{4\pi\varepsilon_0}\frac{\boldsymbol{\mu}\cdot\mathbf{x}}{x^3}.$$

4.6. Find the spherical polar components of the force exerted on a charge q by the dipole of problem 4.5. Where must the charge be for this force to be perpendicular to the dipole axis?

4.7. Show that the potential energy of a charge q' at the position $\mathbf{x} + \mathbf{b}'$ relative to the centre of an electric dipole $\boldsymbol{\mu}$ may be written as

$$\mathscr{V}' = \frac{q'}{4\pi\varepsilon_0 x^3}\boldsymbol{\mu}\cdot(\mathbf{x} + \mathbf{b}')\left[1 - \frac{3}{2}\frac{2\mathbf{x}\cdot\mathbf{b}' + \mathbf{b}'^2}{x^2} + \dots\right],$$

provided

$$b' < x.$$

Now form another dipole by adding a charge $-q'$ at the position $\mathbf{x} - \mathbf{b}'$, and derive a similar expression for its potential energy. By adding these and taking the case

$$b' \ll x,$$

show that the mutual potential energy of dipoles $\boldsymbol{\mu}$ and $\boldsymbol{\mu}'$ is

$$\mathscr{V} = \frac{1}{4\pi\varepsilon_0}\left[\frac{\boldsymbol{\mu}\cdot\boldsymbol{\mu}'}{x^3} - 3\frac{(\boldsymbol{\mu}\cdot\mathbf{x})(\boldsymbol{\mu}'\cdot\mathbf{x})}{x^5}\right],$$

where $\mathbf{x}$ is the displacement vector joining their centres.

4.8. The earth may be regarded as a uniform sphere of mass 6×10^{24} kg and radius $6{\cdot}4 \times 10^3$ km, and the moon as a uniform sphere of mass $7{\cdot}4 \times 10^{22}$ kg, the distance between their centres being $3{\cdot}8 \times 10^5$ km. Given that in the absence of the moon the gravitational acceleration would be $9{\cdot}8$ m s^{-2}, calculate the deviations from this caused by the moon at a point on the earth's surface (a) nearest the moon, (b) furthest from the moon, (c) at a point on the 'equator' midway between these two points. What is the gravitational acceleration at the centre-of-mass of the two bodies?

Chapter 5

Dynamics of Many Particles

We arrived at the concept of the point particle by removing ourselves sufficiently far from the macroscopic objects we are familiar with for them all to appear as points. The laws describing the motion of these point particle equivalents of real objects then turned out to be very simple. Can we exploit this knowledge to deal with objects that are near us and do have a finite size?

We can, provided we imagine such a body itself to be divided into many parts, each small enough to appear as a point. To be precise, we can think of these as the molecules comprising the object; the sort of apparatus and phenomena involved in most mechanics experiments would certainly not treat a molecule as anything other than a point particle. It is not necessary, however, to subdivide (mentally) an object to such an extent, for groups of many millions of molecules would still behave as point particles as far as this discussion is concerned.

It does not matter whether the 'object' is a rigid body (in which the point particles remain at fixed distances from each other) or a volume of gas (in which they are free to move in any direction relative to each other) or a liquid (in which such freedom of movement is restricted by the fact that the average distance between neighbouring particles is fixed). Moreover, the particles do not need to be of equal mass. This will enable our discussion to include groups of particles, each of which represents a macroscopic object viewed from afar (for example two masses connected by a string or the planets orbiting the sun).

Applying the force and momentum concepts to a group of particles is straightforward, and leads to very simple results which enable us, from some points of view, still to think of a group of particles as a single particle. The energy considerations, although still valid, lead to rather more complicated results, and show up some of the limitations in the 'single particle' view.

The chief extension of our ideas, however, comes from considering rotations, which leads to the concepts of torque and angular momentum. These can take on great importance even for a single particle, as we shall see in Chapter 6, and are fundamental for analysing the motion of a rigid body, as shown in Chapter 8.

5.1. Force and momentum

Suppose we have a system of N particles, with masses $m_1, m_2, \ldots, m_N$ and with position vectors $\mathbf{X}_1, \mathbf{X}_2, \ldots, \mathbf{X}_N$ relative to the origin O of a fixed frame of reference. The ith particle will have a velocity

$$\mathbf{V}_i = \frac{\mathrm{d}\mathbf{X}_i}{\mathrm{d}t}$$

and a momentum

$$\mathbf{P}_i = m_i \mathbf{V}_i$$

which will change according to the resultant force acting on it.

This force may be composed partly of forces whose origin is outside the system. The (vector) sum of all such forces acting on the ith particle we shall call the *external* force, $\mathbf{F}_i$, acting on the particle. There may also be forces between the particles. The sum of those acting on the ith particle we shall call the *internal* force, $\mathbf{f}_i$, acting on the particle (see Figure 5.1). Then the equation of motion of the particle is

$$\frac{\mathrm{d}\mathbf{P}_i}{\mathrm{d}t} = \mathbf{F}_i + \mathbf{f}_i.$$

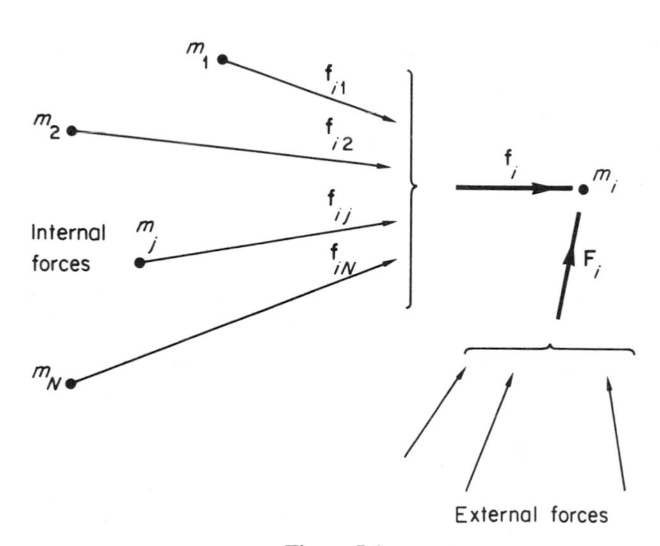

Figure 5.1

Summing over all the particles gives

$$\frac{\mathrm{d}}{\mathrm{d}t}\left(\sum_i \mathbf{P}_i\right) = \sum_i \left(\frac{\mathrm{d}\mathbf{P}_i}{\mathrm{d}t}\right) = \sum_i \mathbf{F}_i + \sum_i \mathbf{f}_i. \tag{5.1}$$

Now since $\mathbf{f}_i$ is the sum of the forces exerted on the ith particle by all the others we may write it as

$$\mathbf{f}_i = \sum_{j \neq i} \mathbf{f}_{ij},$$

where $\mathbf{f}_{ij}$ is the force exerted on particle i by particle j. Hence

$$\sum_i \mathbf{f}_i = \sum \sum_{j \neq i} \mathbf{f}_{ij}. \tag{5.2}$$

This double summation is composed of pairs of forces $\mathbf{f}_{ij} + \mathbf{f}_{ji}$. Since our third basic law states that forces between particles are equal and opposite, each of these pairs is zero. Then

$$\sum_i \mathbf{f}_i = 0. \tag{5.3}$$

Thus if we define

$$\mathbf{P} = \sum_i \mathbf{P}_i \tag{5.4}$$

as the *total momentum* of the system, and

$$\mathbf{F}_{\text{ext}} = \sum_i \mathbf{F}_i \tag{5.5}$$

as the *total external* force acting on the system, equations (5.3), (5.4) and (5.5) enable us to write equation (5.1) as

$$\frac{\mathrm{d}\mathbf{P}}{\mathrm{d}t} = \mathbf{F}_{\text{ext}}. \tag{5.6}$$

(See Section 5.9.1, page 132.)

It should be noted that $\mathbf{F}_{\text{ext}}$ is not the vector sum of a number of forces all necessarily acting at the same point. It is a force, each of whose components is the sum of the corresponding components of the various external forces which act on the system. Thus it is only the magnitudes and directions of these forces which determine $\mathbf{F}_{\text{ext}}$, and *not the points at which they act.*

5.2. Torque and angular momentum

We know that a force can cause a rotation. To gain some insight into this process let us consider how we use a spanner to tighten a nut (Figure 5.2). We apply a force $\mathbf{F}$ at the end of the arm of the spanner in order to rotate the nut in a clockwise direction about its axis, O, and, if the nut has a right-handed thread, it will move along the axis away from the reader.

Several features of this process are familiar to us. The axis of the nut's rotation is perpendicular both to the arm of the spanner and to the direction of the force, and therefore to the plane in which they both lie. We can increase

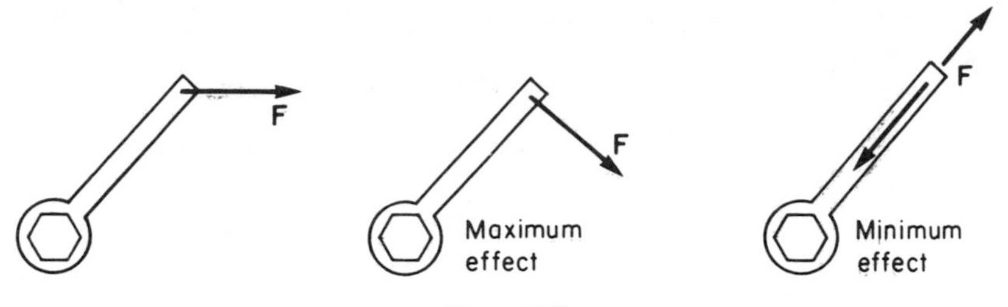

Figure 5.2

the strength of the twist by increasing either the force or the length of the spanner. Moreover, to use a given force with greatest effect we apply it at right angles to the arm of the spanner. The effect diminishes as we move away from this optimum direction until we have no twisting effect at all by the time the force is pointing directly towards, or away from, the axis (Figure 5.2).

It is also true that the effect diminishes if the force has a component along the axis of the screw thread; it has maximum effect when it is perpendicular both to the arm of the spanner and to the screw axis.

We can put these ideas into a definite form by representing the 'twisting' effect of a force as the vector- or cross-product

$$\mathbf{G} = \mathbf{x} \wedge \mathbf{F}.$$

$\mathbf{x}$ is the position vector of the point P at which the force $\mathbf{F}$ acts, relative to O, as shown in Figure 5.3. Then $\mathbf{G}$, which is known as the *torque* or *moment of*

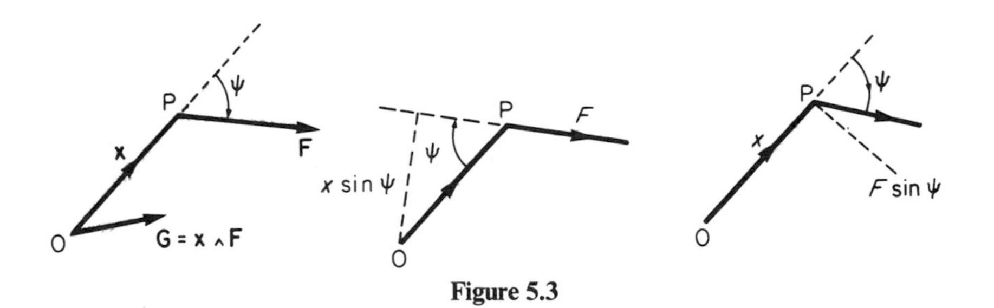

Figure 5.3

the force about 0, is the vector whose direction lies along the perpendicular to $\mathbf{x}$ and $\mathbf{F}$ (that is, the normal to the plane containing $\mathbf{x}$ and $\mathbf{F}$) and whose magnitude is

$$G = xF \sin \psi,$$

where ψ is the positive angle of rotation from $\mathbf{x}$ to $\mathbf{F}$. An angle is counted *positive* when the rotation from $\mathbf{x}$ to $\mathbf{F}$ about the normal is right-handed;

that is, clockwise to a person looking along the direction of the normal. (See Section 5.9.2, page 134.)

The magnitude G may be thought of either as

$$(F) \times (x \sin \psi) = (\text{force}) \times (\text{shortest distance from origin to line of application of force})$$

or as

$$(x) \times (F \sin \psi) = (\text{radial distance}) \times (\text{component of force perpendicular to radius})$$

as shown in Figure 5.3. We may note from the first of these expressions that the torque arising from a force depends only upon the position of the point about which the torque is to be calculated and the line of action of the force; it does not matter where along this line the force is actually applied. (See Section 5.9.3, page 138.)

Using a spanner to rotate a nut is not in practice an example of dynamics, since we usually tighten it against increasing friction until it comes to rest. However, very similar considerations apply if we try to set a wheel rotating about its axis. We do so most easily by using the largest force we can, applying it at the rim rather than near the axis, and along the tangent to the rim (i.e. at right angles to the radius vector from the axis to the point of application). Here, too, the twisting effect of the force is in accord with defining it as the moment of the force.

We have seen already that it is the momentum of a particle that responds to the applied force. There is a corresponding quantity that responds to the moment of a force. This is the *angular momentum*, or *moment of the momentum*,

$$\mathbf{J} = \mathbf{x} \wedge \mathbf{p}.$$

Differentiating this,

$$\frac{d\mathbf{J}}{dt} = \mathbf{x} \wedge \frac{d\mathbf{p}}{dt} + \frac{d\mathbf{x}}{dt} \wedge \mathbf{p}.$$

Now

$$\mathbf{x} \wedge \frac{d\mathbf{p}}{dt} = \mathbf{x} \wedge \mathbf{F} = \mathbf{G}$$

and

$$\frac{d\mathbf{x}}{dt} \wedge \mathbf{p} = \frac{d\mathbf{x}}{dt} \wedge m\frac{d\mathbf{x}}{dt} = 0,$$

since the angle between $d\mathbf{x}/dt$ and itself is zero. Thus

$$\mathbf{G} = \frac{d\mathbf{J}}{dt}, \tag{5.7}$$

which is analogous to

$$\mathbf{F} = \frac{d\mathbf{P}}{dt}$$

for 'ordinary' momentum. The latter is often called *linear* momentum to distinguish it from *angular* momentum. (See Section 5.9.4, page 139.)

As we can see from Figure 5.4, if $\mathbf{b}$ is the vector from O perpendicular to the line of motion of the particle, then, since

$$b = x \sin \psi$$

and $\mathbf{b}$, $\mathbf{x}$ and $\mathbf{v}$ all lie in the same plane, the angular momentum may be written as

$$\mathbf{J} = m\mathbf{b} \wedge \mathbf{v}.$$

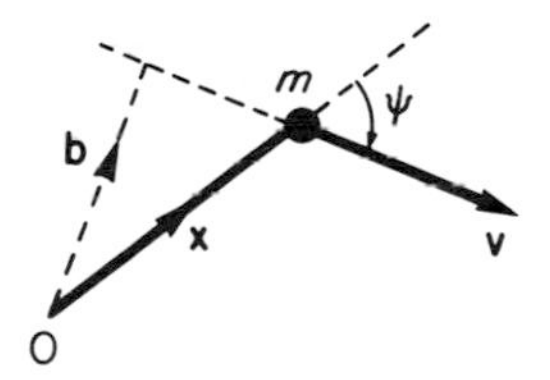

Figure 5.4

Hence, if the particle has a constant momentum $m\mathbf{v}$, which implies that $\mathbf{v}$ and $\mathbf{b}$ are constant, its angular momentum about any fixed point is constant.

As the particle moves, its radius vector will rotate about O with an angular velocity ω, say. Since its end has a velocity component $v \sin \psi$ perpendicular to its length,

$$\omega = \frac{v \sin \psi}{x}, \tag{5.8}$$

and the axis about which the rotation takes place is parallel with $\mathbf{x} \wedge \mathbf{v}$. Thus if we define $\boldsymbol{\omega}$ as a vector with this direction and the magnitude (5.8) the angular momentum has magnitude

$$J = mxv \sin \psi = mx^2\omega,$$

and its direction is the same as that of $\boldsymbol{\omega}$. Thus we may also write

$$\mathbf{J} = mx^2\boldsymbol{\omega}, \tag{5.9}$$

which is a form we shall use in the next section.

The angular momentum of a single particle is very important in analysing such problems as the planetary orbits and the scattering of nuclear particles.

For the present, however, let us carry on with our description of the N-particle system. For the ith particle,

$$\mathbf{G}_i = \frac{\mathrm{d}\mathbf{J}_i}{\mathrm{d}t},$$

and, summing over all particles,

$$\mathbf{G} = \sum_i \mathbf{G}_i = \sum_i \frac{\mathrm{d}\mathbf{J}_i}{\mathrm{d}t} = \frac{\mathrm{d}}{\mathrm{d}t}\left(\sum_i \mathbf{J}_i\right) = \frac{\mathrm{d}\mathbf{J}}{\mathrm{d}t}, \tag{5.10}$$

where $\mathbf{G}$ is the total torque of all the forces acting on the particles and $\mathbf{J}$ is their total angular momentum. Now

$$\begin{aligned}\sum_i \mathbf{G}_i &= \sum_i \mathbf{X}_i \wedge (\mathbf{F}_i + \mathbf{f}_i) \\ &= \sum_i \mathbf{X}_i \wedge \mathbf{F}_i + \sum_i \mathbf{X}_i \wedge \mathbf{f}_i.\end{aligned} \tag{5.11}$$

$\sum_i \mathbf{X}_i \wedge \mathbf{F}_i$ is the moment of the external forces only, and $\sum_i \mathbf{X}_i \wedge \mathbf{f}_i$ is that of the internal forces. Using equation (5.2)

$$\sum_i \mathbf{X}_i \wedge \mathbf{f}_i = \sum_{i \neq j}\sum \mathbf{X}_i \wedge \mathbf{f}_{ij} = \sum_{i<j}\sum (\mathbf{X}_i - \mathbf{X}_j) \wedge \mathbf{f}_{ij},$$

where, in the last summation, we are taking the terms in pairs and using the fact that

$$\mathbf{f}_{ij} = -\mathbf{f}_{ji}.$$

Now $\mathbf{X}_i - \mathbf{X}_j$ is the vector joining particle j to particle i and, according to Newton's third law, this will be parallel with $\mathbf{f}_{ij}$. Therefore

$$(\mathbf{X}_i - \mathbf{X}_j) \wedge \mathbf{f}_{ij} = 0,$$

and the total moment of the internal forces is zero (see Figure 5.5). Thus in the right-hand side of equation (5.11) only external forces need be considered and equation (5.10) can therefore be written as

$$\frac{\mathrm{d}\mathbf{J}}{\mathrm{d}t} = \mathbf{G}_{\mathrm{ext}}. \tag{5.12}$$

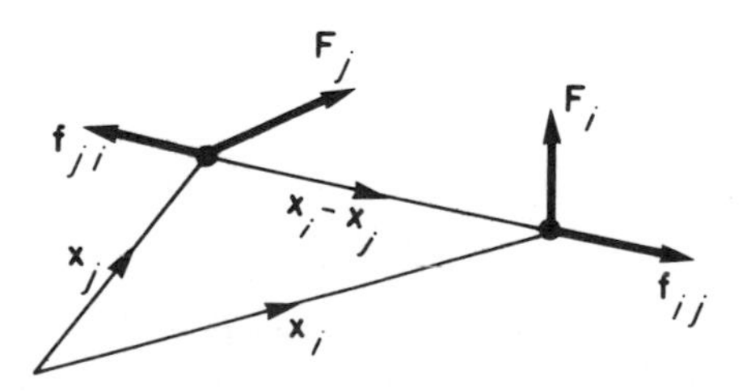

Figure 5.5

5.2.1. *Angular momentum of a rigid planar body; moment of inertia*

Suppose all the particles form a rigid planar body rotating about an axis perpendicular to its plane. Since the body is rigid the angular velocity $\boldsymbol{\omega}$ is the same for all its constituent particles. Then from equation (5.9) the total angular momentum is

$$\mathbf{J} = \sum_i \mathbf{J}_i = \sum_i m_i x_i^2 \boldsymbol{\omega} = I\boldsymbol{\omega}, \tag{5.13}$$

where

$$I = \sum_i m_i x_i^2 \tag{5.14}$$

is the *moment of inertia* of the body about the axis. When the mass is continuously distributed the expression for the moment of inertia becomes

$$I = \int x^2 \, \mathrm{d}m, \tag{5.15}$$

where x is the distance of dm from the axis and the integral is taken over the whole body. (See Section 5.9.5, page 142.) Since all the particles are at fixed distances from the axis, I remains constant as the body rotates. Hence, from equations (5.12) and (5.13),

$$\mathbf{G}_{\text{ext}} = I\frac{\mathrm{d}\boldsymbol{\omega}}{\mathrm{d}t}. \tag{5.16}$$

5.3. Centre-of-mass; momentum, kinetic energy, angular momentum

The motion of a group of particles can often be more easily understood by considering the *centre-of-mass*. This is the point, within the system but not necessarily coinciding with any of the particles, whose position vector, $\mathbf{X}$, is defined by

$$M\mathbf{X} = \sum_i m_i \mathbf{X}_i, \tag{5.17}$$

where

$$M = \sum_i m_i$$

is the total mass of the system.

Then if

$$\mathbf{V} = \frac{\mathrm{d}\mathbf{X}}{\mathrm{d}t}$$

is the velocity of the centre-of-mass, the total momentum of the system is

$$\mathbf{P} = \sum_i \mathbf{P}_i = \sum_i m_i \frac{\mathrm{d}\mathbf{X}_i}{\mathrm{d}t}$$

$$= M\frac{\mathrm{d}\mathbf{X}}{\mathrm{d}t} = M\mathbf{V}.$$

We may therefore write

$$\mathbf{P} = \mathbf{P}_{\text{cm}}, \tag{5.18}$$

and, using equation (5.6),

$$\mathbf{F}_{\text{ext}} = \frac{\mathrm{d}\mathbf{P}_{\text{cm}}}{\mathrm{d}t}, \tag{5.19}$$

where $\mathbf{P}_{\text{cm}}$ is the momentum of a single particle of mass M moving with the velocity of the centre-of-mass.

Since the points of application of the various external forces, $\mathbf{F}_i$, are unimportant in determining their total, $\mathbf{F}_{\text{ext}}$, we could take them all as acting at the centre-of-mass, as shown in Figure 5.6.

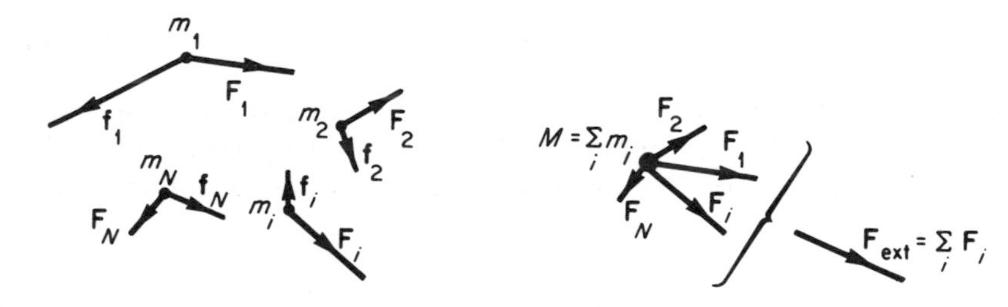

Figure 5.6

It is just the preceding results which are responsible for the observations, noted in Chapter 1, that it is only the force and not its point of application that determines the motion of an object viewed from afar, and that it is only the total mass of the object and not its shape that affects its observed response to the force. What we can now say precisely is that the point particle representing the object at a distance is its total mass concentrated at its centre-of-mass.

When we refer the positions of the particles to their centre-of-mass we obtain some important simplifications. If $\mathbf{x}_i$ is the position vector of particle i relative to the centre-of-mass,

$$\mathbf{X}_i = \mathbf{X} + \mathbf{x}_i.$$

Substituting this into equation (5.17) we have

$$\sum_i m_i \mathbf{x}_i = 0, \tag{5.20}$$

and differentiating this

$$\sum_i m_i \frac{\mathrm{d}\mathbf{x}_i}{\mathrm{d}t} = \sum_i m_i \mathbf{v}_i = 0, \tag{5.21}$$

where $\mathbf{v}_i$ is the velocity of the ith particle relative to the centre-of-mass. Hence if $\mathbf{p}_i$ and $\mathbf{P}_{\text{rel}}$ are the individual and total momenta of the particles arising from their velocities relative to the centre-of-mass,

$$\mathbf{P}_{\text{rel}} = \sum_i \mathbf{p}_i = 0. \tag{5.22}$$

Since

$$\mathbf{V}_i = \mathbf{V} + \mathbf{v}_i,$$

the total kinetic energy of the particles is

$$\begin{aligned}\mathscr{T} &= \tfrac{1}{2}\sum_i m_i \mathbf{V}_i^2 = \tfrac{1}{2}\sum_i m_i(\mathbf{V} + \mathbf{v}_i)^2 \\ &= \tfrac{1}{2}\sum_i m_i \mathbf{V}^2 + \sum_i m_i \mathbf{v}_i \,.\, \mathbf{V} + \tfrac{1}{2}\sum_i m_i \mathbf{v}_i^2.\end{aligned}$$

Now

$$\sum_i m_i \mathbf{V}^2 = \left(\sum_i m_i\right)\mathbf{V}^2 = M\mathbf{V}^2$$

and

$$\sum_i m_i \mathbf{v}_i \,.\, \mathbf{V} = \left(\sum_i m_i \mathbf{v}_i\right) . \, \mathbf{V} = 0,$$

because of equation (5.21). Hence

$$\begin{aligned}\mathscr{T} &= \tfrac{1}{2}M\mathbf{V}^2 + \tfrac{1}{2}\sum_i m_i \mathbf{v}_i^2 \\ &= \mathscr{T}_{\text{cm}} + \mathscr{T}_{\text{rel}},\end{aligned} \tag{5.23}$$

where

$$\mathscr{T}_{\text{cm}} = \tfrac{1}{2}M\mathbf{V}^2 \tag{5.24}$$

is the kinetic energy of the total mass moving with the velocity of the centre-of-mass and

$$\mathscr{T}_{\text{rel}} = \tfrac{1}{2}\sum_i m_i \mathbf{v}_i^2 \tag{5.25}$$

is the total kinetic energy the particles possess by virtue of their velocities *relative* to the centre-of-mass. We can see from the results (5.18) and (5.23) that while the momentum of a system of particles is entirely determined by the motion of its centre-of-mass, its energy content is not and that it is quite possible for a system to possess $\mathscr{T}_{\text{rel}}$, or 'internal' kinetic energy, which would not be apparent to the observer who sees the system as a point particle.

The total angular momentum can be expressed as follows:

$$\begin{aligned}\sum_i \mathbf{J}_i &= \sum_i \mathbf{X}_i \wedge \mathbf{P}_i \\ &= \sum_i (\mathbf{X} + \mathbf{x}_i) \wedge (m_i\mathbf{V} + \mathbf{p}_i) \\ &= \sum_i \mathbf{X} \wedge m_i\mathbf{V} + \sum_i \mathbf{X} \wedge \mathbf{p}_i + \sum_i \mathbf{x}_i \wedge m_i\mathbf{V} + \sum_i \mathbf{x}_i \wedge \mathbf{p}_i.\end{aligned}$$

Now

$$\sum_i \mathbf{X} \wedge m_i\mathbf{V} = \mathbf{X} \wedge \left(\sum_i m_i\right)\mathbf{V} = \mathbf{X} \wedge M\mathbf{V} = \mathbf{X} \wedge \mathbf{P} = \mathbf{J}_{\text{cm}} \tag{5.26}$$

is the angular momentum of the whole mass concentrated at the centre-of-mass and

$$\sum_i \mathbf{x}_i \wedge \mathbf{p}_i = \mathbf{J}_{\text{rel}} \tag{5.27}$$

is the angular momentum relative to the centre-of-mass, while

$$\sum_i \mathbf{X} \wedge \mathbf{p}_i = \mathbf{X} \wedge \left(\sum_i \mathbf{p}_i\right) = 0$$

and

$$\sum_i \mathbf{x}_i \wedge m_i\mathbf{V} = \left(\sum_i m_i\mathbf{x}_i\right) \wedge \mathbf{V} = 0$$

by virtue of equations (5.22) and (5.20). Thus the total angular momentum is

$$\mathbf{J} = \sum_i \mathbf{J}_i = \mathbf{J}_{\text{cm}} + \mathbf{J}_{\text{rel}}, \tag{5.28}$$

and, from equation (5.12),

$$\mathbf{G}_{\text{ext}} = \frac{\mathrm{d}\mathbf{J}_{\text{cm}}}{\mathrm{d}t} + \frac{\mathrm{d}\mathbf{J}_{\text{rel}}}{\mathrm{d}t}. \tag{5.29}$$

In evaluating $\mathbf{G}_{\text{ext}}$ it is important to remember that, in contrast to the case of linear momentum (equation 5.19), the line of action, not merely the direction, of each of the external forces, $\mathbf{F}_i$, must be known.

5.3.1. Centre-of-gravity

The centre-of-mass has a further significance when the whole system lies within a constant gravitational field. The external force on the ith particle is then

$$\mathbf{F}_i = m_i\mathbf{g},$$

where $\mathbf{g}$ is the same for all particles (the gravitational force on unit mass).

The total external force is

$$\mathbf{F}_{\text{ext}} = \sum_i \mathbf{F}_i = \left(\sum_i m_i\right)\mathbf{g} = M\mathbf{g}. \tag{5.30}$$

The external torque on the ith particle is

$$\mathbf{G}_i = \mathbf{X}_i \wedge m_i\mathbf{g} = (\mathbf{X} + \mathbf{x}_i) \wedge m_i\mathbf{g}.$$

The total external torque is

$$\begin{aligned}\mathbf{G}_{\text{ext}} = \sum_i \mathbf{G}_i &= \sum_i \mathbf{X} \wedge m_i\mathbf{g} + \sum_i \mathbf{x}_i \wedge m_i\mathbf{g} \\ &= \left(\sum_i m_i\right)\mathbf{X} \wedge \mathbf{g} + \left(\sum_i m_i\mathbf{x}_i\right) \wedge \mathbf{g}.\end{aligned}$$

The last term vanishes according to equation (5.20), leaving

$$\mathbf{G}_{\text{ext}} = \mathbf{X} \wedge M\mathbf{g}. \tag{5.31}$$

Equations (5.30) and (5.31) show that the total force and torque may be calculated by assuming that gravity is acting upon a single particle of mass M at the centre-of-mass. From this point of view the centre-of-mass may be called the *centre-of-gravity*, although the former concept is the more fundamental since it does not depend upon the existence of any forces.

5.4. Conservation of energy

Suppose all the forces, external and internal, can be derived from potential energies. Thus for the ith particle

$$\mathscr{V}_i = \mathscr{V}_{i\,\text{ext}} + \mathscr{V}_{i\,\text{int}},$$

and, summing over all the particles,

$$\mathscr{V} = \sum_i \mathscr{V}_i = \sum_i \mathscr{V}_{i\,\text{ext}} + \sum_i \mathscr{V}_{i\,\text{int}} = \mathscr{V}_{\text{ext}} + \mathscr{V}_{\text{int}}. \tag{5.32}$$

Energy will be conserved for the whole system:

$$\mathscr{T} + \mathscr{V} = \mathscr{E} = \text{constant}.$$

Therefore, from equations (5.23) and (5.32),

$$(\mathscr{T}_{\text{cm}} + \mathscr{V}_{\text{ext}}) + (\mathscr{T}_{\text{rel}} + \mathscr{V}_{\text{int}}) = \mathscr{E}. \tag{5.33}$$

5.5. Isolated system of particles

When no external forces act on the system,

$$\mathbf{F}_{\text{ext}} = \mathbf{G}_{\text{ext}} = 0, \qquad \mathscr{V}_{\text{ext}} = \text{constant}. \tag{5.34}$$

Then, from equations (5.18), (5.19) and (5.34), total linear momentum will be conserved,

$$\mathbf{P} = \mathbf{P}_{cm} = \text{constant}, \tag{5.35}$$

and, from equations (5.28), (5.29) and (5.34), so will total angular momentum,

$$\mathbf{J} = \mathbf{J}_{cm} + \mathbf{J}_{rel} = \text{constant}. \tag{5.36}$$

However, since the centre-of-mass will move with constant velocity, $\mathbf{J}_{cm}$ will be constant, so that both parts of equation (5.36) will be separately conserved:

$$\mathbf{J}_{cm} = \text{constant}, \qquad \mathbf{J}_{rel} = \text{constant}. \tag{5.37}$$

Since

$$\mathscr{T}_{cm} = \frac{1}{2}M\mathbf{V}^2 = \frac{(M\mathbf{V})^2}{2M} = \frac{\mathbf{P}_{cm}^2}{2M},$$

equation (5.35) shows that the centre-of-mass kinetic energy will be conserved:

$$\mathscr{T}_{cm} = \text{constant}. \tag{5.38}$$

Therefore, from equation (5.33), the total internal energy will be conserved:

$$\mathscr{T}_{int} + \mathscr{V}_{int} = \mathscr{E}_{int} = \text{constant}. \tag{5.39}$$

Although the results (5.35), (5.37) and (5.39) are strictly true only in the absence of external forces, they are often very good approximations even when finite external forces are present. The conditions for this to be true are that the distances travelled by thc particles of a system and the times taken should be short enough for the external forces to cause only a negligible change in the energies and momenta of the particles during the processes under consideration. For example, the application of conservation principles to any normal collision phenomena in an earth laboratory will not be affected by the earth's gravity.

5.6. Moving frame of reference

It is often more convenient to make physical measurements in a moving laboratory. How does this affect the mechanics of a system of particles? In Chapter 8 this problem is treated at some length. Here we shall deal simply with the case where the origin of the moving reference frame, O′, has a constant velocity, **U**, relative to the fixed origin, O, that we have been using so far:

$$\overrightarrow{OO'} = \mathbf{X}_{O'} = \mathbf{A} + \mathbf{U}t,$$

where **A** is the displacement of O′ at time zero (see Figure 5.7).

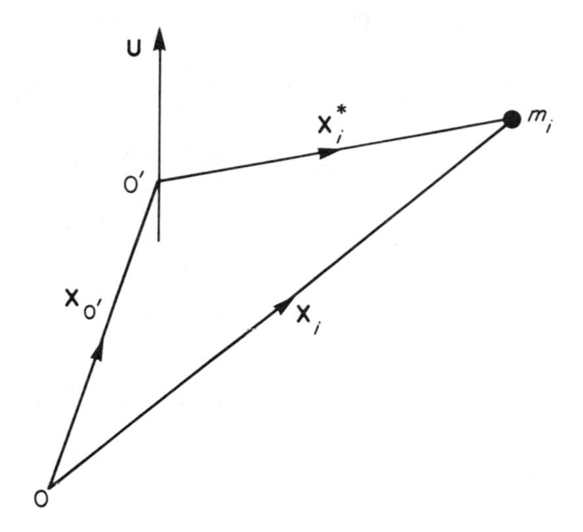

Figure 5.7

Then if starred quantities refer to the position velocity, etc., of a particle relative to O′,

$$\mathbf{X}_i = \mathbf{X}_{O'} + \mathbf{X}_i^* = \mathbf{A} + \mathbf{U}t + \mathbf{X}_i^*$$

$$\mathbf{V}_i = \frac{\mathrm{d}\mathbf{X}_i}{\mathrm{d}t} = \mathbf{U} + \frac{\mathrm{d}\mathbf{X}_i^*}{\mathrm{d}t} = \mathbf{U} + \mathbf{V}_i^*.$$

Hence the total momentum is

$$\mathbf{P} = \sum_i m_i\mathbf{V}_i = \sum_i m_i\mathbf{U} + \sum_i m_i\mathbf{V}_i^*$$

$$= M\mathbf{U} + \sum \mathbf{P}_i^* = M\mathbf{U} + \mathbf{P}^*.$$

Since $M\mathbf{U}$ is a constant,

$$\frac{\mathrm{d}\mathbf{P}}{\mathrm{d}t} = \frac{\mathrm{d}\mathbf{P}^*}{\mathrm{d}t},$$

and substitution into equation (5.6) then gives

$$\frac{\mathrm{d}\mathbf{P}^*}{\mathrm{d}t} = \mathbf{F}_{\text{ext}}. \tag{5.40}$$

(See Section 5.9.6, page 142.) The angular momentum is

$$\begin{aligned}\mathbf{J} = \sum_i \mathbf{X}_i \wedge m_i\mathbf{V}_i &= \sum_i (\mathbf{X}_{O'} + \mathbf{X}_i^*) \wedge m_i(\mathbf{U} + \mathbf{V}_i^*) \\ &= \sum_i \mathbf{X}_{O'} \wedge m_i\mathbf{U} + \sum_i \mathbf{X}_{O'} \wedge m_i\mathbf{V}_i^* \\ &\quad + \sum_i \mathbf{X}_i^* \wedge m_i\mathbf{U} + \sum_i \mathbf{X}_i^* \wedge m_i\mathbf{V}_i^* \\ &= M\mathbf{X}_{O'} \wedge \mathbf{U} + \mathbf{X}_{O'} \wedge \left(\sum_i \mathbf{P}_i^*\right) \\ &\quad + \left(\sum_i m_i\mathbf{X}_i^*\right) \wedge \mathbf{U} + \sum_i \mathbf{X}_i^* \wedge \mathbf{P}_i^*.\end{aligned}$$

Then, since

$$\frac{\mathrm{d}\mathbf{X}_{O'}}{\mathrm{d}t} = \mathbf{U}$$

and is constant,

$$\begin{aligned}\frac{\mathrm{d}\mathbf{J}}{\mathrm{d}t} &= M\mathbf{U} \wedge \mathbf{U} + \mathbf{X}_{O'} \wedge \frac{\mathrm{d}\mathbf{P}^*}{\mathrm{d}t} + \mathbf{U} \wedge \mathbf{P}^* + \mathbf{P}^* \wedge \mathbf{U} + \frac{\mathrm{d}}{\mathrm{d}t}\left(\sum_i \mathbf{J}_i^*\right) \\ &= \mathbf{X}_{O'} \wedge \frac{\mathrm{d}\mathbf{P}^*}{\mathrm{d}t} + \frac{\mathrm{d}\mathbf{J}^*}{\mathrm{d}t}.\end{aligned} \tag{5.41}$$

The total external torque is

$$\begin{aligned}\mathbf{G}_{\mathrm{ext}} = \sum_i \mathbf{X}_i \wedge \mathbf{F}_i &= \sum_i (\mathbf{X}_{O'} + \mathbf{X}_i^*) \wedge \mathbf{F}_i \\ &= \mathbf{X}_{O'} \wedge \mathbf{F}_{\mathrm{ext}} + \mathbf{G}_{\mathrm{ext}}^*.\end{aligned} \tag{5.42}$$

Hence, from equations (5.12), (5.40), (5.41) and (5.42),

$$\frac{\mathrm{d}\mathbf{J}^*}{\mathrm{d}t} = \mathbf{G}_{\mathrm{ext}}^*. \tag{5.43}$$

Comparing the results (5.40) and (5.43) with (5.6) and (5.12), we see that the laws they describe have exactly the same form in the two cases. The further results (5.19), (5.23) and (5.29), which are derived from (5.6) and (5.12), will likewise retain the same form. Thus we can choose any point moving with constant velocity as the reference point for the analysis of a problem, and use all the equations of motion and conservation laws so far derived as if it were at rest. In the absence of external forces the centre-of-mass itself is such a point and is frequently the best one to use. (See Section 5.9.7, page 144.)

5.7. Elastic and inelastic collisions

The discussion in Sections 2.6.3 and 2.6.4 led us to class as *elastic collisions* those collisions between the particles of a system in which the internal potential energy is the same before and after the collision and in which no other forms of energy, such as heat and radiation, are involved. This concept can be extended to three dimensions. For example two hardened steel balls colliding with a relative velocity of around $1\ \mathrm{m\,s^{-1}}$ would generate very little heat and their constitution and structure would remain the same whatever the directions of their motion before and after the collision. The collision would therefore be an elastic one. The collision time (the period they are in contact and slightly deformed) is so small and the distance travelled during this is so short that any finite external forces would have no effect upon the momenta and energies of the system. Thus $\mathscr{V}_{\text{ext}}$ is constant throughout the collision, and since $\mathscr{V}_{\text{int}}$ is also unchanged at the end of the collision,

$$\mathscr{T} = \mathscr{T}_{\text{cm}} + \mathscr{T}_{\text{rel}} = \mathscr{E} - \mathscr{V}_{\text{ext}} - \mathscr{V}_{\text{int}} = \text{constant}.$$

Since the external forces leave $\mathbf{P}_{\text{cm}}$ and therefore $\mathscr{T}_{\text{cm}}$ unchanged, it is also true that

$$\mathscr{T}_{\text{rel}} = \mathscr{E} - \mathscr{T}_{\text{cm}} - \mathscr{V}_{\text{ext}} - \mathscr{V}_{\text{int}} = \text{constant}. \tag{5.44}$$

Conservation of energy in these cases thus simply implies conservation of *kinetic* energy. However, it is easy to think of collisions for which this cannot be true. For example the model of Figure 5.8 consists of two masses, one of which has a ratchet spring to which the other becomes attached on colliding.

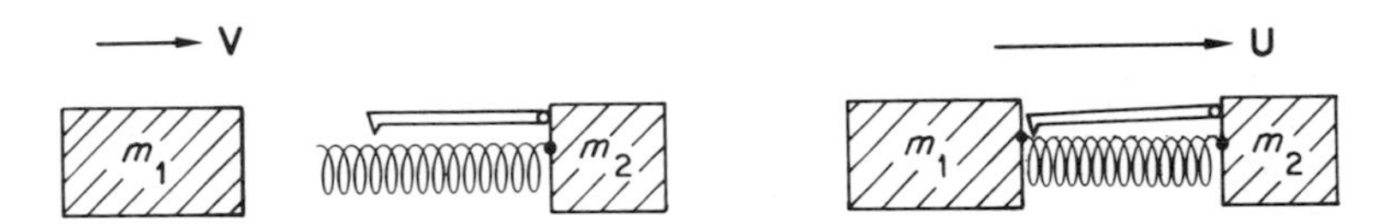

Figure 5.8

Initially one particle is stationary, the other approaching it with velocity **V**. During the collision the spring is compressed and is kept in this condition by the ratchet, the composite particle then moving off with velocity **U**. When no external forces act, $\mathbf{P}_{\text{cm}}$ and therefore $\mathscr{T}_{\text{cm}}$ will remain constant. After the collision there is no relative motion and therefore $\mathscr{T}_{\text{cm}}$ is the *total* kinetic energy. However, before the collision there will be some (positive) $\mathscr{T}_{\text{rel}}$ as well. Thus

$$\mathscr{T}_{\text{initial}} = \mathscr{T}_{\text{rel}} + \mathscr{T}_{\text{cm}} > \mathscr{T}_{\text{cm}} = \mathscr{T}_{\text{final}}. \tag{5.45}$$

Kinetic energy therefore decreases as a result of the collision. In a similar way, if we were to start with the moving compound particle and release the trigger so that the separate particles flew apart, there would be a gain in total kinetic energy.

It is clear in these cases how the energy balance is made up. In the first, the loss of kinetic energy is accounted for by the gain in potential energy of the compressed spring; in the second, this exchange is reversed. Since we know how to calculate the energy of a spring we can use conservation of energy as well as of momentum to analyse the motion, and can determine the final velocity and the compression of the spring.

However, energy need not be stored in such a simple way. For example, when a rocket ejects its burning fuel products it is, in effect, disintegrating, and the total kinetic energy of the rocket casing and the fuel is increasing. This increase is at the expense of the chemical energy of the fuel, which decreases during combustion. Energy overall is still conserved, but this principle is of no use for analysing the motion unless we can calculate chemical energies.

The situation is even more complicated when heat, radiation and other forms of energy can take part. Although we believe conservation of energy to be one of the most fundamental of physical laws, and still to be true in such complicated cases, its calculation now demands even more information about the characteristics of the various forms that energy can take.

All collisions or interactions between particles in which there is no simple conservation of total kinetic energy are called *inelastic*. As we have seen, any compound formation or disintegration is necessarily inelastic.

Thus conservation of energy is a law that must be applied with caution and often cannot be used for lack of information. Conservation of momentum, on the other hand, is usually much more easily formulated and generally applicable.

5.8. Impulsive forces

Insofar as the results of this chapter have been based upon the application of Newton's second law to each particle,

$$\frac{d\mathbf{P}_i}{dt} = m_i\frac{d\mathbf{V}_i}{dt} = \mathbf{F}_i + \mathbf{f}_i, \tag{5.46}$$

with its implication of finite masses, accelerations and forces, we might have some doubts about their validity when impulsive forces are involved. These appear to give sudden changes of velocity and therefore infinite accelerations which equation (5.46) would not properly describe.

In fact, no forces or accelerations are truly infinite. They may be very large and last for a very short time, but they are not discontinuous and *can* be described by equation (5.46). Nevertheless, the deductions to be made must be slightly modified. For a single particle we can write, from equation (5.46),

$$m_i \int_{(1)}^{(2)} d\mathbf{V}_i = \int_{(1)}^{(2)} (\mathbf{F}_i + \mathbf{f}_i)\, dt.$$

As in Section 5.1, terms involving internal forces cancel when this is summed over all particles, leaving only

$$\int_{(1)}^{(2)} \sum_i m_i \, \mathrm{d}\mathbf{V}_i = \int_{(1)}^{(2)} \left(\sum_i \mathbf{F}_i \right) \mathrm{d}t,$$

or

$$\mathbf{P}_{(2)} - \mathbf{P}_{(1)} = \int_{(1)}^{(2)} \mathbf{F}_{\text{ext}} \, \mathrm{d}t, \tag{5.47}$$

where $\mathbf{P}_{(1)}, \mathbf{P}_{(2)}$ are the total momenta of the system at the beginning and end of the period under consideration.

Suppose, because of the superposition of a large number of impulsive forces, the total external force has a variation in time typified by the component F_1 in Figure 5.9a. Then the total momentum will have a corresponding variation like that of Figure 5.9b.

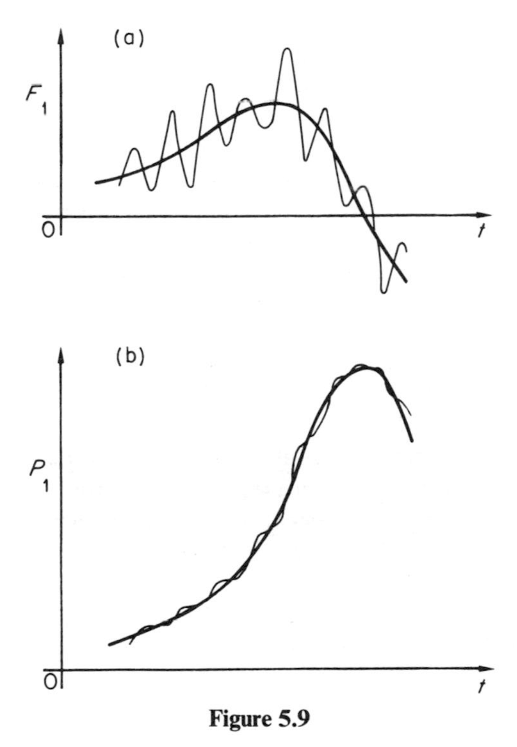

Figure 5.9

F_1 will have a 'spikey' shape because of the impulses which are acting and P_1 will have a correspondingly uneven shape, but if we cannot observe such detail they will appear as the smooth heavy curves shown in these figures. Equation (5.47) will then apply to these 'smoothed-out' components of force and momentum, and likewise to the other components. If we now take a small time interval,

$$t_{(2)} - t_{(1)} = \delta t,$$

the corresponding change in the smoothly varying momentum will be

$$\delta \mathbf{P} = \mathbf{P}_{(2)} - \mathbf{P}_{(1)} = \int_{t_{(1)}}^{t_{(1)}+\delta t} \mathbf{F}_{\text{ext}}\, \mathrm{d}t,$$

and in the limit

$$\frac{\mathrm{d}\mathbf{P}}{\mathrm{d}t} = \lim_{\delta t \to 0} \frac{\delta \mathbf{P}}{\delta t} = \mathbf{F}_{\text{ext}}.$$

By quite similar arguments we can show that

$$\frac{\mathrm{d}\mathbf{J}}{\mathrm{d}t} = \mathbf{G}_{\text{ext}},$$

where **J** and **G** are the 'smoothed-out' angular momentum and total external torque that are observed, rather than the rapidly varying quantities that truly represent the effect of a large number of small impulses. (See Section 5.9.8, page 144.)

It should be noted that the masses appearing in the summations that give $\mathbf{P}_{(1)}$ and $\mathbf{P}_{(2)}$ may not appear to be the same, since some particles may be at rest at $t_{(1)}$ and moving at $t_{(2)}$, and vice versa. In this sense $\mathrm{d}\mathbf{P}/\mathrm{d}t$ will involve changes of mass as well as of velocity. However, to describe this mathematically in all circumstances as

$$\frac{\mathrm{d}\mathbf{P}}{\mathrm{d}t} = \frac{\mathrm{d}}{\mathrm{d}t}(M\mathbf{V}) = M\frac{\mathrm{d}\mathbf{V}}{\mathrm{d}t} + \frac{\mathrm{d}M}{\mathrm{d}t}\mathbf{V}$$

can be very misleading unless it is quite clear what $\mathrm{d}\mathbf{V}/\mathrm{d}t$ and $\mathrm{d}M/\mathrm{d}t$ refer to. In practice, it is much safer to account for the momentum, $m_i\mathbf{V}_i$, of *every* part of a system before and after an interval δt, even when some of these are zero, as a preliminary to finding $\mathrm{d}\mathbf{P}/\mathrm{d}t$ (see problem 5.14). Similar considerations apply to the calculation of $\mathrm{d}\mathbf{J}/\mathrm{d}t$.

5.9. Comments and worked examples

5.9.1. See Section 5.1

Example 16

A hopper deposits sand at a rate $\mathrm{d}m/\mathrm{d}t$ with zero velocity upon a length of conveyor belt which has mass M and is moving horizontally with velocity v upon smooth supports. The upper surface of the belt is rough and drags the sand along with it.

Classify and discuss the roles played by the various forces which may act on this system. Obtain the equation of motion of the system and show that if the velocity v is to be kept constant the horizontal force that must be exerted on the belt is $v(\mathrm{d}m/\mathrm{d}t)$.

Internal forces. These are the forces within the belt that keep it together and the forces between the sand and the belt. The vertical components of the forces exerted by the belt on the sand balance the gravitational forces acting upon the sand and the horizontal components are the frictional forces that accelerate the sand up to the velocity of the belt. There are also the forces between the stationary sand and the hopper. The vertical components of the forces exerted by the hopper on the sand balance the gravitational forces on the sand and their horizontal components are zero in total since this sand is stationary.

External forces. There are gravitational forces on the belt and the moving sand, and upon the hopper and the stationary sand. There are also forces exerted on the belt by its supports and by any mechanism pulling the belt along.

In considering the changes of momentum of the whole system the internal forces all cancel each other. There is no vertical motion and therefore no change in the vertical component of momentum of any part of the system. Thus the total gravitational force on the moving sand and belt is cancelled by the vertical component of the forces exerted by the supports beneath (although both are increasing at the rate $g(\mathrm{d}m/\mathrm{d}t)$). Similarly, the total gravitational force on the stationary sand and hopper is cancelled by the total forces exerted by the supports of the hopper, which will therefore be vertical. (These forces will decrease at the rate $g(\mathrm{d}m/\mathrm{d}t)$.)

The only non-zero component of momentum of the whole system is the horizontal one which, as we can see from Figure 5.10, is

$$P = (M + m)v,$$

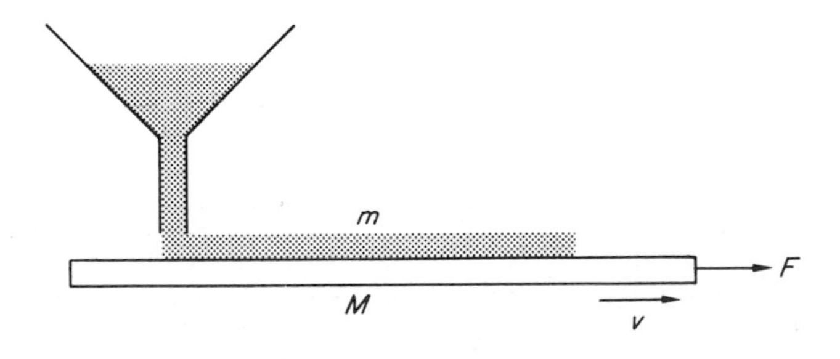

Figure 5.10

where m is the total mass of sand that has been deposited on the belt up to the time considered. Hence if F is the horizontal component of the external forces (i.e. the pull on the belt),

$$F = \frac{\mathrm{d}P}{\mathrm{d}t} = \frac{\mathrm{d}}{\mathrm{d}t}(M + m)v.$$

This is the required equation of motion. When v is constant it simplifies to

$$F = \frac{\mathrm{d}m}{\mathrm{d}t}v.$$

5.9.2. See Section 5.2

Note that although two vectors, **a** and **b**, may define a plane uniquely, there is an ambiguity about what is meant by the normal to that plane. As Figure 5.11 shows, we can choose either the unit vector $\hat{\mathbf{n}}$ to represent the

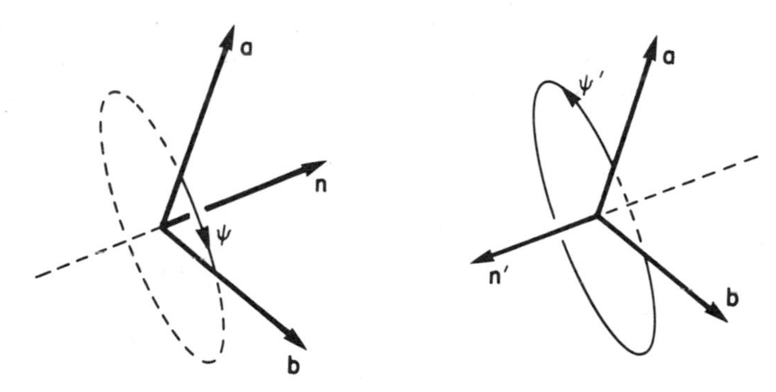

Figure 5.11

normal direction or the opposite unit vector $\hat{\mathbf{n}}'$. There is a similar ambiguity in measuring the *positive* angle of rotation from **a** to **b**, ψ corresponding to a positive rotation about $\hat{\mathbf{n}}$, and ψ' to a positive rotation about $\hat{\mathbf{n}}'$.

However,

$$\psi + \psi' = 2\pi,$$

and therefore

$$\sin\psi = -\sin\psi'.$$

Thus the vector with magnitude $ab\sin\psi$ and direction $\hat{\mathbf{n}}$ is the *same* as the vector with magnitude $ab\sin\psi'$ and direction $\hat{\mathbf{n}}'$. The vector-product as defined in Section 5.2 is therefore unambiguous. By considering in Figure 5.11 the angles of rotation involved in the definitions of $\mathbf{a} \wedge \mathbf{b}$ and $\mathbf{b} \wedge \mathbf{a}$, it is also clear that

$$\mathbf{a} \wedge \mathbf{b} = -\mathbf{b} \wedge \mathbf{a}. \tag{5.48}$$

Now the area of a triangle of sides a and b, with included angle ψ, is $\frac{1}{2}ab \sin \psi$. Hence if $\mathbf{\Delta}(\mathbf{a}, \mathbf{b})$ denotes a vector whose magnitude is the area of the triangle defined by $\mathbf{a}$ and $\mathbf{b}$, and whose direction is the normal to the plane containing these vectors, in the sense previously defined,

$$\mathbf{a} \wedge \mathbf{b} = 2\mathbf{\Delta}(\mathbf{a}, \mathbf{b}). \tag{5.49}$$

This geometrical interpretation enables us to prove the distribution law for the vector-product. Note first that if we project the triangle, $\mathbf{\Delta}$, in a direction $\hat{\mathbf{s}}$ (i.e. view it from a direction defined by the unit vector $\hat{\mathbf{s}}$, at an angle θ, say, to the normal $\mathbf{n}$ to the triangle), the area of the projection will be

$$\Delta \cos \theta = \mathbf{\Delta} \,.\, \hat{\mathbf{s}},$$

as shown in Figure 5.12a.

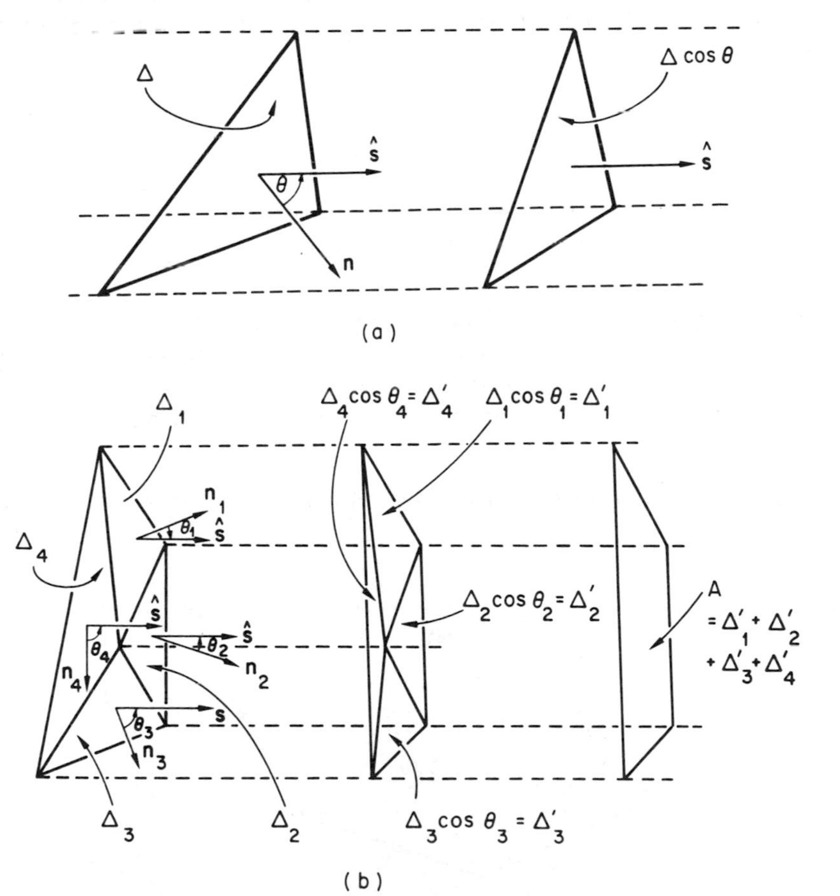

Figure 5.12

Any surface, not necessarily plane, can be projected in a chosen direction, $\hat{\mathbf{s}}$. The area, A, of this projection can be divided into a network of triangles Δ_i', each of which is the projection of a triangle Δ_i forming part of the original surface. Hence, summing over the surface,

$$A = \sum_i \Delta_i' = \sum_i \Delta_i \cos\theta_i = \sum_i \mathbf{\Delta}_i \,.\, \hat{\mathbf{s}}.$$

An example of a surface divided into four triangles is shown in Figure 5.12b.

Now if we view any closed non-reentrant surface from any direction we shall see a boundary curve or polygon dividing it into two parts. One part presents its *outside* to the viewer, with all the directions θ_i between 0 and $\pi/2$, while the other (hidden by the first) presents its *inside*, with all the directions θ_i between $\pi/2$ and π. Then if the former has a total projected area A the latter will clearly have a total projected area $-A$. Hence, summing over the whole surface,

$$\sum_i \mathbf{\Delta}_i \,.\, \hat{\mathbf{s}} = \left(\sum_i \mathbf{\Delta}_i\right) .\, \hat{\mathbf{s}} = 0.$$

Since this is true for any direction $\hat{\mathbf{s}}$ it follows that

$$\sum_i \mathbf{\Delta}_i = 0. \tag{5.50}$$

Let us apply this result to the polyhedron defined by the vectors **a**, **b** and **c**, shown in Figure 5.13. Two of the faces are already triangles, and we divide each of the three parallelograms into two equal coplanar triangles. Then, using equation (5.50) and making sure that the normal to each area is directed outwards,

$$2\mathbf{\Delta}(\mathbf{b}, \mathbf{a}) + 2\mathbf{\Delta}(\mathbf{c}, \mathbf{a}) + 2\mathbf{\Delta}(\mathbf{a}, \mathbf{b} + \mathbf{c}) + \mathbf{\Delta}(\mathbf{b}, \mathbf{b} + \mathbf{c}) + \mathbf{\Delta}(\mathbf{b} + \mathbf{c}, \mathbf{b}) = 0.$$

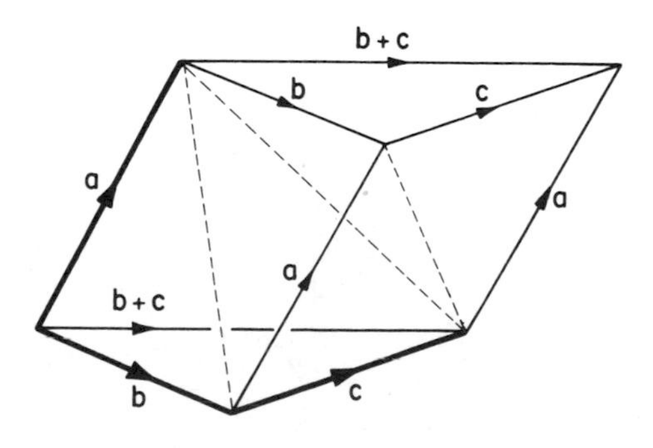

Figure 5.13

The last two vectors are equal and opposite, and applying the results (5.48) and (5.49) to the first three, we obtain

$$\mathbf{a} \wedge (\mathbf{b} + \mathbf{c}) = \mathbf{a} \wedge \mathbf{b} + \mathbf{a} \wedge \mathbf{c}. \tag{5.51}$$

Applying result (5.48) again to this gives

$$(\mathbf{b} + \mathbf{c}) \wedge \mathbf{a} = \mathbf{b} \wedge \mathbf{a} + \mathbf{c} \wedge \mathbf{a}. \tag{5.52}$$

Repeated applications of (5.51) and (5.52) then give the general distribution law:

$$\begin{aligned}(\mathbf{a} + \mathbf{a}' + \mathbf{a}'' + \ldots) \wedge (\mathbf{b} + \mathbf{b}' + \mathbf{b}'' + \ldots)\\ = \mathbf{a} \wedge \mathbf{b} + \mathbf{a} \wedge \mathbf{b}' + \mathbf{a}' \wedge \mathbf{b} + \mathbf{a}'' \wedge \mathbf{b} + \ldots.\end{aligned} \tag{5.53}$$

One immediate result of this law is that if two vectors are expressed in terms of the unit vectors **i**, **j**, **k** defining a right-handed Cartesian frame,

$$\mathbf{a} = a_1\mathbf{i} + a_2\mathbf{j} + a_3\mathbf{k}, \qquad \mathbf{b} = b_1\mathbf{i} + b_2\mathbf{j} + b_3\mathbf{k},$$

then

$$\begin{aligned}\mathbf{a} \wedge \mathbf{b} = a_1 b_1 \mathbf{i} \wedge \mathbf{i} + a_2 b_2 \mathbf{j} \wedge \mathbf{j} + a_3 b_3 \mathbf{k} \wedge \mathbf{k}\\ + a_2 b_3 \mathbf{j} \wedge \mathbf{k} + a_3 b_2 \mathbf{k} \wedge \mathbf{j} + \ldots.\end{aligned} \tag{5.54}$$

Now it is obvious from the definition of the vector-product and the fact that **i**, **j**, **k** are a right-handed orthogonal set of unit vectors that

$$\begin{aligned}&\mathbf{i} \wedge \mathbf{i} = \mathbf{j} \wedge \mathbf{j} = \mathbf{k} \wedge \mathbf{k} = 0;\\ &\mathbf{j} \wedge \mathbf{k} = -\mathbf{k} \wedge \mathbf{j} = \mathbf{i}, \qquad \mathbf{k} \wedge \mathbf{i} = -\mathbf{i} \wedge \mathbf{k} = \mathbf{j},\\ &\mathbf{i} \wedge \mathbf{j} = -\mathbf{j} \wedge \mathbf{i} = \mathbf{k}.\end{aligned} \tag{5.55}$$

Substituting the results (5.55) into the expression (5.54),

$$\mathbf{a} \wedge \mathbf{b} = (a_2 b_3 - a_3 b_2)\mathbf{i} + (a_3 b_1 - a_1 b_3)\mathbf{j} + (a_1 b_2 - a_2 b_1)\mathbf{k}.$$

This form of the vector-product may be written as the determinant

$$\mathbf{a} \wedge \mathbf{b} = \begin{vmatrix} \mathbf{i} & \mathbf{j} & \mathbf{k} \\ a_1 & a_2 & a_3 \\ b_1 & b_2 & b_3 \end{vmatrix}. \tag{5.56}$$

If we consider the gradient as a vector-operator

$$\nabla = \frac{\partial}{\partial x_1}\mathbf{i} + \frac{\partial}{\partial x_2}\mathbf{j} + \frac{\partial}{\partial x_3}\mathbf{k},$$

taking the vector-product of this with a vector **F** then gives, according to (5.56),

$$\nabla \wedge \mathbf{F} = \left(\frac{\partial F_3}{\partial x_2} - \frac{\partial F_2}{\partial x_3}\right)\mathbf{i} + \left(\frac{\partial F_1}{\partial x_3} - \frac{\partial F_3}{\partial x_1}\right)\mathbf{j} + \left(\frac{\partial F_2}{\partial x_1} - \frac{\partial F_1}{\partial x_2}\right)\mathbf{k}. \quad (5.57)$$

The Cartesian components of this are the same as those derived for **curl F**. This is the justification for the notation used in Section 4.5 (equation 4.36):

$$\mathbf{curl\ F} \equiv \nabla \wedge \mathbf{F}. \quad (5.58)$$

5.9.3. See Section 5.2

Example 17

G is the torque of a force **F** about a point O. $\hat{\mathbf{s}}$ is a unit vector defining an arbitrary direction and OL is a line through O in the direction $\hat{\mathbf{s}}$. MN is the common perpendicular to the line of action of **F** and to OL (see Figure 5.14).

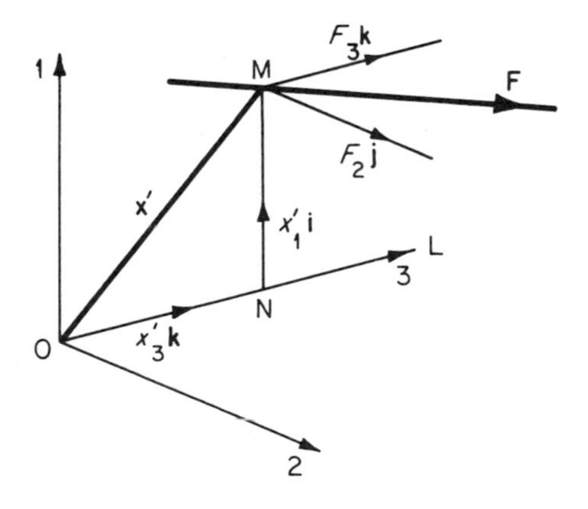

Figure 5.14

Show that the component of **G** in the direction $\hat{\mathbf{s}}$ is the product of MN with the component of **F** that is normal to the plane OLM.

We saw in Section 5.2 that in calculating the torque the force may be considered to act at any point along its line of action. Let us therefore take M as the point of action. Denoting its position vector by

$$\mathbf{x}' = \overrightarrow{\mathrm{OM}},$$

the torque is then

$$\mathbf{G} = \mathbf{x}' \wedge \mathbf{F}.$$

Now choose Cartesian axes 1 in the direction of NM, 3 along OL and 2 to complete the right-handed set. $\mathbf{x}'$ lies in the plane (3, 1) and may therefore be expressed as

$$\mathbf{x}' = x_1'\mathbf{i} + x_3'\mathbf{k}.$$

Since $\mathbf{F}$ is perpendicular to $\overrightarrow{NM}$ and therefore to $\mathbf{i}$, it may be written as

$$\mathbf{F} = F_2\mathbf{j} + F_3\mathbf{k}.$$

Hence the torque is

$$\mathbf{G} = \mathbf{x}' \wedge \mathbf{F} = \begin{vmatrix} \mathbf{i} & \mathbf{j} & \mathbf{k} \\ x_1' & 0 & x_3' \\ 0 & F_2 & F_3 \end{vmatrix},$$

from which we can see that the component in the direction of $\hat{\mathbf{s}}$ is

$$G_3 = x_1'F_2.$$

Since x_1' is MN, the length of the common perpendicular, and F_2 is the component of F in the direction perpendicular to both $\overrightarrow{MN}$ and $\hat{\mathbf{s}}$, this proves the required relationship.

Note that this result precisely relates the basic definition of the moment of a force about a point to the more common practical notion of the twisting effect of a force about an axis, with which the discussion of Section 5.2 opened.

5.9.4. See Section 5.2

Example 18

Consider the conveyor belt of Example 16 as continuous and wrapped round two end rollers, as it would be in practice. One of the rollers is driven by a belt transmitting a torque G such that the speed of the belt is maintained at constant value v (see Figure 5.15). The mass of that part of the belt in contact with the rollers, and of the rollers themselves, may be neglected in comparison with the mass of the rest of the belt, M, and of the sand being carried, m.

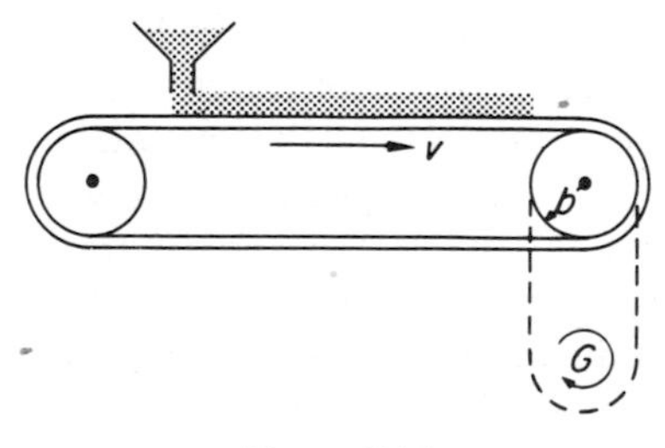

Figure 5.15

Show that

$$G = \frac{dm}{dt} vb,$$

where b is the radius of the driven roller, and that the forces holding the axle on which the driven roller rotates must have a horizontal component of at least $(dm/dt)v$.

Suppose S_1, S_2 are the vertical and horizontal components of the external force acting at the axis of the idling roller and R_1, R_2 the corresponding components for the driven roller. The latter will also be subject to the forces F, F' in the driving belt. In addition, there will be external gravitational forces and forces supporting the horizontal parts of the belt. Since the belt is flexible and has no vertical motion except at the rollers, the vertical support force on each element of the belt will be equal and opposite *and act at the same point* as the gravitational force on that element of the belt and any sand it may be supporting (see Figure 5.16a). These forces therefore cancel in their contributions both to the total external force and to the total external torque acting on the system and need not be considered further.

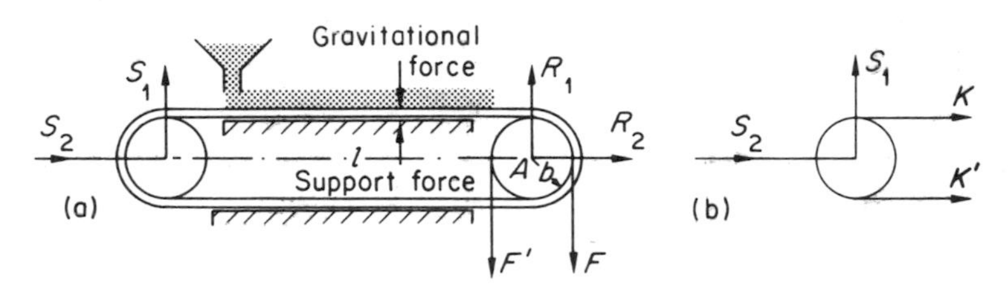

Figure 5.16

The momentum from left to right of the upper part of the belt plus sand is $(\frac{1}{2}M + m)v$ and of the lower part is $-\frac{1}{2}Mv$. There are no vertical components of momentum. Thus the total momentum is horizontal and of value

$$P_2 = mv. \tag{5.59}$$

The angular momentum is clearly perpendicular to the plane of the diagram. If it is calculated about the axis, A, of the driven roller the momentum of each element of the upper part of the belt plus sand is at the same perpendicular distance b from that axis, and their total angular momentum is therefore $(\frac{1}{2}M + m)vb$, directed into the plane of the diagram. The angular momentum of the lower part, similarly, is $\frac{1}{2}Mvb$. (Note that the two angular momentum contributions are in the *same* direction, unlike the linear momentum.) Thus the total angular momentum is

$$J_3 = (M + m)vb. \tag{5.60}$$

The horizontal component of the total external force is

$$F_{\text{ext}\,2} = R_2 + S_2, \tag{5.61}$$

and the component into the plane of the diagram of the total external torque about A is

$$G_{\text{ext}\,3} = Fb - F'b + S_1 l, \tag{5.62}$$

where l is the distance between the centres of the rollers.

We can proceed no further by considering only forces external to the whole system. It is necessary to consider the left-hand roller by itself, as shown in Figure 5.16b, where K, K' are the horizontal forces exerted on it by the conveyor belt. Now the roller is considered to have zero mass, and can therefore have no change in its linear or angular momentum. The forces external to the roller (which will now include K and K') must therefore total zero and exert zero torque about its axis. Hence,

$$\begin{aligned} S_1 &= 0, \\ S_2 + K + K' &= 0, \\ Kb - K'b &= 0, \end{aligned} \tag{5.63}$$

from which

$$K = K', \qquad S_2 = -2K. \tag{5.64}$$

From equations (5.62) and (5.63),

$$G_{\text{ext}\,3} = (F - F')b = G,$$

the torque exerted by the driving belt. Hence, from equations (5.7) and (5.60),

$$G = \frac{\mathrm{d}J_3}{\mathrm{d}t} = \frac{\mathrm{d}m}{\mathrm{d}t}vb. \tag{5.65}$$

From equations (5.61) and (5.64),

$$R_2 = F_{\text{ext}\,2} + 2K.$$

Hence, from equations (5.6) and (5.59),

$$R_2 = \frac{\mathrm{d}m}{\mathrm{d}t}v + 2K.$$

Since the least value of K is zero, the least value of R_2 is

$$R_{2\,\text{min}} = \frac{\mathrm{d}m}{\mathrm{d}t}v. \tag{5.66}$$

This last result is a very clear example of the fact that a force may change momentum without the point of its application moving. $R_{2\,\text{min}}$ is necessary in order to account for the change of momentum as the sand is given the velocity of the belt, but since the axis of the roller on which it is acting does not move this force does no work, and therefore cannot provide the gain in energy which the sand is also acquiring. This energy comes from the torque supplied by the driving belt.

5.9.5. See Section 5.2.1

It must not be assumed that the result (5.13) for a planar body implies that $\boldsymbol{\omega}$ and $\mathbf{J}$ are necessarily parallel vectors in all cases. When the body is not planar but is still rotating about a fixed axis with angular velocity $\boldsymbol{\omega}$, we can still define the moment of inertia about the axis in the more general form

$$I = \sum_i m_i \rho_i^2 \left(= \int \rho^2 \, \mathrm{d}m \right), \tag{5.67}$$

where ρ_i is the (perpendicular) distance of m_i from the axis. The *component* of angular momentum along this axis, J_{axial}, is then

$$J_{\text{axial}} = I\omega, \tag{5.68}$$

but there may be components in other directions as well. The angular momentum of rigid bodies is dealt with more fully in Chapter 9.

5.9.6. See Section 5.6

Example 19

The exhaust of a rocket is ejected backwards with a constant relative speed u. In the absence of any external forces, the speed $v_{(1)}$ and mass $m_{(1)}$ of the rocket (plus fuel) at one instant are related to the speed $v_{(2)}$ and mass $m_{(2)}$ at any other instant by

$$\log_e \frac{m_{(1)}}{m_{(2)}} = \frac{v_{(2)} - v_{(1)}}{u}.$$

Show this to be true using (a) a frame in which the rocket is initially at rest, (b) a frame which is travelling with the instantaneous velocity of the rocket.

Since there are no external forces the centre-of-mass of the rocket and its fuel (including the ejected exhaust gases) will move with constant velocity. Provided the rocket motors fire either directly forward or backward the problem then reduces, with proper choice of axes, to one-dimensional motion.

(a) Suppose that in the frame in which the rocket was initially at rest it has at some later time a velocity v and total mass m. A short time later the rocket will have ejected a small mass, m', of exhaust with speed u backward relative to itself, and therefore with a velocity $v - u$ in the frame being used. As a result its mass will have decreased to $m - m'$ and its velocity increased to $v + \mathrm{d}v$ (see Figure 5.17a).

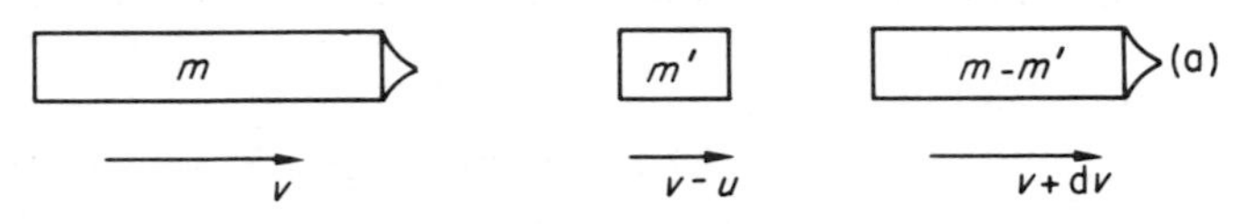

Figure 5.17(a)

Since no external forces act total momentum is conserved during this change:

$$(m - m')(v + \mathrm{d}v) + m'(v - u) = mv. \tag{5.69}$$

(The momentum of the gases exhausted earlier will be the same before and after and therefore need not be added to each side of this equation.) Neglecting the second-order quantity $m'\,\mathrm{d}v$, equation (5.69) reduces to

$$m\,\mathrm{d}v - m'u = 0. \tag{5.70}$$

If we denote by $\mathrm{d}m$ the *increment* in m (corresponding to $\mathrm{d}v$) then

$$\mathrm{d}m = -m'.$$

Hence

$$m\mathrm{d}v + \mathrm{d}mu = 0$$

or

$$\frac{\mathrm{d}m}{m} = -\frac{\mathrm{d}v}{u}.$$

On integration this gives the required result,

$$\int_{(1)}^{(2)} \frac{\mathrm{d}m}{m} = -\frac{1}{u}\int_{(1)}^{(2)} \mathrm{d}v; \ \log_e \frac{m_{(1)}}{m_{(2)}} = \frac{v_{(2)} - v_{(1)}}{u}. \tag{5.71}$$

(b) Now take instead a frame moving with constant velocity v relative to the first so that at the instant the rocket and remaining fuel have mass m they will be at rest in this frame and therefore have zero momentum. After the mass m' of exhaust has been ejected the rocket will be moving with velocity $\mathrm{d}v$, and the exhaust backward with speed u and therefore with velocity $-u$ (see Figure 5.17b).

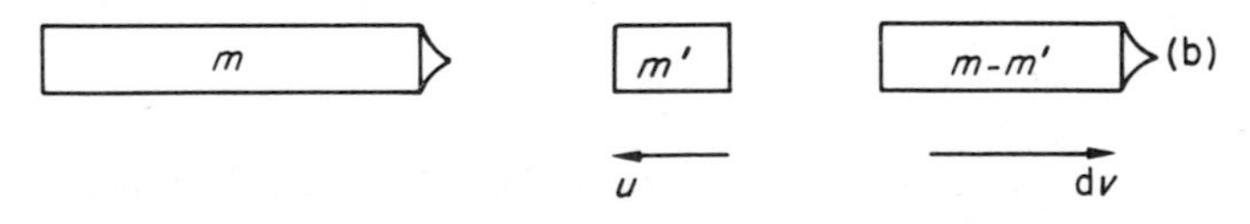

Figure 5.17(b)

In this frame also the momentum will be conserved (and is zero):

$$(m - m')\mathrm{d}v - m'u = 0.$$

From this equation (5.70) again follows and from it the result (5.71).

5.9.7. See Section 5.6

The fact that the various results concerning the motion of groups of particles all appear the same in frames moving with constant velocities with respect to each other is inherent in the observation made in Chapter 1 that Newton's laws for a single particle could be based upon any inertial frame. What we have demonstrated in Section 5.6 is that this starting point is a self-consistent one.

An alternative approach is to establish Newton's laws (experimentally) for only one frame and then to show that they will be true for any other moving with constant relative velocity. Whichever view we adopt, we arrive at the conclusion that, as far as classical mechanics is concerned, there is no criterion for picking out one inertial frame as having special significance. In particular, we *cannot* choose one which is at *absolute rest*; we can always regard as 'stationary' the one most convenient for describing any particular problem.

5.9.8. See Section 5.8

It may be noted that in the motion of a rocket (Example 19) and a conveyor belt (Examples 16 and 18) we are really dealing with impulsive effects. In one case the individual molecules experience sudden changes in momentum as they collide with each other and with the rocket itself during the combustion process, and in the other each grain of sand is jerked forward as it reaches the belt.

The mathematical descriptions used refer to the smoothed-out motions that we normally consider for practical purposes. Sometimes this is inadequate. For example if a conveyor belt were transporting large rocks we might, even in practice, need to consider the irregular motion that would arise as each rock fell from the hopper and was given an impulsive change of momentum by the belt.

5.10. Problems

5.1. Determine $\mathbf{x} \wedge \mathbf{y}$, where $\mathbf{x}$ and $\mathbf{y}$ are the following pairs of vectors:

(a) $3\mathbf{i} + 2\mathbf{j} + 2\mathbf{k}$, $2\mathbf{i} - 2\mathbf{j} + \mathbf{k}$

(b) $3\mathbf{i} + 2\mathbf{j} + 6\mathbf{k}$, $2\mathbf{i} + 6\mathbf{j} - 3\mathbf{k}$

(c) $\mathbf{i} + 2\mathbf{j} + 2\mathbf{k}$, $2\mathbf{i} + \mathbf{j} + 2\mathbf{k}$

(d) the unit vector making angles 30° and 60° with axes 1 and 2, the unit vector making angles 60° and 30° with axes 2 and 3.

5.2. A force of magnitude 10 N makes equal angles with all three axes and acts through a point in the 2,3 plane 2 m distant from the origin and equidistant from the axes 2 and 3. Find the torque exerted by the force about the origin and show that it has magnitude $20/\sqrt{3}$ Nm.

5.3. Show that

$$\mathbf{a} \wedge (\mathbf{b} \wedge \mathbf{c}) = (\mathbf{a} \cdot \mathbf{c})\mathbf{b} - (\mathbf{a} \cdot \mathbf{b})\mathbf{c}.$$

(*Hint:* use the expression (5.56) twice to find the first component of $\mathbf{a} \wedge (\mathbf{b} \wedge \mathbf{c})$ and show that this may be written as $b_1(a_1c_1 + a_2c_2 + a_3c_3) - c_1(a_1b_1 + a_2b_2 + a_3b_3)$.)

5.4. If equal and opposite (*but not colinear*) forces, $\mathbf{F}$, $-\mathbf{F}$, act upon a system, show that their combined torque is independent of the point about which it is calculated, and may be written as $\mathbf{h} \wedge \mathbf{F}$ where $\mathbf{h}$ is a common perpendicular to the two forces. Hence show that such a pair of forces cannot affect the centre-of-mass motion of a system but only the rotation about it. (The torque arising from a pair of equal and opposite forces is called a *couple*.)

5.5. With what interpretation of the various velocities is the result (5.71) valid for the case in which the rocket is ejecting the exhaust gases *forward* in order to slow down? Eight-ninths of the total mass of a rocket is fuel. It starts from rest and uses some of the fuel for acceleration in a field-free space. What fraction of the fuel must remain available to enable to rocket to be brought to rest again?

5.6. Eighty per cent of the total mass of a rocket is fuel. The rocket accelerates from rest, in a field-free space, until half the fuel is consumed. At this stage part of the casing is separated, with negligible relative velocity, leaving the remaining rocket again with eighty per cent of its total mass as fuel. This fuel is then used up in achieving a final velocity v. What would have been the final velocity if the rocket had functioned entirely as a single stage?

5.7. A rocket of total initial mass m_0 ejects exhaust at a constant relative speed u and at a constant (time) rate αm_0, until its mass is reduced to βm_0. Then, with the rocket motor shut off, it continues to cruise at constant velocity. Show that in travelling a total distance s the time taken by the rocket is

$$t = \frac{1}{\alpha} - \frac{s}{u \log_e \beta} + \frac{1 - \beta}{\alpha \log_e \beta}$$

($\int \log_e x \, dx = x(\log_e x - 1)$).

5.8. A yo-yo consists of two heavy discs connected by a short narrow spindle around which a string can wind. Initially the string is fully wound on the spindle, and it is then pulled vertically upwards at such a rate as to cause the yo-yo to remain at a constant height as it rotates. When the string is fully unwound the upper end is held stationary so that the yo-yo climbs up the string. Show that it comes to rest again at the top of the string.

5.9. A uniform circular disc, of mass M and radius a, is free to roll *without slipping* along the edge of a fixed horizontal surface. A light rigid rod of length x is fixed to the disc with one end at its centre, and is initially pointing vertically downwards. A horizontal force R is applied to the end of this rod, as shown in Figure 5.18.

(a) What is the initial angular acceleration of the disc, and in which direction relative to R does the rod begin to move, when $x > a$? Show that in the initial small rotation of

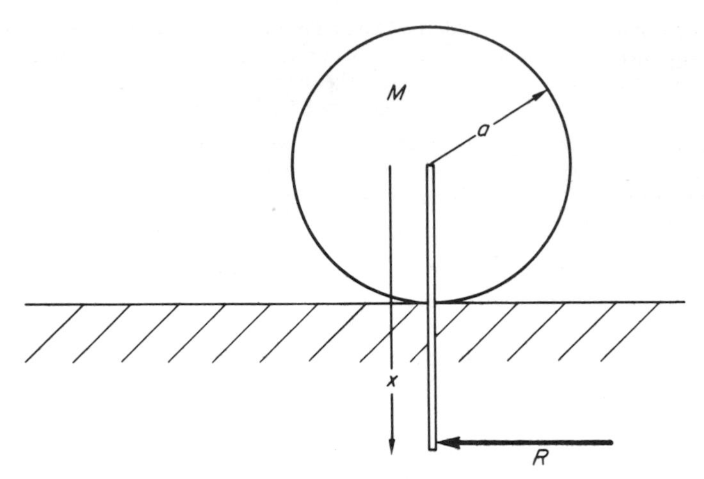

Figure 5.18

the disc the work done by R accounts for the gain in kinetic energy, but the impulse it delivers does not account for the change in momentum. Which other force must be included for the latter purpose?

(b) Where, if at all, can a horizontal force be applied to the disc to account for the initial changes in kinetic energy *and* momentum?

5.10. A cylinder, whose mass is M, radius b and moment of inertia I, can rotate, not necessarily freely, about an axle fixed at the centre of a horizontal beam of length $2a$ $(a > b)$ resting on supports at its two ends, as shown in Figure 5.19. The cylinder is made to rotate by a downward force F applied to a rope wound on the cylinder, its angular velocity at any time being ω. R and S are the support forces. Show that consideration of the linear and angular momentum of the whole system will give only two equations relating R, S and $d\omega/dt$ to the force F. How would you explain this indeterminacy?

Determine $d\omega/dt$ when the axle (a) is smooth, (b) exerts frictional forces such that if

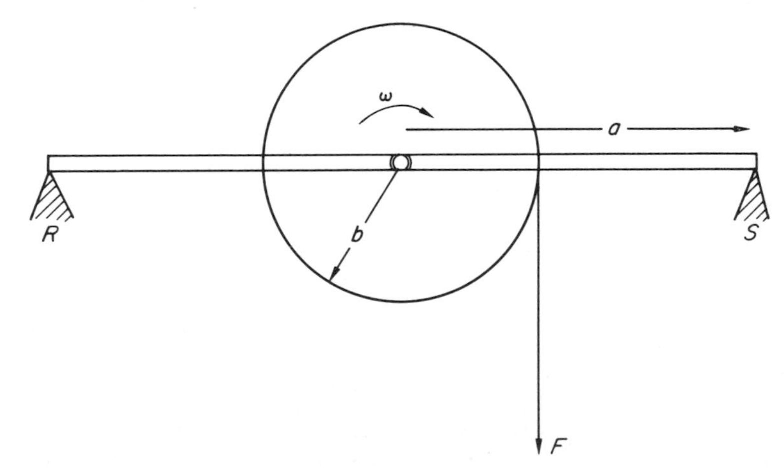

Figure 5.19

K is the total load on the axle the torque about the centre is αK. Show that in the latter case the support forces are

$$R = \tfrac{1}{2}(F + Mg)(1 - \alpha/a), \qquad S = \tfrac{1}{2}(F + Mg)(1 + \alpha/a).$$

According to this result R would vanish or become negative for a sufficiently large value of α, apparently indicating that when the friction is large enough the left-hand end of the bar would rise! Show that this is impossible and that α has a maximum value that ensures a positive value for R.

5.11. Show that the power delivered by the force moving the conveyor belt of Example 16 is *twice* the rate at which the kinetic energy of the system is increasing. What is happening to the excess energy being delivered to the system?

5.12. Suppose that in Example 16 (page 132) a small pump were fixed at the bottom of the hopper, which projected the sand forward so that, as it landed, the sand was already moving forward with the velocity of the belt. Show that although the momentum of the system still increases at the rate $v(\mathrm{d}m/\mathrm{d}t)$, no force is required to keep the belt moving at constant velocity. Which external force provides this increase? Does this force provide the kinetic energy acquired by the sand and, if not, what is the source of this energy?

5.13. Suppose that in Example 16 the belt were *smooth*, but had a series of fins projecting from it, to carry the material along. Suppose, moreover, that the material were not sand, but consisted of larger pieces of matter, deposited at such a rate that at most one piece was trapped between each pair of fins and bounced back and forth elastically between them. By considering in detail the motion of one such piece, show that if $\mathrm{d}m/\mathrm{d}t$ is the mean rate (averaged over a large number of pieces) at which the hopper deposits mass, then the force required to maintain the belt at a mean speed v is $v(\mathrm{d}m/\mathrm{d}t)$ and that, in this case, the work done by this force *does* account for the rate at which the kinetic energy increases.

5.14. A tank of water is free to run along a smooth horizontal track. The water escapes from the tank at a constant rate, so that the total mass of tank and water at time t is

$$m(t) = m_0 - \alpha t,$$

where α is a positive constant. If $v(t)$ is the velocity, the momentum of the tank and the water in it is

$$p = m(t)v(t).$$

Show, nevertheless, that although m and v are both functions of time, the equation of motion, when a force F is applied to the tank, is

$$F = m\frac{\mathrm{d}v}{\mathrm{d}t}$$

and *not*

$$F = m\frac{\mathrm{d}v}{\mathrm{d}t} + \frac{\mathrm{d}m}{\mathrm{d}t}v.$$

If the tank is initially at rest and the force is maintained at a constant value F_0 while the total mass changes from m_0 to m, show that the distance travelled during this change is $F_0[m_0 - m - m\log(m_0/m)]/\alpha^2$.

Find the momentum and energy gained by the tank with its remaining water, and by the water which leaks away, and hence show that the whole system gains a momentum equal to the impulse delivered by the force, and energy equal to the work done by it.

Chapter 6

Motion Under a Central Force

Both the gravitational and electrostatic forces between point particles act along the line joining them. Indeed, if a force depends only upon the relative position of the particles, and not upon their velocities, no other direction is possible. Point particles have no intrinsic direction associated with them, and it therefore follows that the only direction they can define, apart from that of their velocities, is that of the line joining them. Thus in three dimensions, such forces, as well as having great practical importance, are the simplest forces that can be associated with point particles.

Since, as we shall shortly see, motion under a central force is two-dimensional, and the common notation for two-dimensional polar co-ordinates is (r, θ), we shall use $\mathbf{r}$ for defining the position of a particle. It will also be convenient at times to use the abbreviations

$$\dot{r} = \frac{\mathrm{d}r}{\mathrm{d}t}, \qquad \dot{\theta} = \frac{\mathrm{d}\theta}{\mathrm{d}t}, \qquad \text{etc.}$$

6.1. Central forces; constant angular momentum

When one of the particles is fixed at the origin, while the position of the other is given by the vector $\mathbf{r}$, the force on it, $\mathbf{F}$, will be parallel or antiparallel with $\mathbf{r}$. It is then called a *central force*, and its moment about the origin will be zero:

$$\mathbf{G} = \mathbf{r} \wedge \mathbf{F} = 0.$$

The angular momentum, $\mathbf{J}$, of the moving particle about the origin will therefore be constant. Constant $\mathbf{J}$ implies a *fixed direction*, which we shall denote by the fixed unit vector $\mathbf{k}$, as well as a *constant magnitude* J:

$$\mathbf{r} \wedge \mathbf{p} = m\mathbf{r} \wedge \mathbf{v} = \mathbf{J} = J\mathbf{k}. \tag{6.1}$$

The initial position $\mathbf{r}_0$ and velocity $\mathbf{v}_0$ lie in a plane through the origin perpendicular to $\mathbf{k}$. The particle cannot move out of this plane, for if it did its position vector, $\mathbf{r}$, would then not be perpendicular to $\mathbf{k}$, in contradiction to equation (6.1). Thus *the motion is planar.*

In this case we need only two-dimensional polar coordinates (r, θ) to define the position of the particle. From Figure 6.1 we can see that when it

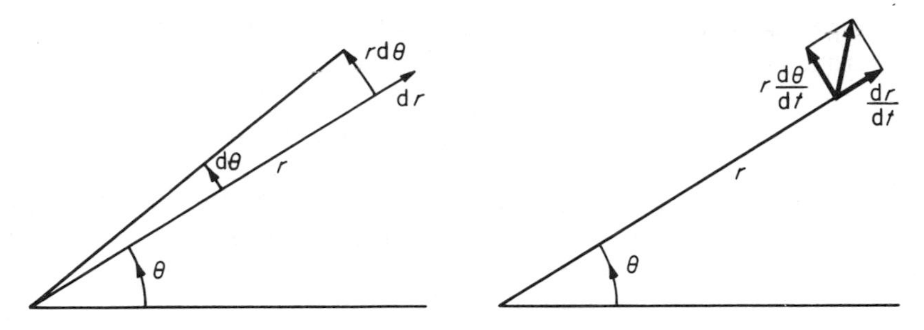

Figure 6.1

moves from (r, θ) to $(r + \mathrm{d}r, \theta + \mathrm{d}\theta)$ the radial component of displacement is $\mathrm{d}r$ and that perpendicular to the radius is $r\,\mathrm{d}\theta$, neglecting higher order infinitesimals. Thus the corresponding components of velocity are

$$v_r = \frac{\mathrm{d}r}{\mathrm{d}t} = \dot{r}, \qquad v_\theta = r\frac{\mathrm{d}\theta}{\mathrm{d}t} = r\dot{\theta}. \tag{6.2}$$

Hence

$$J = mrv_\theta = mr^2\dot{\theta} = \text{constant.} \tag{6.3}$$

6.2. Conservative central forces

If the central force is also conservative,

$$\mathscr{T}(r, \theta) + \mathscr{V}(r) = \mathscr{E} = \text{constant.}$$

From equations (6.2) and (6.3),

$$\begin{aligned}\mathscr{T}(r, \theta) &= \frac{m}{2}v^2 = \frac{m}{2}(v_r^2 + v_\theta^2)\\ &= \frac{m}{2}(\dot{r}^2 + r^2\dot{\theta}^2)\\ &= \frac{m\dot{r}^2}{2} + \frac{J^2}{2mr^2}.\end{aligned} \tag{6.4}$$

Hence we may rewrite the conservation of total energy as

$$\frac{1}{2}m\dot{r}^2 + \frac{J^2}{2mr^2} + \mathscr{V}(r) = \mathscr{E} = \text{constant.} \tag{6.5}$$

The corresponding equation for linear, one-dimensional motion is

$$\tfrac{1}{2}m\dot{x}^2 + \mathscr{V}(x) = \mathscr{E} = \text{constant.} \tag{6.6}$$

Comparing these two we can see that, *as far as radial distance is concerned*, the particle moves as if it had a potential energy

$$\mathscr{U}(r) = \frac{J^2}{2mr^2} + \mathscr{V}(r) \tag{6.7}$$

(see Figure 6.2). Radial distance alone does not specify the position of the particle. For given values of $\mathscr{E}$ and J equation (6.5) can give $r(t)$, but we need to use the angular momentum equation (6.3) to determine $\theta(t)$.

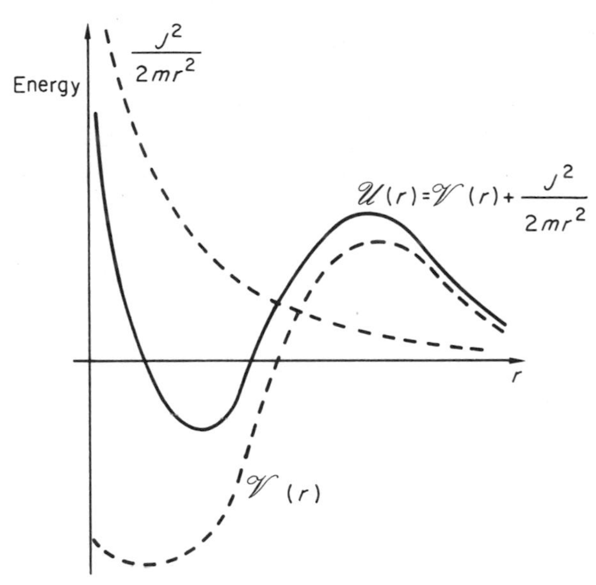

Figure 6.2

Nevertheless, by considering the pseudo-potential energy $U(r)$ we can discuss bound and unbound motion in the manner of Sections 2.5.1 and 2.5.2. Thus, for the case of Figure 6.3, a particle with total energy $\mathscr{E}$ as shown will have a motion either bound, with

$$r_1 \leqslant r \leqslant r_2,$$

or unbound, with

$$r \geqslant r_3.$$

There are important differences between this and one-dimensional motion. Radial distances can never be negative. Moreover the term $J^2/2mr^2$, which is equivalent to a strong repulsive force and, because it depends upon J^2, is sometimes called the *angular momentum barrier*, usually imposes a lower limit for the radial distance. Only a (true) *attractive* radial force varying more rapidly than $1/r^2$ can counteract this.

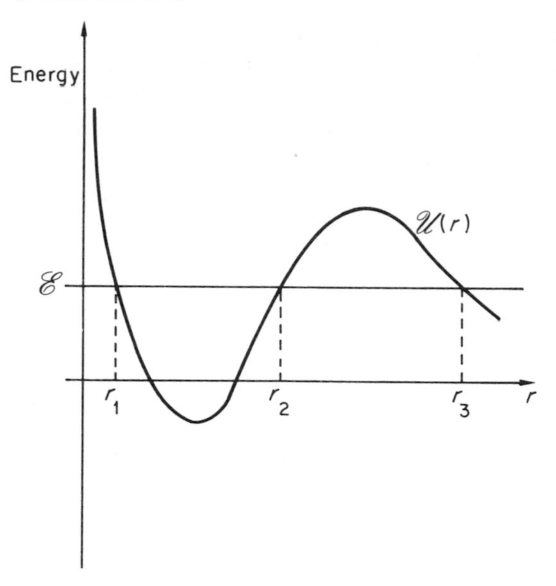

Figure 6.3

Except for the special case of zero angular momentum, or when the particle moves to infinity, there will, from equation (6.3), always be some angular velocity. Thus the particle does not move linearly between r_1 and r_2 or between r_3 and infinity. Instead, it has an orbit which lies either between the circles of radii r_1 and r_2 or outside the circle of radius r_3, as shown in Figure 6.4a and b. At points such as A, B and C where the orbit touches the bounding circles the velocity will be perpendicular to the radius vector,

$$v_r = \dot{r} = 0, \qquad v_\theta = r\dot{\theta} \neq 0.$$

(See Section 6.6.1, page 168.)

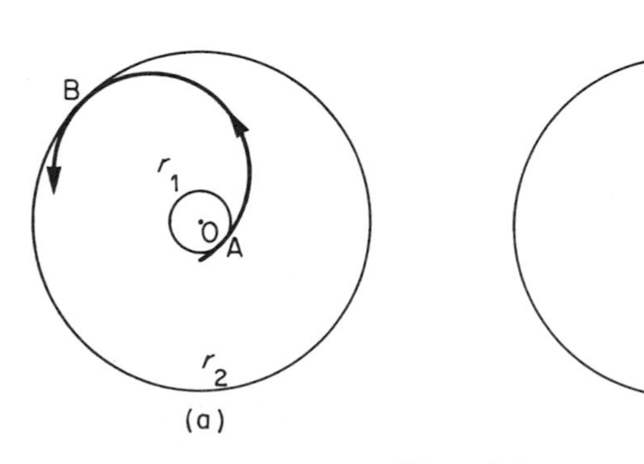

Figure 6.4

Points of equilibrium in one dimension, where $\mathscr{V}(x)$ has a minimum or maximum, are possible positions of stable or unstable rest. Where $\mathscr{U}(r)$ is a minimum or maximum it is possible for r to remain constant, but, from equation (6.3), θ will then vary at a constant rate. Thus the minima and maxima of the pseudo-potential energy $\mathscr{U}(r)$ give possible radii for stable and unstable *circular orbits*, traversed with constant angular velocity.

6.3. 'Inverse square' motion; general properties

Since both the gravitational and electrostatic forces have an inverse square dependence, motions with the potential energy

$$\mathscr{V}(r) = \frac{k}{r} \tag{6.8}$$

have great historical and practical importance. The energy equation is then

$$\frac{m}{2}\dot{r}^2 + \frac{J^2}{2mr^2} + \frac{k}{r} = \mathscr{E}. \tag{6.9}$$

6.3.1. Repulsive force, unbound motion

Positive k implies a repulsive force and negative k an attractive force. For a repulsive force the pseudo-potential energy,

$$\mathscr{U}(r) = \frac{J^2}{2mr^2} + \frac{k}{r}, \quad k > 0, \tag{6.10}$$

will be of the form shown in Figure 6.5. For real motion the total energy is necessarily positive. In consequence, the motion will always be unbound,

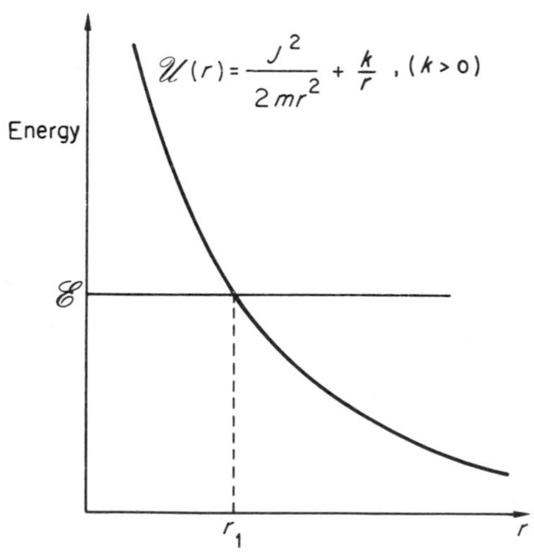

Figure 6.5

with a nearest approach distance r_1. As the particle recedes from the force centre,

$$r \to \infty,$$

equation (6.5) shows that

$$v_r = \dot{r} \to \left(\frac{2E}{m}\right)^{1/2},$$

while equations (6.2) and (6.3) show that

$$v_\theta = r\dot{\theta} = \frac{J}{mr} \to 0.$$

Thus the particle's trajectory 'at infinity' approaches a straight line at a fixed value of θ, called an *asymptote*. By reversing the motion, so that the particle moves towards the force centre, gaining angular velocity as it does so, and then out again towards infinity, we can see that there will be two asymptotes, as shown in Figure 6.6. It should be noted that the asymptotes

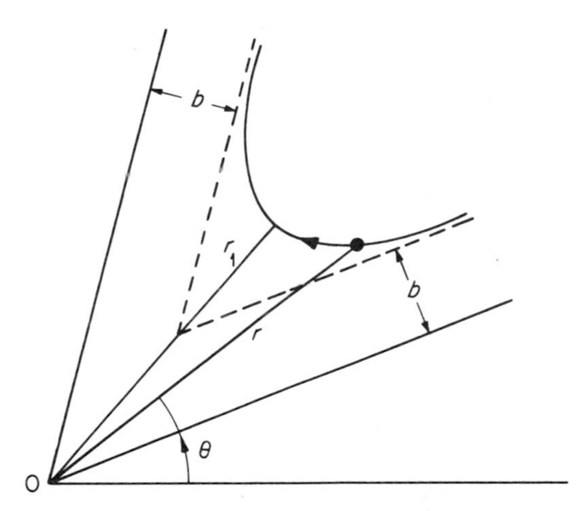

Figure 6.6

do not pass through the force centre. They are displaced by a distance b in accordance with the relationship

$$J = mbv_r(\infty) = b(2m\mathscr{E})^{1/2}. \tag{6.11}$$

6.3.2. Attractive force, unbound motion

For an attractive inverse square force the pseudo-potential energy,

$$\mathscr{U}(r) = \frac{J^2}{2mr^2} + \frac{k}{r}, \quad k < 0, \tag{6.12}$$

will have the form shown in Figure 6.7. When the total energy is positive the motion is essentially the same as for a repulsive force with two asymptotes

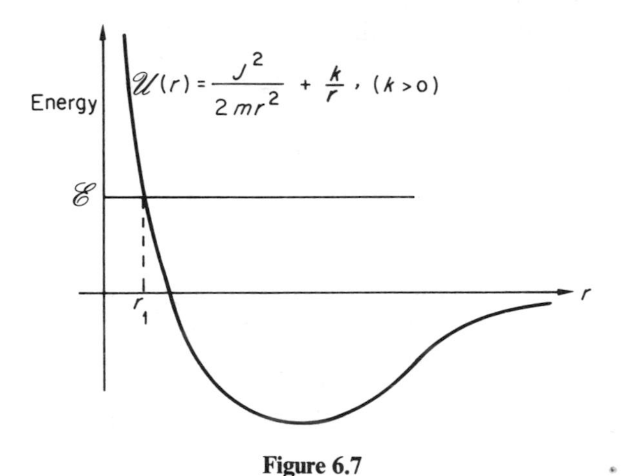

Figure 6.7

for the trajectory at infinity. The main difference is that because the force is now attractive the particle is pulled from the asymptote *towards* the origin, as in Figure 6.8, rather than pushed away from it, as in Figure 6.6. Equation (6.11) will be unchanged, thus showing that the asymptotes in this case are displaced from the origin the same distance as before.

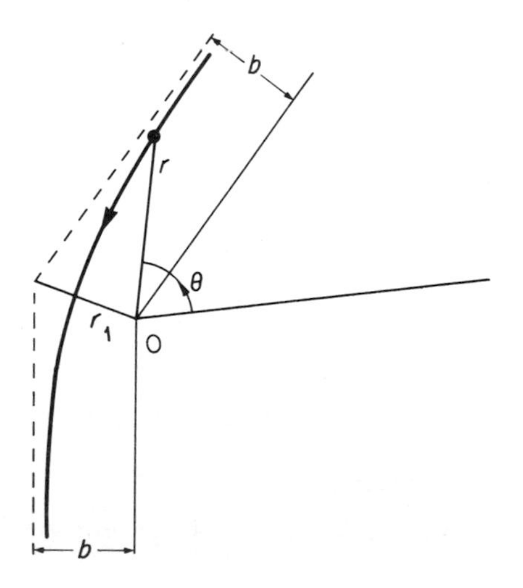

Figure 6.8

6.3.3. Attractive force, bound motion

When the total energy is negative, motion is bound between two circles of radii r_1 and r_2, as shown in Figures 6.9 and 6.10. Since at these radii

$$\dot{r} = 0,$$

equation (6.9) shows that r_1 and r_2 are the roots of the quadratic equation

$$\frac{J^2}{2mr^2} + \frac{k}{r} - \mathscr{E} = 0. \tag{6.13}$$

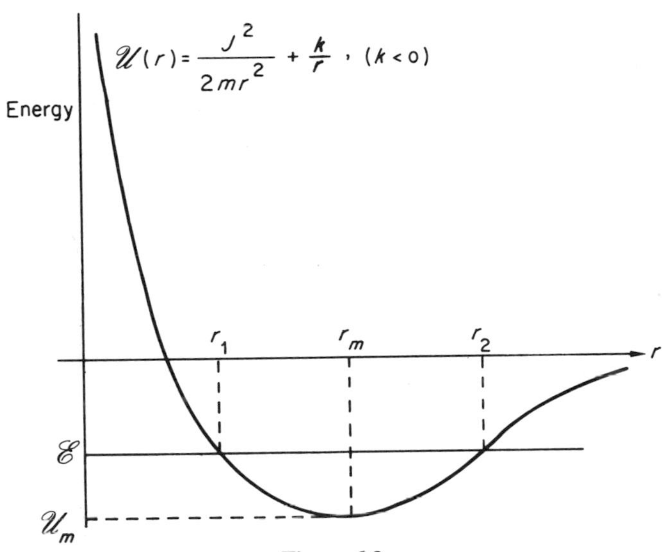

Figure 6.9

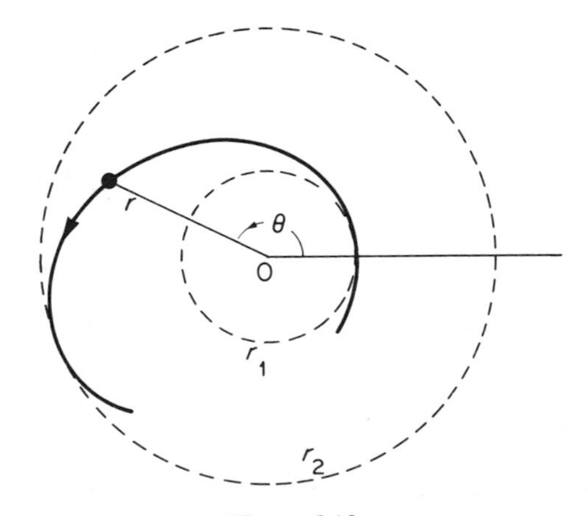

Figure 6.10

Solving this,

$$r_1 = \frac{k}{2\mathscr{E}}\left[1 - \left(1 + \frac{2\mathscr{E}J^2}{mk^2}\right)^{1/2}\right],$$

$$r_2 = \frac{k}{2\mathscr{E}}\left[1 + \left(1 + \frac{2\mathscr{E}J^2}{mk^2}\right)^{1/2}\right]. \tag{6.14}$$

One special case arises when

$$\mathscr{E} = \mathscr{U}_m,$$

where $\mathscr{U}_m$ is the minimum value of $\mathscr{U}(r)$. At this minimum

$$\frac{d\mathscr{U}}{dr} = 0,$$

which, from equation (6.12), occurs at the values

$$r = r_m = -\frac{J^2}{km}, \qquad \mathscr{U} = \mathscr{U}_m = -\frac{mk^2}{2J^2}. \tag{6.15}$$

$\mathscr{U}_m$ is the lowest energy for a given angular momentum. The motion is then circular,

$$r = r_m = r_1 = r_2 = -\frac{J^2}{mk} = \frac{k}{2\mathscr{U}_m}, \tag{6.16}$$

with constant angular velocity,

$$\omega_m = \dot{\theta} = \frac{J}{mr^2}. \tag{6.17}$$

Combining equations (6.16) and (6.17),

$$\omega_m^2 r_m^3 = -\frac{k}{m}. \tag{6.18}$$

(See Section 6.6.2, page 169.) Real motion is not possible for

$$\mathscr{E} < \mathscr{U}_m.$$

The other special case is at the boundary between unbound and bound motion,

$$\mathscr{E} = 0.$$

Equations (6.14) give an indefinite (0/0) value for one root and infinity for the other in this case. We may expect difficulties here since we have multiplied equation (6.13) by r^2 to give the results (6.14), and this is valid only if r is finite. Going back to the original equation (6.13) we have, when $\mathscr{E} = 0$,

$$\frac{1}{r}\left(k + \frac{J^2}{2mr}\right) = 0,$$

from which

$$k + \frac{J^2}{2mr} = 0 \quad \text{or} \quad \frac{1}{r} = 0.$$

Hence

$$r_1 = -\frac{J^2}{2mk}, \qquad r_2 = \infty.$$

In this case both velocity components approach zero at infinity

$$v_r = \dot{r} \to \left(\frac{2\mathscr{E}}{m}\right)^{1/2} = 0,$$

$$v_\theta = r\dot{\theta} = \frac{J^2}{mr} \to 0,$$

and there is, therefore, no asymptotic motion. We might say that the particle 'just reaches infinity', in contrast to the unbound case where it still has a finite velocity 'at infinity'.

6.4. 'Inverse square' motion; exact path

Although many properties of 'inverse square' motion can be obtained directly from the conservation of energy,

$$\frac{m}{2}\left(\frac{\mathrm{d}r}{\mathrm{d}t}\right)^2 + \frac{J^2}{2mr^2} + \frac{k}{r} = \mathscr{E} = \text{constant}, \tag{6.19}$$

and from conservation of angular momentum,

$$mr^2\frac{\mathrm{d}\theta}{\mathrm{d}t} = J = \text{constant}, \tag{6.20}$$

it is sometimes necessary to calculate the exact path (i.e. to find $r(\theta)$). From these two equations,

$$\left(\frac{\mathrm{d}r}{\mathrm{d}t}\right)^2 = \frac{2\mathscr{E}}{m} - \frac{2k}{mr} - \frac{J^2}{m^2r^2}$$

and

$$\left(\frac{\mathrm{d}\theta}{\mathrm{d}t}\right)^2 = \frac{J^2}{m^2r^4}.$$

Hence

$$\left(\frac{dr}{d\theta}\right)^2 = \left(\frac{dr/dt}{d\theta/dt}\right)^2 = r^2\left(\frac{2\mathscr{E}mr^2}{J^2} - \frac{2kmr}{J^2} - 1\right).$$

The substitution

$$u = \frac{1}{r}, \qquad \frac{du}{d\theta} = -\frac{1}{r^2}\frac{dr}{d\theta}$$

gives

$$\left(\frac{du}{d\theta}\right)^2 = \frac{2m\mathscr{E}}{J^2} - \frac{2kmu}{J^2} - u^2$$

$$= \frac{2m\mathscr{E}}{J^2} + \frac{m^2k^2}{J^4} - \left(\frac{mk}{J^2} + u\right)^2. \tag{6.21}$$

Since

$$\mathscr{E} \geqslant \mathscr{U}_m = -\frac{mk^2}{2J^2}$$

for real motion, whether k is positive or negative, we can put

$$W^2 = \frac{2m\mathscr{E}}{J^2} + \frac{m^2k^2}{J^4} = \frac{m^2k^2}{J^4}\left(1 + \frac{2\mathscr{E}J^2}{mk^2}\right), \qquad W \text{ real.}$$

We can also substitute

$$w = u + \frac{mk}{J^2}, \qquad \frac{dw}{d\theta} = \frac{du}{d\theta}$$

in equation (6.21) to give the simpler form

$$\left(\frac{dw}{d\theta}\right)^2 = W^2 - w^2. \tag{6.22}$$

This has the general solution

$$w = W\cos(\theta - \theta_0),$$

where θ_0 is an adjustable constant.

Hence, remembering that r and u are necessarily positive,

$$\frac{1}{r} = u = \frac{m|k|}{J^2}\left(1 + \frac{2\mathscr{E}J^2}{mk^2}\right)^{1/2}\cos(\theta - \theta_0) - \frac{mk}{J^2}. \tag{6.23}$$

Depending upon the value of the *eccentricity*,

$$e = \left(1 + \frac{2\mathscr{E}J^2}{mk^2}\right)^{1/2}, \tag{6.24}$$

equation (6.23) represents a hyperbola, parabola or ellipse with the origin as focus, and with the direction

$$\theta = \theta_0$$

as the major axis. The main features of these trajectories have already been discussed above. Equation (6.23) now enables us to consider them in greater detail. One important deduction can be made from equation (6.23) without knowing anything about the precise type of trajectory it describes. Since the right-hand side is unchanged when θ is increased by 2π, a particle that completely encircles the force centre once will follow the same path in subsequent revolutions. In other words, if a complete orbit is possible it will be a *closed orbit*.

6.4.1. Repulsive force ($k > 0$), *positive energy* ($\mathscr{E} > 0$)

The eccentricity has the value

$$e = \left(1 + \frac{2\mathscr{E}J^2}{mk^2}\right)^{1/2} > 1. \tag{6.25}$$

Hence the trajectory, which, since k is positive, may be written

$$\frac{1}{r} = u = \frac{mk}{J^2}[e\cos(\theta - \theta_0) - 1], \tag{6.26}$$

is an hyperbola. Minimum r implies maximum u, which occurs when $\cos(\theta - \theta_0)$ has its maximum value:

$$r = r_1 \Rightarrow u = u_{\max} \Rightarrow \cos(\theta - \theta_0) = 1 \Rightarrow \theta = \theta_0,$$

$$r_1 = \frac{J^2}{mk(e-1)}.$$

Since u cannot be negative, the least physically permissible value of the term in square brackets (equation 6.26) is zero:

$$r \to \infty \Rightarrow u = 0 \Rightarrow \cos(\theta - \theta_0) = \frac{1}{e} = \cos\alpha, \text{ say.} \tag{6.27}$$

Thus the asymptotes lie in the directions

$$\theta = \theta_a = \theta_0 \pm \alpha$$

and at a distance from the origin

$$b = \frac{J}{(2m\mathscr{E})^{1/2}},$$

as we saw in Section 6.3.1 (see Figure 6.11).

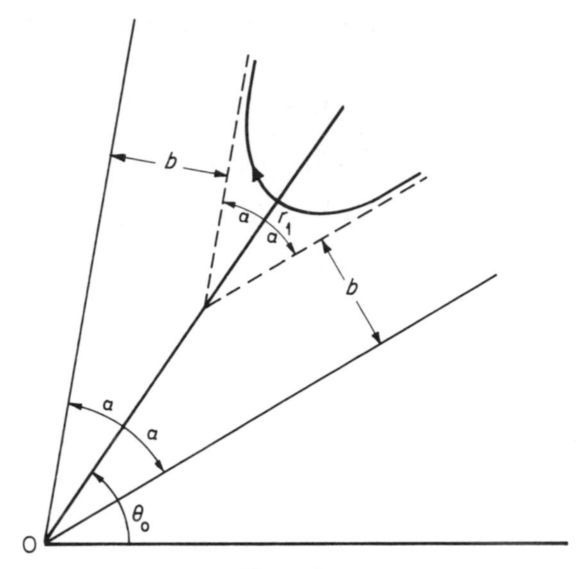

Figure 6.11

6.4.2. Attractive forces ($k < 0$), *positive energy* ($\mathscr{E} > 0$)

The eccentricity still exceeds unity:

$$e = \left(1 + \frac{2\mathscr{E}J^2}{mk^2}\right)^{1/2} > 1. \tag{6.28}$$

The trajectory, which, since k is negative, is now

$$\frac{1}{r} = u = \frac{m|k|}{J^2}[e\cos(\theta - \theta_0) + 1], \tag{6.29}$$

is therefore also an hyperbola.

The direction of minimum approach is again θ_0 although the minimum distance is less:

$$r = r_1 \Rightarrow u = u_{\text{max}} \Rightarrow \cos(\theta - \theta_0) = 1 \Rightarrow \theta = \theta_0,$$

$$r_1 = \frac{J^2}{m|k|(e+1)},$$

while the asymptotic directions are given by

$$r \to \infty \Rightarrow u = 0 \Rightarrow \cos(\theta - \theta_0) = -\frac{1}{e} = -\cos\alpha,$$

or

$$\theta = \theta_0 + \pi \pm \alpha,$$

and the asymptotes are at the same distance from the origin,

$$b = \frac{J}{(2m\mathscr{E})^{1/2}}$$

(see Figure 6.12).

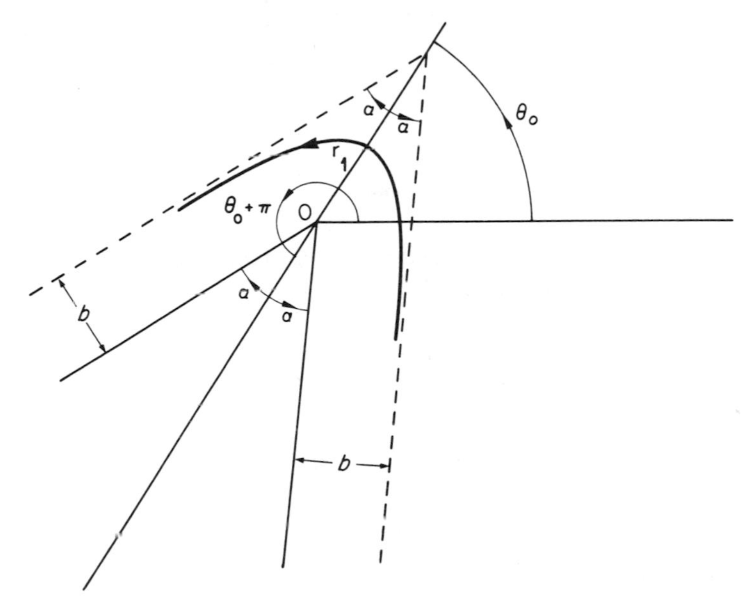

Figure 6.12

6.4.3. Attractive force ($k < 0$), *zero energy* ($\mathscr{E} = 0$)

The eccentricity is

$$e = \left(1 + \frac{2\mathscr{E}J^2}{mk^2}\right)^{1/2} = 1 \tag{6.30}$$

and the trajectory is

$$\frac{1}{r} = u = \frac{m|k|}{J^2}[\cos(\theta - \theta_0) + 1], \tag{6.31}$$

which is a parabola. The direction of nearest approach is still

$$\theta = \theta_0,$$

and

$$r_1 = \frac{J^2}{2m|k|}$$

(see Figure 6.13).

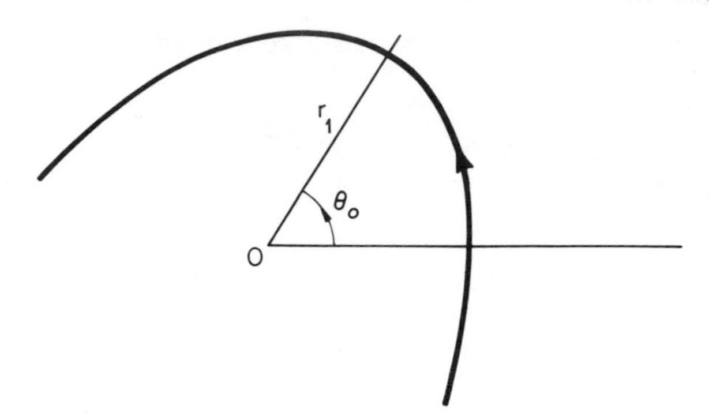

Figure 6.13

6.4.4. Attractive force ($k < 0$), *negative energy* ($\mathcal{U}_m < \mathcal{E} < 0$)

From equations (6.24) and (6.15),

$$e = \left(1 + \frac{2\mathcal{E}J^2}{mk^2}\right)^{1/2} = \left(1 - \frac{\mathcal{E}}{\mathcal{U}_m}\right)^{1/2}. \tag{6.32}$$

Hence the eccentricity has a value

$$0 < e < 1. \tag{6.33}$$

The trajectory

$$\frac{1}{r} = u = \frac{m|k|}{J^2}[e\cos(\theta - \theta_0) + 1] \tag{6.34}$$

is therefore an ellipse. All values of θ are now possible and there is a position of maximum as well as minimum distance from the force centre:

$$r = r_1 \Rightarrow u = u_{\max} \Rightarrow \cos(\theta - \theta_0) = 1 \Rightarrow \theta = \theta_0,$$

$$r_1 = \frac{J^2}{m|k|(1 + e)}; \tag{6.35}$$

$$r = r_2 \Rightarrow u = u_{\min} \Rightarrow \cos(\theta - \theta_0) = -1 \Rightarrow \theta = \theta_0 + \pi,$$

$$r_2 = \frac{J^2}{m|k|(1 - e)}. \tag{6.36}$$

The directions θ_0 and $\theta_0 + \pi$ lie along the major axis of the ellipse (see Figure 6.14). (See Section 6.6.3, page 175.)

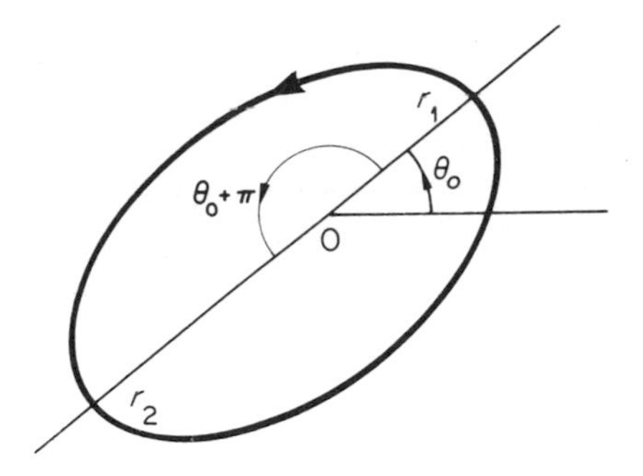

Figure 6.14

6.4.5. Attractive force ($k < 0$), *minimum energy* ($\mathcal{U}_m = \mathcal{E} < 0$)

The eccentricity is now zero:

$$e = \left(1 + \frac{2\mathcal{E}J^2}{mk^2}\right)^{1/2} = \left(1 - \frac{\mathcal{E}}{\mathcal{U}_m}\right)^{1/2} = 0, \tag{6.37}$$

and the orbit becomes a circle of radius

$$r = \frac{J^2}{m|k|}. \tag{6.38}$$

Since

$$J = mr^2\dot{\theta},$$

the result (6.38) is equivalent to the simple relationship for radial acceleration in the case of circular motion:

$$mr\dot{\theta}^2 = \frac{|k|}{r^2}.$$

6.5. Motion of two particles

We have seen in Sections 4.8.2 and 4.8.3 that bodies with density or electric charge distributed with spherical symmetry interact with each other as if they were point particles. This enables us to give approximate analyses, for example, of planetary motion and of the scattering of charged atomic particles based upon the preceding discussion of orbits.

That discussion assumed a fixed force centre. When the two interacting particles have very different masses, as is the case with the earth and the sun or an electron and an atomic nucleus, there is little error in assuming the more massive one to be fixed. The results of Section 6.4 can therefore be applied directly. However, in the case of double stars where the masses are comparable this is clearly invalid, and even in the moon–earth system where the mass ratio is 1:80 we may need to take account of the fact that the earth moves slightly. When an alpha particle is scattered by a light nucleus there is a similar need to allow for the fact that both particles can move.

Suppose the two particles have masses m_1 and m_2 with positions P_1, P_2 determined by the vectors $\mathbf{R}_1$, $\mathbf{R}_2$, relative to a fixed point O. Their centre-of-mass, C, is given by

$$\mathbf{R} = \frac{m_1\mathbf{R}_1 + m_2\mathbf{R}_2}{m_1 + m_2}.$$

Relative to this their positions are

$$\mathbf{r}_1 = \mathbf{R}_1 - \mathbf{R} = \frac{m_2}{m_1 + m_2}(\mathbf{R}_1 - \mathbf{R}_2) = \frac{m_2\mathbf{r}}{M}$$

and

$$\mathbf{r}_2 = \mathbf{R}_2 - \mathbf{R} = \frac{m_1}{m_1 + m_2}(\mathbf{R}_2 - \mathbf{R}_1) = -\frac{m_1\mathbf{r}}{M},$$

where

$$\mathbf{r} = \mathbf{R}_1 - \mathbf{R}_2$$

is the position of particle 1 relative to particle 2 and

$$M = m_1 + m_2$$

is the total mass. We see from this that P_1, C, P_1 are colinear and that

$$\frac{r_2}{r_1} = \frac{P_2C}{CP_1} = \frac{m_1}{m_2} \tag{6.39}$$

(see Figure 6.15).

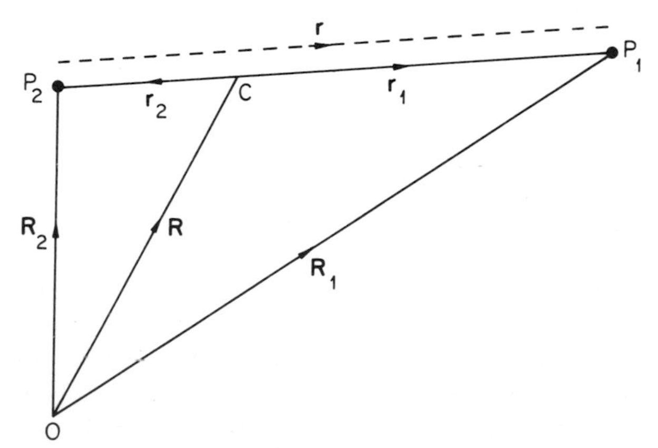

Figure 6.15

If $\mathbf{F}_1$, $\mathbf{F}_2$ are the external forces on the particles and $\mathbf{f}_{12}$ the force exerted on particle 1 by particle 2, the equations of motion for the two particles are

$$m_1 \frac{\mathrm{d}\mathbf{V}_1}{\mathrm{d}t} = m_1 \frac{\mathrm{d}}{\mathrm{d}t}(\mathbf{V} + \mathbf{v}_1) = \mathbf{F}_1 + \mathbf{f}_{12},$$

$$m_2 \frac{\mathrm{d}\mathbf{V}_2}{\mathrm{d}t} = m_2 \frac{\mathrm{d}}{\mathrm{d}t}(\mathbf{V} + \mathbf{v}_2) = \mathbf{F}_2 - \mathbf{f}_{12}.$$

Hence

$$m_1 m_2 \frac{\mathrm{d}}{\mathrm{d}t}(\mathbf{v}_1 - \mathbf{v}_2) = m_2 \mathbf{F}_1 - m_1 \mathbf{F}_2 + M\mathbf{f}_{12}. \tag{6.40}$$

Now

$$\frac{\mathrm{d}}{\mathrm{d}t}(\mathbf{v}_1 - \mathbf{v}_2) = \frac{\mathrm{d}\mathbf{v}}{\mathrm{d}t} = \frac{\mathrm{d}^2\mathbf{r}}{\mathrm{d}t^2}$$

is the acceleration of particle 1 relative to particle 2 and

$$\mathbf{f}_{12} = \mathbf{f}(\mathbf{r})$$

is the force between the particles. Equation (6.40) will reduce to

$$\frac{m_1 m_2}{M} \frac{\mathrm{d}^2\mathbf{r}}{\mathrm{d}t^2} = \mathbf{f}(\mathbf{r})$$

provided

$$m_2 \mathbf{F}_1 = m_1 \mathbf{F}_2.$$

This will certainly hold if there are no external forces. It will also hold if the forces on the particles are parallel and proportional to their masses. This is true of the sun's gravitational force on the earth and moon since their separation is so small compared with their distance from the sun. In such cases the equation of motion simplifies to

$$\mu \frac{\mathrm{d}^2\mathbf{r}}{\mathrm{d}t^2} = \mathbf{f}(\mathbf{r}), \tag{6.41}$$

where

$$\mu = \frac{m_1 m_2}{M} = \frac{m_1 m_2}{m_1 + m_2}$$

is called the *reduced mass*. We see from this that the motion of particle 1 relative to particle 2 is that of a particle of mass μ acted on by the *same* force as particle 1 actually experiences.

For the usual case of a conservative (central) force between the particles, equation (6.41) becomes

$$\mu \frac{d^2\mathbf{r}}{dt^2} = f(r)\hat{\mathbf{r}}. \tag{6.42}$$

(See Section 6.6.4, page 178.) For an observer on one of the particles this equation of relative motion represents his observations; for example we normally observe the position of the moon relative to the earth. However, it is often useful to know the motion of one or both particles relative to their centre-of-mass, since this is either at rest or moves with constant velocity when there are no external forces acting, and even when such forces do act, the centre-of-mass motion may be calculated separately.

Relative to the centre-of-mass particle 1 has position

$$\mathbf{r}_1 = \frac{m_2}{M}\mathbf{r},$$

and since $\mathbf{r}_1$ and $\mathbf{r}$ are in the same direction, unit vector $\hat{\mathbf{r}}$ is the same for both:

$$\hat{\mathbf{r}}_1 = \hat{\mathbf{r}}.$$

Then equation (6.42) may be rewritten:

$$m_1 \frac{d^2\mathbf{r}_1}{dt^2} = f\left(\frac{Mr_1}{m_2}\right)\hat{\mathbf{r}}_1. \tag{6.43}$$

Thus particle 1 will have a motion appropriate to a central conservative force directed to or away from the centre-of-mass, with the law of force suitably adjusted by using Mr_1/m_2 instead of r_1, its actual distance from the centre-of-mass. For the second particle the equation of motion is similar:

$$m_2 \frac{d^2\mathbf{r}_2}{dt^2} = f\left(\frac{Mr_2}{m_1}\right)\hat{\mathbf{r}}_2. \tag{6.44}$$

6.5.1. Conservation of energy and angular momentum

We saw in Section 4.5 that when no external forces act, total energy is conserved within the system. For two particles this may be written

$$\mathscr{T}_{\text{rel}} + \mathscr{V}_{\text{int}} = \mathscr{E}_{\text{rel}} = \text{constant}.$$

Using equation (6.39),

$$\begin{aligned}\mathscr{T}_{\text{rel}} &= \frac{m_1}{2}\left[\left(\frac{dr_1}{dt}\right)^2 + r_1^2\left(\frac{d\theta}{dt}\right)^2\right] + \frac{m_2}{2}\left[\left(\frac{dr_2}{dt}\right)^2 + r_2^2\left(\frac{d\theta}{dt}\right)^2\right] \\ &= \frac{m_1 M}{2m_2}\left[\left(\frac{dr_1}{dt}\right)^2 + r_1^2\left(\frac{d\theta}{dt}\right)^2\right]\end{aligned}$$

and

$$\mathscr{V}_{\text{int}} = \mathscr{V}(r) = \mathscr{V}\left(\frac{M}{m_2} r_1\right).$$

Hence

$$\frac{m_1 M}{2m_2}\left[\left(\frac{\mathrm{d}r_1}{\mathrm{d}t}\right)^2 + r_1^2\left(\frac{\mathrm{d}\theta}{\mathrm{d}t}\right)^2\right] + \mathscr{V}\left(\frac{Mr_1}{m_2}\right) = \mathscr{E}_{\text{rel}}$$

or

$$\frac{m_1}{2}\left[\left(\frac{\mathrm{d}r_1}{\mathrm{d}t}\right)^2 + r_1^2\left(\frac{\mathrm{d}\theta}{\mathrm{d}t}\right)^2\right] + \frac{m_2}{M}\mathscr{V}\left(\frac{Mr_1}{m_2}\right) = \frac{m_2}{M}\mathscr{E}_{\text{rel}}. \tag{6.45}$$

Similarly,

$$\frac{m_2}{2}\left[\left(\frac{\mathrm{d}r_2}{\mathrm{d}t}\right)^2 + r_2^2\left(\frac{\mathrm{d}\theta}{\mathrm{d}t}\right)^2\right] + \frac{m_1}{M}\mathscr{V}\left(\frac{Mr_2}{m_1}\right) = \frac{m_1}{M}\mathscr{E}_{\text{rel}}. \tag{6.46}$$

Equations (6.45) and (6.46) show that the particles separately conserve total energy if we take their velocities relative to the centre-of-mass, and provided we associated with each a fraction of the internal potential energy inversely proportional to its mass.

The angular momentum is also conserved in the absence of external forces. Now

$$\begin{aligned}\mathbf{J}_{\text{rel}} &= m_1\mathbf{r}_1 \wedge \frac{\mathrm{d}\mathbf{r}_1}{\mathrm{d}t} + m_2\mathbf{r}_2 \wedge \frac{\mathrm{d}\mathbf{r}_2}{\mathrm{d}t} \\ &= \left(m_1 + \frac{m_1^2}{m_2}\right)\mathbf{r}_1 \wedge \frac{\mathrm{d}\mathbf{r}_1}{\mathrm{d}t} \\ &= \frac{M}{m_2} m_1\mathbf{r}_1 \wedge \frac{\mathrm{d}\mathbf{r}_1}{\mathrm{d}t} = \frac{M}{m_2}\mathbf{J}_{\text{rel}\,1},\end{aligned}$$

where

$$\mathbf{J}_{\text{rel}\,1} = m_1\mathbf{r}_1 \wedge \frac{\mathrm{d}\mathbf{r}_1}{\mathrm{d}t}$$

is the angular momentum of particle 1 relative to the centre-of-mass. Thus

$$\mathbf{J}_{\text{rel}\,1} = \frac{m_2}{M}\mathbf{J}_{\text{rel}} = \textbf{constant} \tag{6.47}$$

and, similarly,

$$\mathbf{J}_{\text{rel}\,2} = \frac{m_1}{M}\mathbf{J}_{\text{rel}} = \textbf{constant}. \tag{6.48}$$

Equations (6.45), (6.46), (6.47) and (6.48) show that the kinetic energy and angular momentum relative to the centre-of-mass, and the internal potential energy, are all shared by the two particles inversely as their masses, and that either one separately conserves its total energy and angular momentum.

6.6. Comments and worked examples

6.6.1. See Section 6.2

Example 20

A space probe is fired vertically upwards from the earth's surface to reach a maximum height h above that surface. Another probe is fired horizontally to reach the same maximum height. Show that the ratio of the two launching speeds is $\sqrt{h(2R + h)}/(R + h)$, where R is the earth's radius. (The earth is assumed to be a stationary sphere free of any atmosphere.)

Since the gravitational force is a central conservative force directed towards the earth's centre, the total energy, $\mathscr{E}$, and the angular momentum, J, about its centre are conserved. We may write the total energy in either of the forms

$$\mathscr{E} = \mathscr{T} + \mathscr{V} = \frac{1}{2}mv^2 - \frac{GMm}{r}$$

$$= \frac{1}{2}m\dot{r}^2 + \frac{J^2}{2mr^2} - \frac{GMm}{r} = \text{constant}, \tag{6.49}$$

where M is the mass of the earth, m that of the probe, and G is the gravitational constant.

(a) If the vertical firing speed is v_v, then the launching conditions are

$$v = v_v, \qquad r = R \qquad (J = 0).$$

The conditions at maximum height are

$$v = 0, \qquad r = R + h \qquad (J = 0).$$

Substituting these two sets of values into equation (6.49),

$$\frac{1}{2}mv_v^2 - \frac{GMm}{R} = -\frac{GMm}{R + h},$$

$$\frac{1}{2}mv_v^2 = GMm\left(\frac{1}{R} - \frac{1}{R + h}\right). \tag{6.50}$$

(b) If the horizontal firing speed is v_h, then the launching conditions are

$$v = v_h, \qquad r = R \qquad (J = mRv_h).$$

The conditions at maximum height are

$$\dot{r} = 0, \qquad r = R + h \qquad (J = mRv_{\mathrm{h}}).$$

Substituting these two sets of values into equation (6.49),

$$\frac{1}{2}mv_{\mathrm{h}}^2 - \frac{GMm}{R} = \frac{J^2}{2m(R+h)^2} - \frac{GMm}{R+h}$$

$$= \frac{1}{2}m\left(\frac{R}{R+h}\right)^2 v_{\mathrm{h}}^2 - \frac{GMm}{R+h},$$

$$\frac{1}{2}mv_{\mathrm{h}}^2\left[1 - \left(\frac{R}{R+h}\right)^2\right] = GMm\left(\frac{1}{R} - \frac{1}{R+h}\right). \tag{6.51}$$

From equations (6.50) and (6.51) we see that

$$\frac{v_{\mathrm{v}}}{v_{\mathrm{h}}} = \left[1 - \left(\frac{R}{R+h}\right)^2\right]^{1/2} = \frac{\sqrt{h(2R+h)}}{R+h}.$$

6.6.2. See Section 6.3.3

Example 21

A particle of mass m is moving in a circular path of radius r_0 with angular momentum J_0 about a fixed point, towards which it is attracted by an inverse square force. A small impulse, I, of short duration, directed radially outwards, is given to the particle. Show that its perturbed trajectory oscillates about the original circular path with amplitude Ir_0^2/J_0 and period $2\pi mr_0^2/J_0$.

What is the perturbed motion when the impulse is applied in the direction of the original velocity?

Since the force is attractive and inverse square the potential energy will be of the form k/r, $k < 0$. Hence, as shown in Sections 6.2 and 6.3, the radial motion of the particle when it has angular momentum J and total energy E will be governed by the energy equation

$$\tfrac{1}{2}m\dot{r}^2 + \mathscr{U}(r) = \mathscr{E}, \tag{6.52}$$

where

$$\mathscr{U}(r) = \frac{J^2}{2mr^2} + \frac{k}{r} \tag{6.53}$$

is a pseudo-potential energy. $\mathscr{U}(r)$ has the minimum value

$$\mathscr{U}_m = -\frac{mk^2}{2J^2} \tag{6.54}$$

at the radius

$$r_m = -\frac{J^2}{mk}, \tag{6.55}$$

(see equation (6.15)). At this radius

$$\left(\frac{\mathrm{d}^2\mathscr{U}}{\mathrm{d}r^2}\right)_{r_m} = \frac{3J^2}{mr_m^4} + \frac{2k}{r_m^3} = \frac{m^3k^4}{J^6}.$$

Hence $\mathscr{U}(r)$ may be expanded as a Taylor series near r_m:

$$\begin{aligned}\mathscr{U}(r) &= \mathscr{U}_m + (r - r_m)\left(\frac{\mathrm{d}\mathscr{U}}{\mathrm{d}r}\right)_{r_m} + \frac{1}{2}(r - r_m)^2\left(\frac{\mathrm{d}^2\mathscr{U}}{\mathrm{d}r^2}\right)_{r_m} + \ldots \\ &\simeq \mathscr{U}_m + \tfrac{1}{2}K(r - r_m)^2, \end{aligned} \tag{6.56}$$

where

$$K = \frac{m^3k^4}{J^6} > 0. \tag{6.57}$$

Combining equations (6.52) and (6.56) and changing to the displacement

$$x = r - r_m,$$

we have

$$\tfrac{1}{2}m\dot{x}^2 + \tfrac{1}{2}Kx^2 = \mathscr{E} - \mathscr{U}_m. \tag{6.58}$$

By analogy with one-dimensional motion (Section 3.1.3) we can see from equation (6.58) that when the particle is slightly perturbed from its circular path its *radial* motion will consist of simple harmonic motion about the radius, r_m, at which the pseudo-potential energy, $\mathscr{U}(r)$, is a minimum. Equation (6.58) shows that the period of the oscillations will be

$$\tau = 2\pi\left(\frac{m}{k}\right)^{1/2} = 2\pi\frac{J^3}{mk^2}, \tag{6.59}$$

and their amplitude will be

$$x_0 = \left[\frac{2(\mathscr{E} - \mathscr{U}_m)}{K}\right]^{1/2} = \frac{J^3}{mk^2}\left[\frac{2(\mathscr{E} - \mathscr{U}_m)}{m}\right]^{1/2}. \tag{6.60}$$

It is important to realize that the values of J, $\mathscr{E}$ and $\mathscr{U}_m$ that occur in the expressions (6.59) and (6.60) refer to the conditions that pertain to the *perturbed motion*. In the present problem we must therefore be careful to calculate them *after the impulse has been applied.*

(a) In the initial circular path defined by r_0 and J_0 the speed v_0 of the particle is the same as its angular component of velocity v_θ, and is given by

$$J_0 = mr_0v_\theta = mr_0v_0,$$

The force constant k is given by

$$-\frac{k}{r_0^2} = \frac{mv_0^2}{r_0}.$$

Hence

$$v_0 = \frac{J_0}{mr_0} \tag{6.61}$$

and

$$k = -\frac{J_0^2}{mr_0}. \tag{6.62}$$

The total initial energy is therefore

$$\mathscr{E}_0 = \frac{1}{2}mv_0^2 + \frac{k}{r_0} = -\frac{1}{2}mv_0^2 = -\frac{J_0^2}{2mr_0^2}, \tag{6.63}$$

which is also the minimum of the initial pseudo-potential energy function (see Figure 6.16).

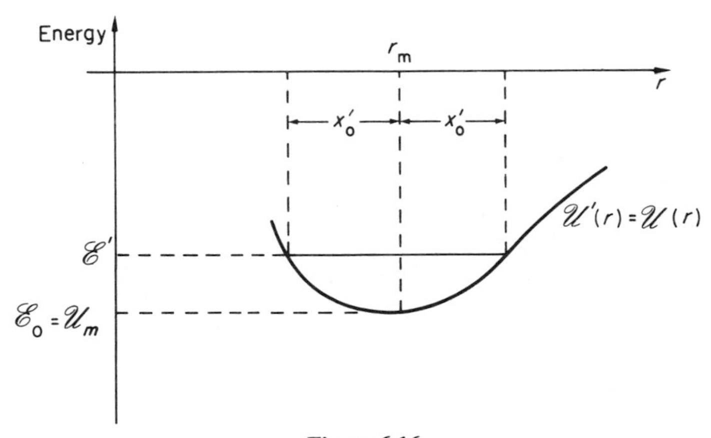

Figure 6.16

Suppose that the radial and angular components of velocity, angular momentum, etc., immediately after the application of the radial impulse are denoted by v_r', v_θ', J', etc. Since the impulse is short the position of the particle will not change during the impulse,

$$r' = r_0,$$

and since the impulse is *radial* the angular component of velocity also will not change,

$$v'_\theta = v_0.$$

However, the radial component will change from zero to

$$v'_r = \frac{I}{m},$$

since the impulse changes the radial component of momentum by an amount I.

Hence the angular momentum is unchanged,

$$J' = mr'v'_\theta = mr_0v_0 = J_0,$$

but the total energy becomes

$$\begin{aligned}\mathscr{E}' &= \frac{m}{2}(v_r'^2 + v_\theta'^2) + \frac{k}{r'} \\ &= \frac{m}{2}\left[\left(\frac{I}{m}\right)^2 + v_0^2\right] + \frac{k}{r_0},\end{aligned}$$

which, from equation (6.63), may be written

$$\mathscr{E}' = \mathscr{E}_0 + \frac{I^2}{2m}. \tag{6.64}$$

Since J remains constant the pseudo-potential energy function will be the same after the impulse as before. The only change will be that described by equation (6.64), which shows that the total energy is raised by an amount $I^2/2m$ above the minimum. Hence, we may use equations (6.59) and (6.60) to show that the radial motion is simple harmonic about r_0 with amplitude

$$x'_0 = \frac{J_0^3}{mk^2}\left[\frac{2}{m}\left(\frac{I^2}{2m}\right)\right]^{1/2} = \frac{J_0^3 I}{m^2k^2},$$

and with period

$$\tau' = \frac{2\pi mr_0^3}{mk^2}.$$

Using equation (6.62), these may be written

$$x'_0 = \frac{Ir_0^2}{J_0}, \qquad \tau' = \frac{2\pi J_0^3}{mk^2}. \tag{6.65}$$

[We may note that since

$$J_0 = mr_0^2\omega_0,$$

where ω_0 is the angular velocity of the unperturbed circular motion, then

$$\tau' = \frac{2\pi}{\omega_0} = \tau_0,$$

where τ_0 is the period of that circular motion. The perturbed motion therefore completes one cycle of oscillation about the circular path at the same time as the orbit itself is completed. The perturbed path therefore joins itself or is closed. In Section 6.4 we see that the path is closed, even when it is no longer a small perturbation about a circle.]

(b) Suppose that immediately after the application of the tangential impulse, the various dynamical quantities are denoted by v_r'', v_θ'', J'', etc. It will still be true that the position of the particle is unchanged,

$$r'' = r_0,$$

but now it will be the radial component of velocity that is also unchanged,

$$v_r'' = 0.$$

It is an angular velocity change that now accounts for the tangential impulse:

$$m(v_\theta'' - v_0) = I,$$

which gives

$$v_\theta'' = v_0 + \frac{I}{m}.$$

The total energy therefore becomes

$$\begin{aligned}
\mathscr{E}'' &= \frac{m}{2}(v_r''^2 + v_\theta''^2) + \frac{k}{r''} \\
&= \frac{m}{2}\left(v_0 + \frac{I}{m}\right)^2 + \frac{k}{r_0} \\
&= \frac{m}{2}v_0^2 + \frac{k}{r_0} + v_0 I + \frac{I^2}{2m}.
\end{aligned}$$

Using equations (6.61) and (6.63) this may be written

$$\mathscr{E}'' = \mathscr{E}_0 + \frac{J_0 I}{m r_0} + \frac{I^2}{2m}. \tag{6.66}$$

The angular momentum changes to

$$\begin{aligned}
J'' &= m r'' v_\theta'' \\
&= m r_0\left(v_0 + \frac{I}{m}\right) \\
&= J_0 + \delta J,
\end{aligned} \tag{6.67}$$

where

$$\delta J = r_0 I. \tag{6.68}$$

Thus, not only is the total energy of the system raised, but, because of the increase in angular momentum, the pseudo-potential energy function is raised as a whole, and the radius at which its minimum occurs is thereby increased, as shown in Figure 6.17.

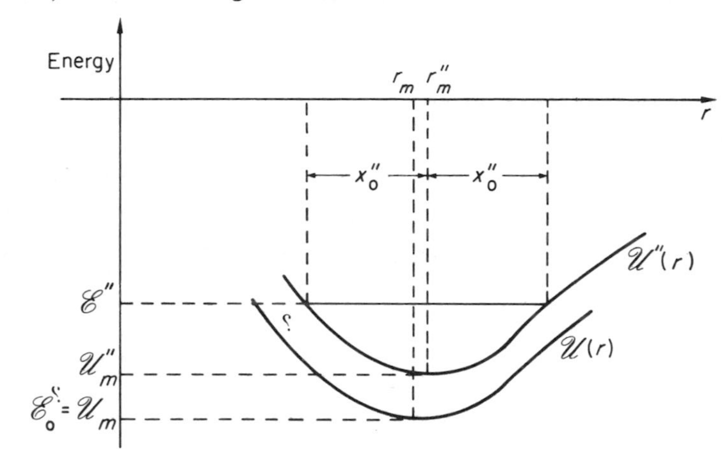

Figure 6.17

From equation (6.54) this minimum energy becomes

$$\mathscr{U}''_m = -\frac{mk^2}{2J''^2}.$$

Since I is small δJ will be small, and we may therefore expand $\mathscr{U}''_m$ as the Taylor series

$$\mathscr{U}''_m = -\frac{mk^2}{2J_0^0} + \frac{mk^2}{J_0^3}\delta J - \frac{3mk^2}{2J_0^2}(\delta J)^2 + \dots,$$

which, using equations (6.62), (6.63) and (6.68), may be written

$$\mathscr{U}''_m = \mathscr{E}_0 + \frac{J_0 I}{mr_0} - \frac{3I^2}{2m} + \dots. \tag{6.69}$$

The position of minimum energy is, from equation (6.55),

$$r''_m = -\frac{J''^2}{mk} = -\frac{J_0^2}{mk} - \frac{2J_0}{mk}\delta J + \dots,$$

or, using equations (6.62) and (6.68),

$$r''_m \simeq r_0 + \frac{2r_0^2 I}{J_0}. \tag{6.70}$$

Thus, the perturbed motion oscillates about the enlarged radius (6.70) with the period given by equation (6.59):

$$\tau'' = 2\pi \frac{J''^3}{mk^2}$$

$$= 2\pi \frac{J_0^3}{mk^2}\left(1 + \frac{3\delta J}{J_0} + \ldots\right),$$

or, using equations (6.62) and (6.65),

$$\tau'' \simeq \tau\left(1 + \frac{3r_0 I}{J_0}\right).$$

The amplitude of oscillation about r''_m is given by equation (6.60), with the values (6.66), (6.67) and (6.69):

$$x''_0 = \frac{J''^3}{mk^2}\left[\frac{2(\mathscr{E}'' - \mathscr{U}''_m)}{m}\right]^{1/2}$$

$$= \frac{J_0^3}{mk^2}\left(1 + \frac{3\delta J}{J_0} + \ldots\right)\left(\frac{4I^2}{m^2} + \ldots\right)^{1/2}$$

$$= \frac{2J_0^3 I}{m^2k^2} + \ldots.$$

Using equations (6.62) and (6.65) this may be written:

$$x''_0 \simeq \frac{2r_0^2 I}{J_0} = 2x'_0. \tag{6.71}$$

[Equations (6.66) and (6.69) show the importance of not discarding too soon the second-order quantities. If only the first-order terms had been kept $\mathscr{E}'' - \mathscr{U}''_m$ would have been zero. This would have implied that the path was circular, but at the larger radius (6.70) corresponding to the larger angular momentum. It is clearly impossible, simply by giving a single impulse tangential to the initial circular path, suddenly to move the particle out to a larger radius while keeping its velocity parallel with the original direction.]

6.6.3. See Section 6.4.4

Example 22

Prove the validity of Kepler's laws of planetary motion:

(a) All the planets travel in ellipses with the sun at a focus.
(b) The radius vector from the sun to a planet sweeps out equal areas in equal times.
(c) The squares of the periods of the planets' orbits are proportional to the cube of the major diameters of their orbits.

(a) Assuming the sun to be a stationary spherical mass M and a planet to be a spherical mass m, the gravitational force on the planet will be a conservative central force with potential energy

$$\mathscr{V}(r) = -\frac{GMm}{r},$$

where G is the gravitational constant. This is of the form discussed in Sections 6.3 and 6.4 with

$$k = -GMm. \tag{6.72}$$

Any bound motion, as shown in Sections 6.4 and 6.4.4, will therefore be elliptical with the force centre (the sun) as a focus.

(b) As the radius vector r moves through a small angle $\delta\theta$, the approximately triangular shape it sweeps out will have an area

$$\delta A = \tfrac{1}{2}r^2\,\delta\theta$$

(see Figure 6.18). Hence

$$\frac{\mathrm{d}A}{\mathrm{d}t} = \frac{1}{2}r^2\frac{\mathrm{d}\theta}{\mathrm{d}t}. \tag{6.73}$$

Now the angular momentum has magnitude

$$J = mr^2\frac{\mathrm{d}\theta}{\mathrm{d}t}, \tag{6.74}$$

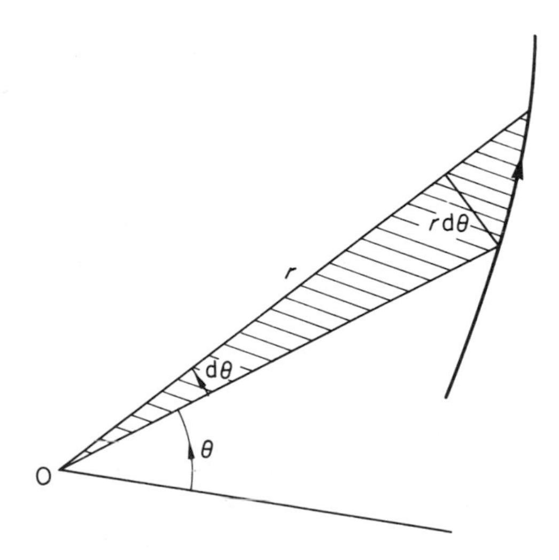

Figure 6.18

which is constant because the gravitational force is central. Equations (6.73) and (6.74) then show that the area swept out by the radius vector increases at the constant rate

$$\frac{\mathrm{d}A}{\mathrm{d}t} = \frac{J}{2m}. \tag{6.75}$$

Note that neither the conservative nature of the force nor its form of radial dependence need be involved for this second law.

(c) An ellipse with major diameter $2a$ and eccentricity e may be shown to have area

$$A = \pi a^2(1 - e^2)^{1/2}.$$

Since the period, τ, is the time for the radius vector to sweep out the complete area of the ellipse, we have, using equation (6.75),

$$\tau = \frac{A}{\mathrm{d}A/\mathrm{d}t} = \frac{2m\pi a^2(1 - e^2)^{1/2}}{J}.$$

Hence

$$\frac{\tau^2}{a^3} = \frac{4m^2\pi^2 a(1 - e^2)}{J^2}. \tag{6.76}$$

Now the minimum and maximum radii r_1 and r_2 (which are in the directions θ_0 and $\theta_0 + \pi$) span the major diameter. Then, using equations (6.35) and (6.36),

$$a = \tfrac{1}{2}(r_1 + r_2) = \frac{J^2}{m|k|(1 - e^2)}. \tag{6.77}$$

From equations (6.76) and (6.77) we have

$$\frac{\tau^2}{a^3} = \frac{4\pi^2 m}{|k|}. \tag{6.78}$$

Using equation (6.72) we see that this ratio is independent of the mass of the planet,

$$\frac{\tau^2}{a^3} = \frac{4\pi^2}{GM}. \tag{6.79}$$

[Kepler's laws, formulated in 1609 and 1619, and themselves based upon Tycho Brahe's earlier observations, are of great historical importance because Newton deduced from them the universal laws of gravitation.]

6.6.4. See Section 6.5

Example 23

How are Kepler's laws (Example 22) modified when the sun is no longer assumed to be fixed?

For a fixed sun the equation of motion of a planet is

$$m\frac{\mathrm{d}^2\mathbf{r}}{\mathrm{d}t^2} = -\frac{GMm}{r^2}\hat{\mathbf{r}} = \frac{k}{r^2}\hat{\mathbf{r}}, \tag{6.80}$$

while, according to equation (6.42), when the sun is free to move the equation becomes

$$\mu\frac{\mathrm{d}^2\mathbf{r}}{\mathrm{d}t^2} = -\frac{GMm}{r^2}\hat{\mathbf{r}} = \frac{k}{r^2}\hat{\mathbf{r}}. \tag{6.81}$$

In both equations (6.80) and (6.81) $\mathbf{r}$ gives the vectorial position of the planet relative to the sun.

Although the apparent mass of the planet changes, the force in both cases is a central inverse square force. Hence the first two of Kepler's laws remain true. However, in the result (6.81) m changes to μ, while k remains unchanged. Hence if τ', a' now refer to the orbit of the planet *relative to the sun*, the result (6.78) becomes

$$\frac{\tau'^2}{a'^3} = \frac{4\pi^2\mu}{|k|}.$$

Inserting the values

$$\mu = \frac{Mm}{M + m}, \qquad k = -GMm,$$

we obtain the result

$$\frac{\tau'^2}{a'^3} = \frac{4\pi^2}{G(M + m)} \tag{6.82}$$

instead of (6.79).

Hence if a given value

$$a = a'$$

were used in the results (6.79) and (6.82), the corresponding periods would have the ratio

$$\frac{\tau}{\tau'} = \left(1 + \frac{m}{M}\right)^{1/2} = 1 + \frac{1}{2}\frac{m}{M} + \dots .$$

Thus if

$$\Delta\tau = \tau - \tau'$$

is the error in the period which would arise from the assumption of a fixed sun,

$$\frac{\Delta\tau}{\tau'} = \frac{1}{2}\frac{m}{M}.$$

In the case of the earth, with a period of a year,

$$\tau' \simeq 3{\cdot}15 \times 10^7 \text{ s}, \qquad \frac{m}{M} \simeq 3{\cdot}01 \times 10^{-6}.$$

Hence

$$\Delta\tau \simeq 47{\cdot}4 \text{ s}.$$

[At the time Kepler's laws were enunciated clocks used in astronomy and navigation were commonly in error by this amount in the course of *one day*.]

6.7. Problems

6.1. Show that if a central force is conservative then its magnitude can depend only upon radial distance from the force centre and that the corresponding potential energy must therefore be of the form $\mathscr{V}(r)$.

6.2. A mass m is attached by a light flexible string of length a to a light spring also of length a, with strength constant K. The other end of the spring is pivoted at a fixed point O. Show that the radial equation of motion of the mass for

$$r \geqslant 2a$$

is

$$\frac{1}{2}m\dot{r}^2 + \frac{J^2}{2mr^2} + \frac{1}{2}K(r - 2a)^2 = \mathscr{E},$$

where J is the angular momentum and $\mathscr{E}$ the total energy. What is the corresponding equation for

$$r \leqslant 2a?$$

The mass is pulled out a distance $3a$ from O and given a velocity $\mathbf{V}$ perpendicular to the string.

(a) For what value of V will the mass move in a circular path?

(b) What is the least value, V_m, of V that will keep the string taut in the subsequent motion?

(c) Show that when V is less than V_m the distance of nearest approach to O, r_m, is given by

$$r_m = \frac{3a}{(1 + Ka^2/mV^2)^{1/2}}.$$

6.3. The simplest model of the hydrogen atom is a fixed proton of charge e around which an electron of charge $-e$ and mass m moves in a circular orbit of radius r. Show that the total energy of such an atom is

$$\mathscr{E} = -\frac{e}{8\pi\varepsilon_0 r}$$

and that it should therefore be capable of an inexhaustible supply of energy.

If the circular orbits are restricted to those whose angular momenta are (non-zero) multiples of $\hbar$, where $\hbar$ is $(1/2\pi) \times$ Planck's constant, show that the energy can then only have the values

$$\mathscr{E} = -\frac{\mathscr{E}_0}{n^2},$$

and determine $\mathscr{E}_0$ in electron volts ($e = 1{\cdot}6 \times 10^{-19}$ C, $m = 9{\cdot}1 \times 10^{-31}$ kg, $\hbar = 1{\cdot}05 \times 10^{-34}$ J s, $\varepsilon_0 = 10^{-9}/36\pi$ F m^{-1}, 1 eV $= 1{\cdot}6 \times 10^{-19}$ J).

6.4. A mass m, free to slide on a smooth horizontal plane, is attached to a long string which passes through a hole in the plane and then hangs vertically downwards with a mass M attached to the lower end. Initially the mass m is at a distance a from the hole and has a velocity $\mathbf{V}$ in the horizontal plane. What is the maximum distance from the hole reached by m in the subsequent motion when $\mathbf{V}$ is (a) parallel with, (b) perpendicular to, the string leading from it to the hole? In the latter case, find the maximum velocity of the mass M.

6.5. A sun probe is launched from the earth, which is travelling at a speed v around the sun in what may be assumed to be a circular orbit of radius R. At launching the velocity of the probe is αv and is at an angle θ to the line joining the sun to the earth (in the sun's frame of reference). If the distance of nearest approach is to be γR,($\gamma < 1$), show that the launching speed and direction are linked by the relationship

$$\alpha^2 = \frac{2\gamma(1 - \gamma)}{\sin^2 \theta - \gamma^2}.$$

Discuss qualitatively the main features of this relationship and their physical interpretation. (The effect of the *earth's* gravity on the probe may be neglected.)

6.6. A spherical satellite of mass m and radius a, initially in circular orbit with period T about the sun, meets a dust cloud also in circular orbit but travelling in the opposite direction. For a length L of its path (small compared with the radius of its orbit) the satellite travels through the cloud, collecting all the dust particles with which it collides. Show that the period of its new orbit is diminished by approximately $12\pi a^2 L\rho T/m$, where ρ is the density of matter in the cloud.

6.7. A meteorite of mass m, which may be assumed to have started from infinity at rest, is moving round the sun. At the point of nearest approach to the sun it collides with, and sticks to, an artificial planet of mass M which is moving in the same direction round the sun in a circular orbit of radius R.

Show that in the new orbit resulting from the collision the planet's greatest deviation from its former circular path is approximately $4R(\sqrt{2} - 1)m/M$, provided that $m \ll M$.

6.8. An earth satellite of mass m and the shell of its final rocket stage of mass αm are attached to each other and cruising in circular orbit about the earth at a height h above its surface. An explosive charge projects the shell backwards so that it separates with a speed u relative to the satellite. Show that if the shell is not to collide with the earth, the separation speed must not lie within the limits

$$u = (1 + \alpha)\sqrt{\frac{g}{R + h}}R\left(1 \mp \sqrt{\frac{2R}{2R + h}}\right).$$

Which limitation would this imply *in practice*?

6.9. Two space stations are moving, in the same sense, in coplanar circular orbits about the earth with a small difference h in their heights. An object is projected forward from the inner station with a velocity that just enables it to reach the outer one. Show that its relative speed of impact on arrival is approximately $\pi h/(2T)$, where T is the period of orbit for the outer station.

Chapter 7

Scattering; Cross-section

Some properties of the planets, their moons, comets and other heavenly bodies can be deduced from a detailed study of their paths, using the results of Sections 6.4 and 6.5. In the case of atomic particles we are never able to observe their complete trajectories. When they are bound together, as electrons are to a nucleus to form an atom, we cannot observe their motion at all in any direct sense. This is why it is wrong to assume that such small particles so close together can necessarily be described by concepts and a mathematical framework that are based upon macroscopic observations. In fact they cannot. The basic idea of describing position as a function of time, $\mathbf{r}(t)$, has to be given up. Quantum mechanics, which describes the *state* of a physical system instead of the motion of its various parts, replaces classical mechanics as a proper description.

Nevertheless, isolated atomic and nuclear particles can have quite well-defined trajectories, as we can see from the tracks they leave behind in bubble chambers or nuclear emulsions. Typically we might see the tracks shown in Figure 7.1, which we could interpret as a collision at O, with particle

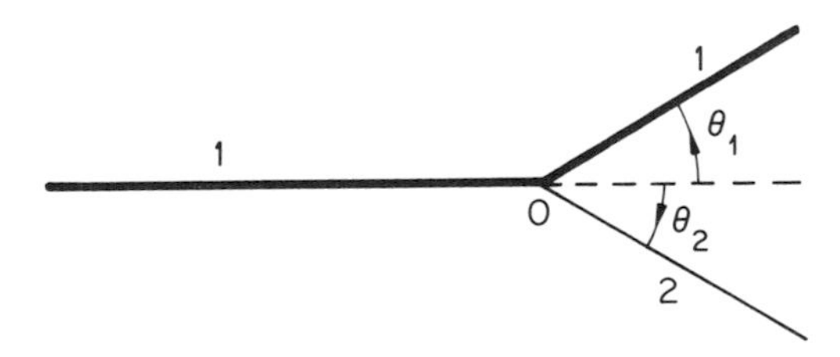

Figure 7.1

1 being scattered through an angle θ_1 and particle 2, initially stationary, setting off at an angle θ_2 to the initial velocity of particle 1. The exact trajectory in the neighbourhood of O, or, indeed, whether it is correct to talk of one, is unimportant. All that we can observe are the angles θ_1 and θ_2. Such observations from scattering experiments are often all we have as a clue to the nature of the particles involved.

7.1. Elastic scattering from a fixed sphere

We start by considering a simple classical collision in which particle 2 is a sphere of radius a and of such great mass that it can be considered stationary, with its centre at O. A particle, 1, of very small mass and dimensions comes from the left in a direction which would pass O at a distance b. If this is less than the radius of the sphere it will strike it and be deflected through an angle θ, as shown in Figure 7.2.

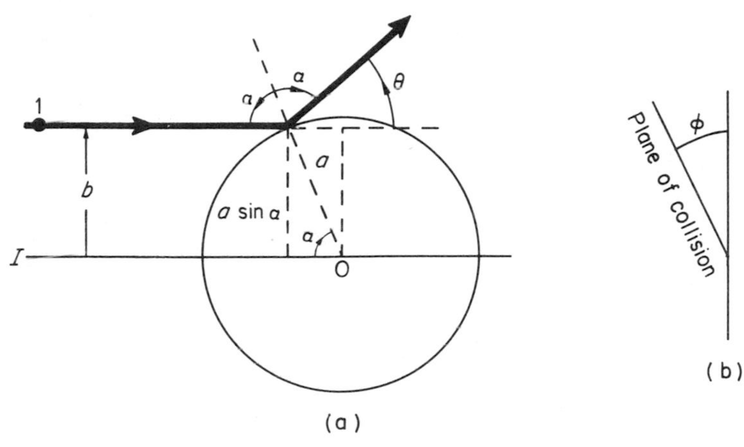

Figure 7.2

We take an axis of symmetry IO through O parallel with the initial velocity of particle 1. If the plane containing this axis and particle 1 makes an azimuthal angle ϕ with some reference plane through OI, b and ϕ give the initial trajectory, while θ and ϕ (the latter being unchanged) give the final trajectory of particle 1. If the collision is elastic the angles of incidence and reflection (α) will be equal. Consequently,

$$\theta = \pi - 2\alpha.$$

Since

$$b = a \sin \alpha,$$

the relationship between b, the *impact parameter*, and θ, the *scattering angle*, is

$$b = a \cos \frac{\theta}{2}. \tag{7.1}$$

A typical scattering experiment does not involve detailed observation of individual particles. Instead of a single particle with a known impact parameter we would have a uniform flux of particles, S (unit area)$^{-1}$ (unit time)$^{-1}$, coming from the left, as in Figure 7.3a. All these particles with impact parameter in the range $(b, b + \mathrm{d}b)$ and azimuthal angle in the range $(\phi, \phi + \mathrm{d}\phi)$ will pass through the shaded area of Figure 7.3b. Since this has area $b\,\mathrm{d}b\,\mathrm{d}\phi$, the rate at which they pass through will be $Sb\,\mathrm{d}b\,\mathrm{d}\phi$ (unit time)$^{-1}$.

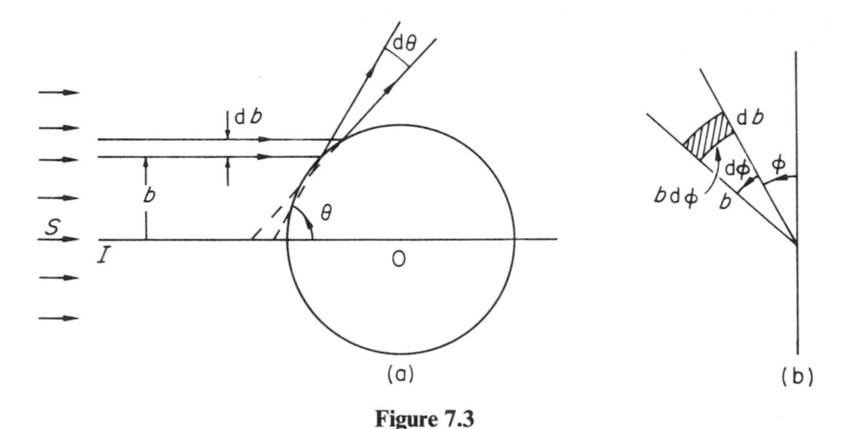

Figure 7.3

After colliding with the sphere, these particular particles will emerge with scattering angle in the range $(\theta, \theta + \mathrm{d}\theta)$ and azimuthal angle in the (unchanged) range $(\phi, \phi + \mathrm{d}\phi)$. If we were to measure the rate at which they emerge by placing a counting device that covers these angular ranges, we would, provided $\mathrm{d}\theta$ and $\mathrm{d}\phi$ are small, expect the counting rate to be proportional to $\mathrm{d}\theta$ and to $\mathrm{d}\phi$, and possibly to depend upon θ and ϕ. It will therefore be of the form $N(\theta, \phi)\,\mathrm{d}\theta\,\mathrm{d}\phi$, where $N(\theta, \phi)$ is the *distribution function*. Since the rate at which particles are counted emerging in the angular ranges $(\theta, \theta + \mathrm{d}\theta)$, $(\phi, \phi + \mathrm{d}\phi)$ must equal the rate at which particles cross the area $b\,\mathrm{d}b\,\mathrm{d}\phi$,

$$N(\theta, \phi)\,\mathrm{d}\theta\,\mathrm{d}\phi = Sb\,\mathrm{d}b\,\mathrm{d}\phi, \tag{7.2}$$

whence

$$N(\theta, \phi) = Sb\frac{\mathrm{d}b}{\mathrm{d}\theta}. \tag{7.3}$$

In this relationship θ is linked with b by equation (7.1). It is clear from that equation, or from Figure 7.3a, that θ decreases as b increases and therefore $\mathrm{d}b$ and $\mathrm{d}\theta$ will have opposite signs. If we are always to consider counting rates as *positive* we must take absolute values in equation (7.2) and replace equation (7.3) by

$$N(\theta, \phi) = Sb\left|\frac{\mathrm{d}b}{\mathrm{d}\theta}\right|. \tag{7.4}$$

Now θ is determined by b, according to equation (7.1), and

$$\frac{db}{d\theta} = -\frac{a}{2}\sin\frac{\theta}{2}.$$

Hence, equation (7.4) becomes

$$N(\theta, \phi) = \frac{Sa^2}{2}\cos\frac{\theta}{2}\sin\frac{\theta}{2} = \frac{Sa^2}{4}\sin\theta. \tag{7.5}$$

In this case we can predict the variation of counting rate with θ and ϕ because we know the scattering object is an elastic sphere. However, if we do not know the nature of the scattering object, a measurement of $N(\theta, \phi)$ over the whole range $0 \leqslant \theta \leqslant \pi$ and $0 \leqslant \phi \leqslant 2\pi$ may give us clues to the shape and properties of the scattering object.

For example, a flat disc would scatter particles only into a single (θ, ϕ) direction. A cone whose axis was parallel with the beam direction would scatter particles uniformly in ϕ but only in a single θ direction. Both these shapes of scattering object could therefore be ruled out if the counting rate were proportional to $\sin\theta$ and independent of ϕ, as in equation (7.5). An elastic spherical scatterer would fit the observations and would be preferred as an explanation of the process, although we have not, of course, proved that only an elastic sphere would give the result (7.5).

Thus although measuring S and $N(\theta, \phi)$ alone cannot determine uniquely the shape and size of the scattering object, they can help to decide between various hypotheses.

7.2. Cross-section

The usual way of presenting scattering phenomena is to note that since S is a flux we can define a small (second-order infinitesimal) area $d^2\sigma$ which, if placed perpendicularly to the incoming beam, would just intercept particles at the rate $N(\theta, \phi)\, d\theta\, d\phi$ at which they leave within the angular ranges $(\theta, \theta + d\theta)$, $(\phi, \phi + d\phi)$. Then

$$S d^2\sigma = N(\theta, \phi)\, d\theta\, d\phi \tag{7.6}$$

or

$$\frac{d^2\sigma}{d\theta\, d\phi} = \frac{N(\theta, \phi)}{S}, \tag{7.7}$$

$d^2\sigma/d\theta\, d\phi$ being known as the *differential scattering cross-section* for, or with respect to, θ, ϕ. If we are interested only in how the scattering varies with one angle, θ say, we sum or integrate over all physically possible values of ϕ, thus defining

$$\frac{d\sigma}{d\theta} = \int_{\phi=0}^{2\pi} \left(\frac{d^2\sigma}{d\theta\, d\phi}\right) d\phi = \frac{1}{S}\int_0^{2\pi} N(\theta, \phi)\, d\phi, \tag{7.8}$$

which is known as the differential scattering cross-section for θ. Finally, by integrating over θ (which varies from 0 to π) we can find the rate of scattering for all directions:

$$\sigma_T = \int_{\theta=0}^{\pi} \left(\frac{d\sigma}{d\theta}\right) d\theta = \frac{1}{S}\int_0^{\pi}\int_0^{2\pi} N(\theta, \phi)\, d\phi\, d\theta, \tag{7.9}$$

σ_T being known as the *total scattering cross-section.* (See Section 7.4.1, page 193.)

It is important to note that these definitions of cross-sections are based solely on counting rates before and after the scattering, since these alone determine S and $N(\theta, \phi)$. They are, therefore, applicable whenever an experiment along these lines is possible, whether or not a theory exists that directly relates the final direction of a particular particle to its initial path.

When a relationship

$$b \equiv b(\theta)$$

is known or predicted, the area $d^2\sigma$ will have a precise location and shape, as in Figure 7.3b, and we may write

$$d^2\sigma = b\, db\, d\phi = b\left|\frac{db}{d\theta}\right| d\theta\, d\phi. \tag{7.10}$$

Then equation (7.7) becomes

$$\frac{d^2\sigma}{d\theta\, d\phi} = \frac{N(\theta, \phi)}{S} = b\left|\frac{db}{d\theta}\right|. \tag{7.11}$$

(For nearly all types of collision θ decreases as b increases, so we must use $|db/d\theta|$ rather than $db/d\theta$ to give positive values for the counting rate and the cross-section.)

For the elastic sphere scattering of Section 7.1 equation (7.1) gives the result:

$$b\left|\frac{db}{d\theta}\right| = \frac{a^2}{4}\sin\theta.$$

Hence

$$\frac{d^2\sigma}{d\theta\, d\phi} = \frac{a^2}{4}\sin\theta, \tag{7.12}$$

$$\frac{d\sigma}{d\theta} = \int_0^{2\pi} \frac{a^2}{4}\sin\theta\, d\phi = \frac{\pi a^2}{2}\sin\theta,$$

and the total cross-section is

$$\sigma_T = \int_0^{\pi} \frac{\pi a^2}{2}\sin\theta\, d\theta = \pi a^2.$$

The total cross-section is therefore just the geometrical area of the sphere as seen by the beam. This is what we would expect since it is only those beam particles within this area which will strike the sphere and suffer any deflexion.

When the target is not symmetrical about the beam axis the scattering angle θ may depend upon ϕ as well as upon b, and the plane of the scattering (defined by ϕ) may even change. In this case, a theory relating the final direction, (θ, ϕ), uniquely to the initial path, $(b, \phi_{\text{initial}})$, will yield a relationship of the form

$$d^2\sigma = b\, db\, d\phi_{\text{initial}} = |f(\theta, \phi)|\, d\theta\, d\phi. \tag{7.13}$$

We would then replace equation (7.11) by the more general form:

$$\frac{d^2\sigma}{d\theta\, d\phi} = f(\theta, \phi). \tag{7.14}$$

7.2.1. Cross-section per unit solid angle

Two practical points that arise in a scattering experiment should be noted. We can see from Figure 7.3a that a particle scattered through an angle θ appears to come from a point on the beam axis that depends upon θ. Thus as we vary the inclination of the counting device we should also vary the point at which it is directed. However, provided the counter is very far from the collision centre compared with the maximum radius at which scattering takes place, this variation is negligible and the counter can remain pointed towards the centre. In atomic and nuclear physics experiments, this condition is always very well satisfied since the counter distance is certain to be many orders of magnitude greater than an atomic radius. In such experiments, a volume of material or target containing many scattering centres (atoms or nuclei) may be used to scatter the beam. The counter distance must then be large compared with the dimensions of the target if the counter is to measure the same scattering angle from all the centres within the target.

A particle counter usually has a fixed aperture of area A. At a distance R from the target this will subtend a solid angle

$$d^2\Omega = \frac{A}{R^2}.$$

When the counter is moved about the target at a fixed value of R it will therefore subtend a constant $d^2\Omega$ at any inclination θ, ϕ, but *not* a constant $d\theta\, d\phi$, since

$$d^2\Omega = \sin\theta\, d\theta\, d\phi.$$

It is therefore common to use the differential scattering cross-section *per unit solid angle*, defined as

$$\frac{d^2\sigma}{d^2\Omega} = \frac{1}{\sin\theta}\frac{d^2\sigma}{d\theta\, d\phi}. \tag{7.15}$$

For scattering from a massive sphere equation (7.12) gives

$$\frac{d^2\sigma}{d^2\Omega} = \frac{a^2}{4} = \text{constant}.$$

Since this shows that the counting rate for a fixed aperture counter is the same at all orientations, such scattering is said to be *isotropic*. (See Section 7.4.2, page 194.)

7.3. Laboratory and centre-of-mass descriptions

When the two particles involved in a collision have comparable masses the description of a collision must take account of the motions of both. Initially the target particle 2 of mass m_2 is at rest at the origin, O, of a co-ordinate framework which we shall call the *laboratory frame*. In this frame the initial momentum of particle 1 of mass m_1 is $\mathbf{P}_{1i}$. After the collision the final momenta of the particles will be $\mathbf{P}_{1f}$ and $\mathbf{P}_{2f}$. Particle 1 will be scattered through an angle θ_1 and particle 2 through θ_2, relative to the initial direction of particle 1. This is shown in Figure 7.4. Since experimentally we can

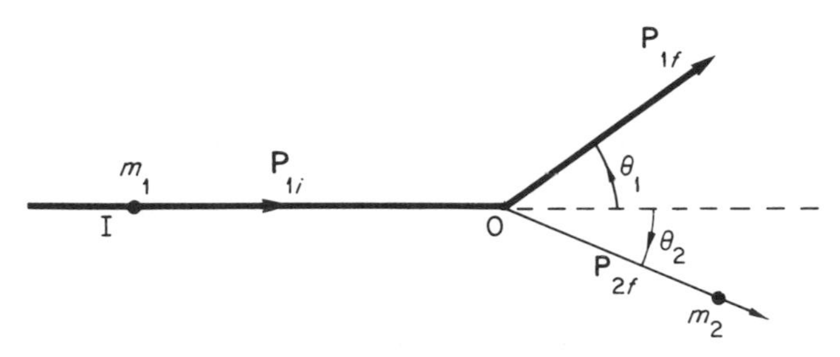

Figure 7.4

normally determine only the initial and final momenta and directions, the scale is such that the detail of the collision is condensed into the region close to O and will not concern us further.

Assuming that the collision is free of external forces, we know from Section 6.5 that it is simpler to analyse it in a coordinate framework centred on C, the centre-of-mass of the two particles. C will move with a constant velocity $\mathbf{V}$, while relative to this the total momentum of the particles will remain zero throughout the collision. In this *centre-of-mass frame* the collision will appear as in Figure 7.5. Since the momenta of the particles are always equal and opposite, one momentum, $\mathbf{p}_i$, before and one, $\mathbf{p}_f$, after are enough to specify the collision, with a single scattering angle, θ_c.

The azimuthal angle, ϕ, is also required to specify the plane in which the collision takes place. However, if this is the same before and after the collision, it is clearly unaltered by changing between the laboratory and centre-of-mass frames, and is therefore omitted from Figures 7.4 and 7.5.

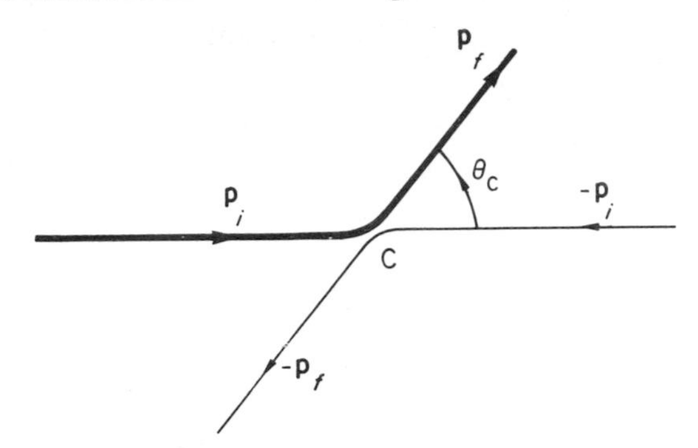

Figure 7.5

By considering the motion of the particles well before the collision, it is clear that the centre-of-mass moves along the axis IO with the constant velocity **V** in the laboratory frame given by

$$\mathbf{P}_{1i} = \mathbf{P}_{\text{cm}} = (m_1 + m_2)\mathbf{V}.$$

Now

$$\begin{aligned} \mathbf{p}_i = m_1\mathbf{v}_{1i} &= m_1(\mathbf{V}_{1i} - \mathbf{V}) \\ &= \mathbf{P}_{1i} - \frac{m_1}{m_1 + m_2}\mathbf{P}_{\text{cm}} = \frac{m_2}{m_1 + m_2}\mathbf{P}_{1i}, \end{aligned} \tag{7.16}$$

and

$$\begin{aligned} \mathbf{p}_f = m_1\mathbf{v}_{1f} &= m_1(\mathbf{V}_{1f} - \mathbf{V}) \\ &= \mathbf{P}_{1f} - \frac{m_1}{m_1 + m_2}\mathbf{P}_{\text{cm}} = \mathbf{P}_{1f} - \frac{m_1}{m_1 + m_2}\mathbf{P}_{1i}. \end{aligned} \tag{7.17}$$

Since momentum is conserved in the laboratory frame

$$\mathbf{P}_{1i} = \mathbf{P}_{1f} + \mathbf{P}_{2f}. \tag{7.18}$$

The relationships (7.16), (7.17) and (7.18) can all be shown in the single Figure 7.6. Note that since, from equation (7.16),

$$\frac{\text{CB}}{\text{AB}} = \frac{p_i}{P_{1i}} = \frac{m_2}{m_1 + m_2},$$

the point C divides AB in the ratio

$$\frac{\text{AC}}{\text{CB}} = \frac{m_1}{m_2}.$$

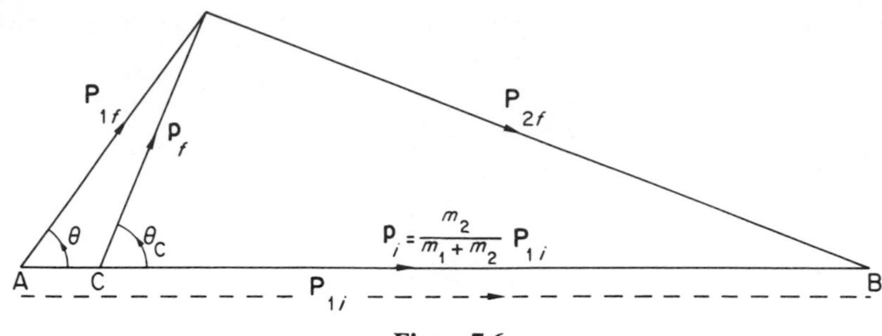

Figure 7.6

7.3.1. Elastic collisions

Only conservation of momentum has been involved so far and diagram 7.6 is true whether or not the collision is elastic. If it is, energy is conserved in the centre-of-mass frame:

$$\mathscr{T}_{\text{rel}} + \mathscr{V}_{\text{int}} = \mathscr{E}_{\text{rel}} = \text{constant}.$$

Since the potential energies of the two particles when they are far apart initially and finally will be the same, their total kinetic energies will also be the same:

$$\mathscr{T}_{\text{rel}\,i} = \mathscr{T}_{\text{rel}\,f}$$

or

$$\frac{p_i^2}{2m_1} + \frac{p_i^2}{2m_2} = \frac{p_f^2}{2m_1} + \frac{p_f^2}{2m_2}.$$

Hence

$$|\mathbf{p}_i| \equiv p_i = p_f \equiv |\mathbf{p}_f| = p, \quad \text{say.}$$

Thus as the scattering angles change, the point Q moves round a circle of radius p centred on C where

$$p = \frac{m_2}{m_1 + m_2} P_{1i}.$$

Figure 7.7 shows that in this case the scattering angles θ_c, θ_l ($\equiv \theta_1$) of the projectile in the two frames are given by

$$\sin\theta_c = \frac{\text{QL}}{\text{CQ}}, \qquad \tan\theta_l = \tan\theta_1 = \frac{\text{QL}}{\text{AL}}.$$

Now

$$\text{AL} = \text{AC} + \text{CL} = \frac{m_1}{m_2}\text{CB} + \text{CQ}\cos\theta_c$$

$$= \left(\frac{m_1}{m_2} + \cos\theta_c\right)\text{CQ}.$$

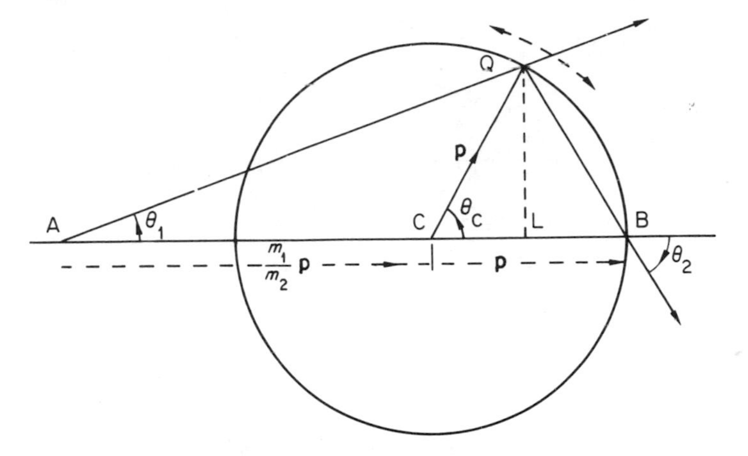

Figure 7.7

Hence

$$\tan \theta_l = \frac{\sin \theta_c}{m_1/m_2 + \cos \theta_c}. \tag{7.19}$$

The differential cross-sections in the two frames will be

$$\frac{d^2\sigma}{d\theta_c \, d\phi} = \left|\frac{b \, db}{d\theta_c}\right|, \qquad \frac{d^2\sigma}{d\theta_l \, d\phi} = \left|\frac{b \, db}{d\theta_l}\right|$$

(b and ϕ are the same for both frames). The relationship between them is therefore

$$\frac{d^2\sigma}{d\theta_l \, d\phi} = \left|\frac{d\theta_c}{d\theta_l}\right| \frac{d^2\sigma}{d\theta_c \, d\phi}. \tag{7.20}$$

For the cross-sections per unit solid angle,

$$\frac{d^2\sigma}{d^2\Omega_c} = \left|\frac{b \, db}{\sin \theta_c \, d\theta_c}\right|, \qquad \frac{d^2\sigma}{d^2\Omega_l} = \left|\frac{b \, db}{\sin \theta_l \, d\theta_l}\right|,$$

the relationship is

$$\frac{d^2\sigma}{d^2\Omega_l} = \left|\frac{\sin \theta_c \, d\theta_c}{\sin \theta_l \, d\theta_l}\right| \frac{d^2\sigma}{d^2\Omega_c}. \tag{7.21}$$

The range of application of these results depends markedly upon the mass ratio of the two particles. The three possibilities,

$$m_2 < m_1, \qquad m_2 = m_1, \qquad m_2 > m_1,$$

give the three diagrams of Figure 7.8.

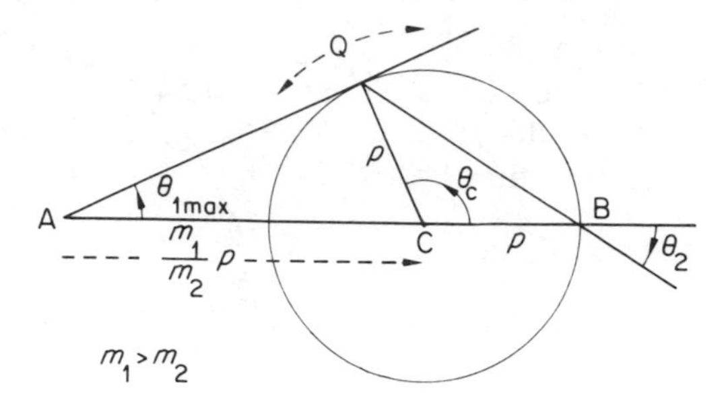

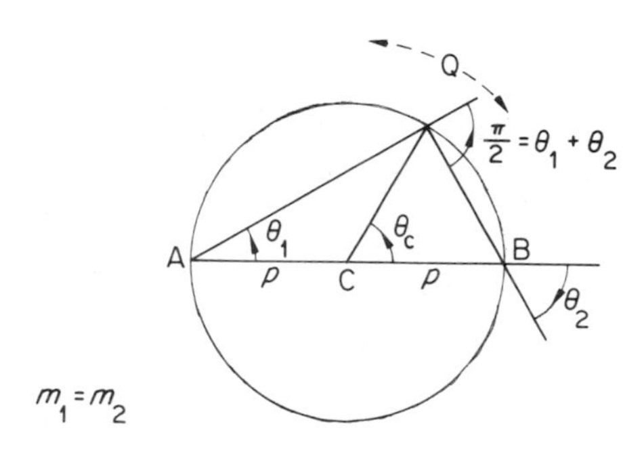

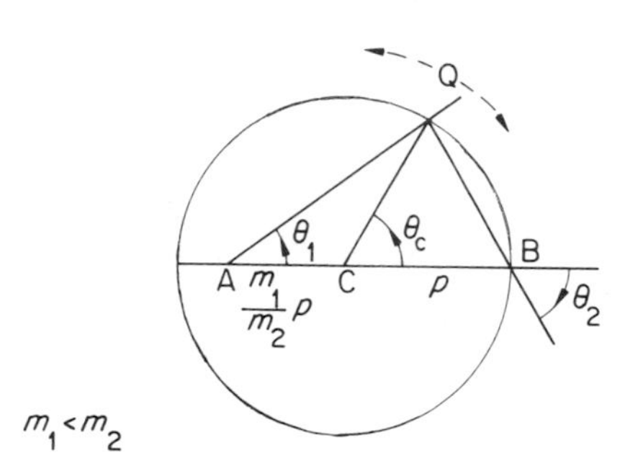

Figure 7.8

In the first case, while θ_c varies from 0 to π, θ_l rises to a maximum given by

$$\sin \theta_{l\,\max} = \frac{m_2}{m_1}$$

and then decreases to zero.

When the masses are equal θ_l just reaches $\pi/2$ as θ_c reaches π (though the final momentum of particle 1 is then zero). Since the diameter AB always subtends a right angle at the circumference, the two particles are always scattered at right angles to each other in the laboratory frame.

Only in the third case does θ_l always increase with θ_c and cover the full range from 0 to π. (See Section 7.4.3, page 196.)

7.4. Comments and worked examples

7.4.1. See Section 7.2

Several notations are used for differential cross-sections. Instead of the $d^2\sigma$ of equations (7.7), (7.11) and (7.15), $d\sigma$ is sometimes used, and sometimes σ is used for either $d^2\sigma/d\theta\, d\phi$ or $d\sigma/d\theta$. As we have defined it, $d^2\sigma$ is a true second-order infinitesimal, $d\sigma$ a first-order infinitesimal and σ a finite cross-section. These preserve a consistent sequence in which each is obtained by a single integration of the previous one. The use of $d^2\sigma/d^2\Omega$ (Section 7.2.1) is consistent with this scheme since it must be remembered that $d^2\Omega$ is also a second-order infinitesimal:

$$d^2\Omega = \sin\theta\, d\theta\, d\phi.$$

The advantage of keeping a strict account of the 'size' of the various cross-sections becomes more important when the counting system depends upon more than the two angles giving the direction of the scattered particle. For example, in an inelastic process there might be, in any given direction, a spread in the values of the final kinetic energy, $\mathscr{T}$, of the scattered particle. If the counting device then selected a range $(\mathscr{T}, \mathscr{T} + d\mathscr{T})$ of energy as well as a range of directions $(\theta, \theta + d\theta)$, $(\phi, \phi + d\phi)$ we could write the counting rate as $N(\theta, \phi, \mathscr{T})\, d\theta\, d\phi\, d\mathscr{T}$ since this rate may depend upon θ, ϕ and $\mathscr{T}$ and will be proportional to $d\theta$, $d\phi$ and $d\mathscr{T}$, provided they are very small. We could therefore define a third-order cross-section $d^3\sigma$ which would intercept just that amount of beam flux S to give the observed counting rate:

$$S\, d^3\sigma = N(\theta, \phi, \mathscr{T})\, d\theta\, d\phi\, d\mathscr{T}.$$

This then gives a differential cross-section with respect to θ, ϕ and $\mathscr{T}$:

$$\frac{d^3\sigma}{d\theta\, d\phi\, d\mathscr{T}} = \frac{N(\theta, \phi, \mathscr{T})}{S}.$$

From this a variety of lower-order differential cross-sections may be derived,

$$\frac{d^2\sigma}{d\theta\, d\phi} = \frac{1}{S}\int N(\theta, \phi, \mathscr{T})\, d\mathscr{T}, \qquad \frac{d^2\sigma}{d\theta\, d\mathscr{T}} = \frac{1}{S}\int N(\theta, \phi, \mathscr{T})\, d\phi, \quad \text{etc.},$$

where the integrations are, in each case, over the whole physically meaningful range of the variable concerned. These integrations correspond experimentally to varying the position or characteristics of the counting device to cover the whole range of the variable and adding the counting rates for each interval of that variable.

More generally, we can imagine the particles scattered from a homogeneous beam to be defined by a range of, say, n parameters, $\alpha, \beta, \ldots, \omega$, and a device counting those in the ranges $(\alpha, \alpha + d\alpha), (\beta, \beta + d\beta), \ldots, (\omega, \omega + d\omega)$. Writing the observed rate in the form $N(\alpha, \beta, \ldots, \omega)\, d\alpha\, d\beta \ldots d\omega$ enables us to define a cross-section $d^n\sigma$:

$$S\, d^n\sigma = N(\alpha, \beta, \ldots, \omega)\, d\alpha\, d\beta \ldots d\omega,$$

from which

$$\frac{d^n\sigma}{d\alpha\, d\beta \ldots d\omega} = \frac{N(\alpha, \beta, \ldots, \omega)}{S}.$$

Integrating this over one or more of the variables gives various lower-order differential cross-sections, leading ultimately to the total cross-section

$$\sigma_T = \frac{1}{S}\int\int \cdots \int N(\alpha, \beta, \ldots, \omega)\, d\alpha\, d\beta \ldots d\omega.$$

7.4.2. See Section 7.2.1

Example 24

A monoenergetic beam of particles each of charge ze is scattered by a heavy nucleus of charge Ze. Find the differential cross-sections for the process.

The heavy nucleus may be regarded as giving a fixed force centre. The potential energy when a particle of charge ze is distant r from the nucleus is

$$\mathscr{V}(r) = \frac{Zze^2}{4\pi\varepsilon_0 r}. \tag{7.22}$$

We may therefore use the results of Section 6.4.1, with

$$k = \frac{Zze^2}{4\pi\varepsilon_0},$$

to determine the relationship between the impact parameter b and the scattering angle θ. Note that θ is not used here to denote the angular position at any point of the particle's trajectory; it is the change from the initial direction at infinity to the final direction at infinity, after the particle has been subjected to the force field around the nucleus. From Figure 6.11 this is

$$\theta = \pi - 2\alpha. \tag{7.23}$$

The impact parameter is the perpendicular distance, b, of the asymptotes from the nucleus at O,

$$b = \frac{J}{(2m\mathscr{E})^{1/2}}. \tag{7.24}$$

Since, from equations (6.28) and (6.29),

$$\cos \alpha = \left(\frac{mk^2}{mk^2 + 2\mathscr{E}J^2} \right)^{1/2} \tag{7.25}$$

we can show the relationships (7.25) and (7.27) in the form of Figure 7.9. From this

$$\cot \frac{\theta}{2} = \left(\frac{2\mathscr{E}J^2}{mk^2} \right)^{1/2}. \tag{7.26}$$

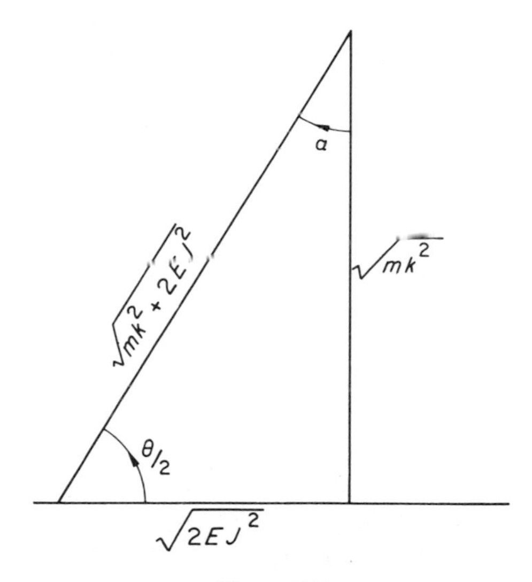

Figure 7.9

Equations (7.24) and (7.26) show that

$$\cot \frac{\theta}{2} = \frac{2\mathscr{E}b}{k}.$$

$\mathscr{E}$ is the same for all the particles in the beam: it is their kinetic energy at infinity, $\mathscr{T}_\infty$, since the potential energy is zero there. Thus

$$\cot \frac{\theta}{2} = \frac{2\mathscr{T}_\infty b}{k}. \tag{7.27}$$

Using this in equation (7.11),

$$\frac{d^2\sigma}{d\theta\, d\phi} = b\left|\frac{db}{d\theta}\right| = \frac{k^2}{8\mathscr{T}_\infty^2}\frac{\cos\theta/2}{\sin^3\theta/2}. \tag{7.28}$$

The cross-section per unit solid angle is

$$\frac{d^2\sigma}{d^2\Omega} = \frac{d^2\sigma}{\sin\theta\, d\theta\, d\phi} = \frac{k^2}{16\mathscr{T}_\infty^2}\frac{1}{\sin^4\theta/2} \tag{7.29}$$

and the differential cross-section with respect to θ is

$$\frac{d\sigma}{d\theta} = \int_0^{2\pi}\frac{d^2\sigma}{d\theta\, d\phi}\, d\phi = \frac{\pi k^2}{8\mathscr{T}_\infty^2}\frac{\sin\theta}{\sin^4\theta/2}. \tag{7.30}$$

[Cross-sections of the forms above constitute the *Rutherford scattering law*. The greater the kinetic energy and the angle of scatter the nearer the beam particles approach the centre of the nucleus. A uniformly charged sphere of finite radius would therefore give deviations from the scattering described above if the energy and/or the angle of scatter of the beam particles were great enough for them to penetrate the charge. In 1913 Rutherford carried out experiments to investigate this possibility, using alpha particles as projectiles. The range of energy and scattering angle over which the expression (7.29) for the cross-section was valid showed that the positive charge of an atom must be concentrated into a region very small compared with the atom itself. The concept of the *nuclear* atom originated in these classic experiments.]

7.4.3. See Section 7.3.1

Example 25

What corrections to the Rutherford scattering formulae are necessary to allow for the motion of the atomic nucleus? Calculate the difference in the differential cross-section per unit solid angle in the backward direction, as observed in the laboratory, between the cases of alpha particles scattered by a fixed and a free aluminium nucleus.

$$\left(\frac{m\,(\alpha\text{-particle})}{m\,(\text{aluminium nucleus})} = 0{\cdot}148.\right)$$

We shall first calculate the cross-sections in the centre-of-mass frame, which moves with constant velocity, v_c, in the beam direction. It is important to realize that the area $d^2\sigma$ that we consider here is exactly the *same* area as the $d^2\sigma$ of Example 24 and that the impact parameter b in the equation

$$d^2\sigma = b\, db\, d\phi$$

is also the same (i.e. it is the distance at which the projectile would pass the target if no forces acted between them). What *is* different here is the need to find the relationship between b and θ_c since we have to calculate $\mathrm{d}^2\sigma/(\mathrm{d}\theta_c\,\mathrm{d}\phi)$. ($\phi$ is the same in all frames moving parallel with the projectile.) We know, from equation (6.43), that the equation of motion of the projectile relative to the centre of mass, C, is

$$m\frac{\mathrm{d}^2\mathbf{r}}{\mathrm{d}t^2} = \frac{Zze^2}{4\pi\varepsilon_0}\left[\frac{M}{(M + m)r}\right]^2 \hat{\mathbf{r}},$$

where m, M are the masses of the projectile (the beam particle) and the target (the nucleus), and ze, Ze their charges. (Note that M is here the mass of the target particle alone, *not* the sum of both masses as in equation (6.43).) This is an 'inverse square' equation corresponding to a potential energy

$$\mathscr{V}(r) = \frac{k_c}{r},$$

where

$$k_c = \frac{Zze^2}{4\pi\varepsilon_0}\left(\frac{M}{M + m}\right)^2 = k\left(\frac{M}{M + m}\right)^2, \tag{7.31}$$

k being the constant relevant to the fixed nucleus case. The scattering angle θ_c is then given by

$$\cot\frac{\theta_c}{2} = \frac{2\mathscr{T}_{\infty c}b_c}{k_c}, \tag{7.32}$$

where $\mathscr{T}_{\infty c}$ is the kinetic energy at infinity, before the collision, in the centre-of-mass frame and b_c is the distance at which the projectile would pass the centre-of-mass. Now the centre-of-mass divides the line joining the projectile to the target in the ratio M/m. By considering the initial condition, in which the projectile is at infinity and moving with speed v_∞ in the laboratory frame, while the target is stationary, we can see that

$$b_c = \frac{M}{M + m}b \tag{7.33}$$

and that

$$v_{c\infty} = \frac{M}{M + m}v_\infty,$$

where $v_{c\infty}$ is the initial projectile speed in the centre-of-mass frame. Hence

$$\mathscr{T}_{c\infty} = \left(\frac{M}{M + m}\right)^2 \mathscr{T}_\infty. \tag{7.34}$$

From equations (7.31), (7.32), (7.33) and (7.34) we obtain the result:

$$b = \frac{k}{2\mathscr{T}_\infty}\frac{M+m}{M}\cot\frac{\theta_c}{2}.$$

Hence

$$\left(\frac{d^2\sigma}{d\theta\, d\phi}\right)_c = \left|b\frac{db}{d\theta_c}\right| = \frac{k^2}{8\mathscr{T}_\infty^2}\left(\frac{M+m}{M}\right)^2\frac{\cos\theta_c/2}{\sin^3\theta_c/2}, \tag{7.35}$$

$$\left(\frac{d^2\sigma}{d^2\Omega}\right)_c = \frac{k^2}{16\mathscr{T}_\infty^2}\left(\frac{M+m}{M}\right)^2\frac{1}{\sin^4\theta_c/2}. \tag{7.36}$$

Thus the laboratory cross-sections are given by

$$\left(\frac{d^2\sigma}{d\theta\, d\phi}\right)_l = \left|\frac{d\theta_c}{d\theta_l}\right|\left(\frac{d^2\sigma}{d\theta\, d\phi}\right)_c, \tag{7.37}$$

$$\left(\frac{d^2\sigma}{d^2\Omega}\right)_l = \left|\frac{\sin\theta_c}{\sin\theta_l}\frac{d\theta_c}{d\theta_l}\right|\left(\frac{d^2\sigma}{d^2\Omega}\right)_c, \tag{7.38}$$

where the right-hand sides of these relationships are given by equations (7.35), (7.36) and the form of equation (7.19) relevant to this problem:

$$\tan\theta_l = \frac{\sin\theta_c}{m/M + \cos\theta_c}. \tag{7.39}$$

The ratios of the laboratory differential cross-sections in the cases of free and fixed nucleus are therefore

$$S = \left.\frac{d^2\sigma}{d\theta\, d\phi}\right)_{l\,\text{free}}\Big/\left.\frac{d^2\sigma}{d\theta\, d\phi}\right)_{l\,\text{fixed}} = \left|\frac{d\theta_c}{d\theta_l}\right|\left(\frac{M+m}{M}\right)^2, \tag{7.40}$$

$$R = \left.\frac{d^2\sigma}{d^2\Omega}\right)_{l\,\text{free}}\Big/\left.\frac{d^2\sigma}{d^2\Omega}\right)_{l\,\text{fixed}} = \left|\frac{d(\cos\theta_c)}{d(\cos\theta_l)}\right|\left(\frac{M+m}{M}\right)^2. \tag{7.41}$$

If we use

$$y = \cos\theta_c, \qquad x = \cos\theta_l,$$

equation (7.39) becomes

$$\frac{\sqrt{1-x^2}}{x} = \frac{\sqrt{1-y^2}}{\beta + y},$$

where

$$\beta = \frac{m}{M}.$$

Hence

$$(\beta + y)^2(1 - x^2) = x^2(1 - y^2).$$

Differentiating this with respect to x,

$$2(\beta + y)(1 - x^2)\frac{dy}{dx} - 2x(\beta + y)^2 = 2x(1 - y^2) - 2x^2y\frac{dy}{dx}.$$

In the backwards direction

$$x = y = -1.$$

Hence

$$\left(\frac{dy}{dx}\right)_{\text{backwards}} = (1 - \beta)^2,$$

or

$$\left(\frac{d(\cos\theta_c)}{d(\cos\theta_l)}\right)_{\text{backwards}} = \left(\frac{M - m}{M}\right)^2. \tag{7.42}$$

Substituting from equation (7.42) into equation (7.41),

$$R_{\text{backwards}} = \left(\frac{M^2 - m^2}{M^2}\right)^2 = (1 - \beta^2)^2.$$

Thus, for the scattering of alpha particles by aluminium nuclei,

$$R_{\text{backwards}} = (1 - 0{\cdot}148^2)^2 = 0{\cdot}96.$$

7.5. Problems

7.1. The following diagrams represent stationary, solid and elastic objects that scatter a beam of small particles travelling from left to right. The problem is to identify the objects from the observed variation of scattering cross-section with direction.

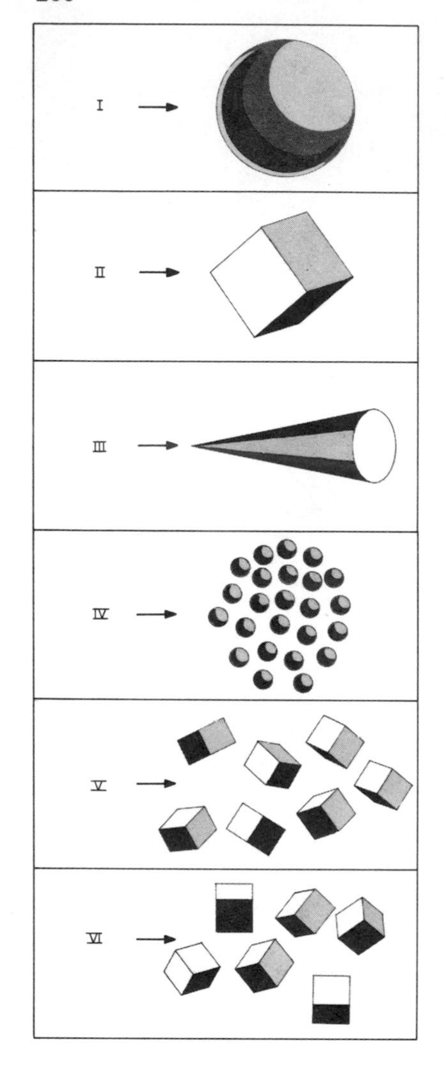

Sphere of radius a.

Cube of side a, having a major diagonal parallel with the beam.

Narrow cone of base radius a, pointing towards the beam.

Composite target, comprising N spheres of radius a.

Composite target, comprising N cubes of side a, each having a major diagonal parallel with the beam, but otherwise randomly oriented. (This is a model for a *polarized* target.)

Composite target, comprising N cubes of side a, completely randomly oriented. (This is a model for an *unpolarized* target.)

In cases IV, V and VI you may assume that N is a large number and that the composite targets are sufficiently thin for the individual cross-sections not to overlap and for secondary scattering to be negligible.

Some differential cross-sections (G1 to G6) and differential cross-sections per unit solid angle (GU1 to GU6) are given below, showing their variation with θ for some fixed value of ϕ. (θ is the scattering angle and ϕ is the azimuthal angle that defines the plane in which the scattering takes place.)

(a) Which of the distributions G1 to G6 would result from each of the targets I to VI?
(b) Which of the distributions GU1 to GU6 would arise from each of the targets I to VI?
(c) How would you expect the cross-sections in each case to vary, if at all, with ϕ?

(d) Which of the following total cross-sections would be appropriate to each of the targets: $\sigma_T = 2a^2$, $\sigma_T = a^2$, $\sigma_T = \pi N a^2$, $\sigma_T = \sqrt{2}Na^2$, $Na^2 < \sigma_T < \sqrt{2}Na^2$, $\sigma_T = \sqrt{3}a^2$, $\pi Na^2 < \sigma_T < 2\pi Na^2$, $\sigma_T = \pi a^2$, $\sigma_T = 3Na^2/2$, $Na^2 < \sigma_T < \sqrt{3}Na^2$, $\sigma_T = \sqrt{3}Na^2$?
(*Note*: some of the suggested results may correspond to more than one of the targets, and some may correspond to none of them.)

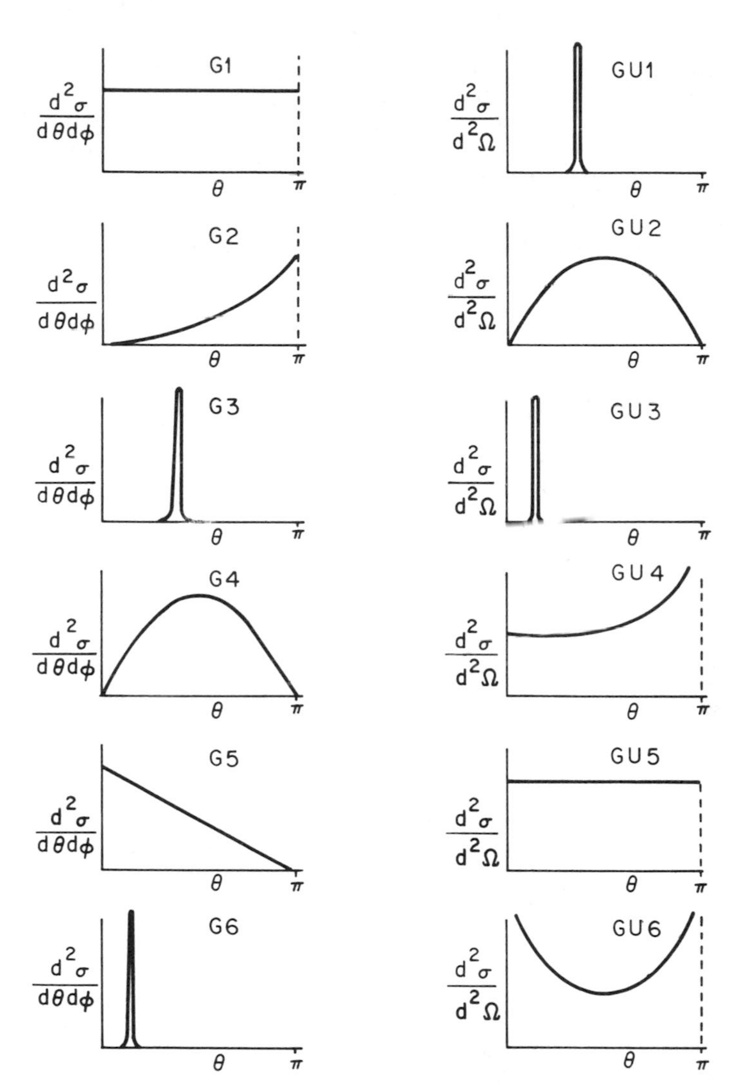

7.2. A beam of particles penetrates a composite target, comprising many scattering centres. Because of collisions with these centres, the flux of the *original* beam (i.e. of particles which have still not made a collision) at a distance x into the target will diminish according to some law $S(x)$. Show that if the density of scattering centres is n and the total scattering cross-section of each centre is σ_T, then

$$S(x) = S_0 \, e^{-x/\lambda},$$

where S_0 is the initial flux, and where

$$\lambda = \frac{1}{\sigma_T n}.$$

λ is therefore the *decay length* of the beam. Show that λ is also equal to

(a) the mean distance travelled by the beam particles before their first collision,

(b) the mean distance, or *mean free path*, travelled by any one beam particle between successive collisions, provided that, as a result of one collision, the cross-section for any subsequent collision is not changed.

7.3. Show that if a particle of mass m_1 is elastically scattered by an initially stationary particle of mass m_2, then the angle, ψ, between their final directions in the laboratory is given by

$$\cot \psi = \frac{m_1 - m_2}{m_1 + m_2} \tan \frac{\theta_c}{2},$$

where θ is the centre-of-mass scattering angle.

The two particles are known to differ in mass by 0·1 per cent, and ψ will therefore have values near $\pi/2$ unless $\tan \theta_c/2$ is large. If ε is the least detectable deviation from $\pi/2$, over what range of values of θ_c will the inequality of the masses be apparent? How many scatterings would need to be observed in order to demonstrate the inequality, given that $\varepsilon = 0{\cdot}01$ radian and that the centre-of-mass scattering is isotropic?

Chapter 8

Frames of Reference

The transforming of the laws of physics from one frame of reference to another is much more than an exercise in mathematical techniques. Very often physical laws are formulated with reference to a hypothetical stationary frame, whereas tests or applications have to be made in laboratories moving with the earth's surface, or in rockets moving with high velocity in outer space. It is essential, therefore, that in some fields of physics and engineering, experimentalists should be familiar with the problem of re-expressing these laws in terms of the observations that are actually made.

The status of physical laws is itself subject to examination from this point of view. A stationary observer will find that a particle sufficiently far away from any others not to be affected by them remains at rest or continues to move in a straight line with constant speed. To an observer spinning round with constant angular velocity on a turntable this same particle will appear to move in a rather complicated curved path. Insofar as each path can be precisely formulated and will accurately describe what is observed by each person, the 'straight line law' and the 'curved path law' are equally valid laws.

Nevertheless, the transformation of the 'straight line law' shows it to have the same simple form for a very large category of observers, whereas the 'curved path law' is not only complicated but its form is special to the particular observer. We have already pointed out that physical laws are always true (or at least tested) only over restricted ranges of the physical quantities concerned (such as mass or velocity), and that the more extensive these ranges, the more basic or fundamental we regard the law. In a similar manner a law is true only for certain categories of observer, and we judge its importance by the extent of these categories. This is why the 'straight line law' of Newton really qualifies as a law, while the 'curved path law' is a curiosity that might appear only as a problem in a book on mechanics.

8.1. Motion of a point particle

The Newtonian laws of mechanics, introduced in Chapter 1, must imply that there is at least one observer S for whom the laws are true. We shall now go carefully over ground already covered to some extent in Chapter 1 in order to show clearly the steps taken by S to describe the motion of a point particle and to establish the validity of any laws this may obey.

For this S needs a frame of reference which he regards as stationary—the S rest frame. It is convenient to use a right-handed orthogonal Cartesian frame defined by the unit vectors **i**, **j**, **k** (see Figure 8.1). The position, P, of the point particle is given by the vector $\overrightarrow{OP}$ or **x**, and S can determine this by constructing perpendiculars PL, PM, PN from P to the three axes. The measured lengths OL, OM, ON, which are the projections of OP on the three axes, are the coordinates x_1, x_2, x_3.

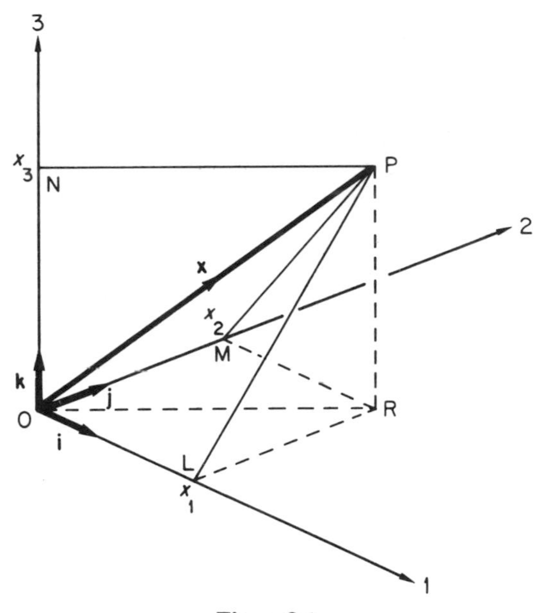

Figure 8.1

Since the vector $\overrightarrow{OL}$ has length OL and has the same direction as the *unit* vector **i**, we write

$$\overrightarrow{OL} = x_1\mathbf{i}.$$

Similarly,

$$\overrightarrow{OM} = x_2\mathbf{j} \quad \text{and} \quad \overrightarrow{ON} = x_3\mathbf{k}.$$

In Figure 8.1 MR is drawn parallel with OL and LR with OM. From the basic law of vector addition we can see that

$$\overrightarrow{OP} = \overrightarrow{OR} + \overrightarrow{ON} = \overrightarrow{OL} + \overrightarrow{OM} + \overrightarrow{ON},$$

or

$$\mathbf{x} = x_1\mathbf{i} + x_2\mathbf{j} + x_3\mathbf{k}.$$

We may note that the physical process of projecting **x** on the axis 1, which S uses in order to measure x_1, is equivalent mathematically to taking the scalar product,

$$\begin{aligned}\mathbf{i}\,.\,\mathbf{x} &= \mathbf{i}\,.\,(x_1\mathbf{i} + x_2\mathbf{j} + x_3\mathbf{k})\\ &= x_1\mathbf{i}\,.\,\mathbf{i} + x_2\mathbf{i}\,.\,\mathbf{j} + x_3\mathbf{i}\,.\,\mathbf{k}\\ &= x_1,\end{aligned}$$

since the vectors **i**, **j**, **k** are orthogonal to one another and each is of unit length. Treating the other two coordinates similarly we find that

$$x_2 = \mathbf{j}\,.\,\mathbf{x} \quad \text{and} \quad x_3 = \mathbf{k}\,.\,\mathbf{x}.$$

Thus

$$\mathbf{x} = (\mathbf{i}\,.\,\mathbf{x})\mathbf{i} + (\mathbf{j}\,.\,\mathbf{x})\mathbf{j} + (\mathbf{k}\,.\,\mathbf{x})\mathbf{k}. \tag{8.1}$$

Any measurements of the motion of P that S makes will be in terms of the three coordinates, which we can group together as the *column matrix* or *vector*

$$(x) \equiv \begin{pmatrix} x_1 \\ x_2 \\ x_3 \end{pmatrix}.$$

Alternatively, we can group them as the row matrix or vector, which is its transpose, denoted by

$$(\tilde{x}) \equiv (x_1 \quad x_2 \quad x_3).$$

(See Section 8.9.1, page 240.)

To emphasize that these three coordinates completely determine **x** for the observer S we shall write

$$\mathbf{x} \underset{S}{\leftrightarrow} (x) \equiv \begin{pmatrix} x_1 \\ x_2 \\ x_3 \end{pmatrix}.$$

By observing the time variations of these coordinates S will obtain what he regards as the velocity (v_s) and acceleration (a_s) of the particle:

$$(v_s) = \left(\frac{\mathrm{d}x}{\mathrm{d}t}\right) \equiv \begin{pmatrix} \mathrm{d}x_1/\mathrm{d}t \\ \mathrm{d}x_2/\mathrm{d}t \\ \mathrm{d}x_3/\mathrm{d}t \end{pmatrix}, \tag{8.2}$$

$$(a_s) = \left(\frac{\mathrm{d}^2x}{\mathrm{d}t^2}\right) \equiv \begin{pmatrix} \mathrm{d}^2x_1/\mathrm{d}t^2 \\ \mathrm{d}^2x_2/\mathrm{d}t^2 \\ \mathrm{d}^2x_3/\mathrm{d}t^2 \end{pmatrix}. \tag{8.3}$$

S can now use these components of velocity and acceleration, *and his rest frame*, to reconstruct in three dimensions the velocity and acceleration vectors **v** and **a**, which we accordingly write as

$$\mathbf{v} \underset{S}{\leftrightarrow} (v_s) = \left(\frac{\mathrm{d}x}{\mathrm{d}t}\right), \tag{8.4}$$

$$\mathbf{a} \underset{S}{\leftrightarrow} (a_s) = \left(\frac{\mathrm{d}^2x}{\mathrm{d}t^2}\right). \tag{8.5}$$

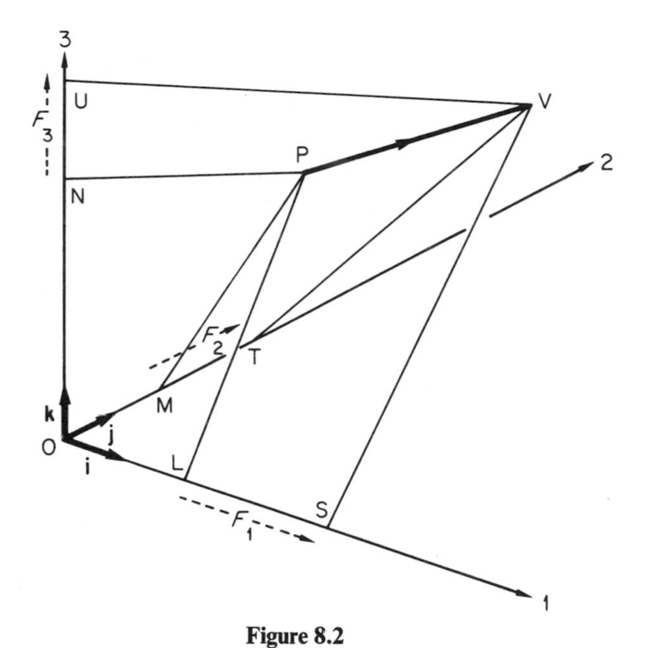

Figure 8.2

Measurement of any other vector by *S* will similarly consist of its projections on the axes of his rest frame. Thus if $\overrightarrow{\mathrm{PV}}$ represents a force **F**, and its projections LS, MT, NU on the three axes have lengths F_1, F_2, F_3 (see Figure 8.2), then

$$\overrightarrow{\mathrm{PV}} = \overrightarrow{\mathrm{LS}} + \overrightarrow{\mathrm{MT}} + \overrightarrow{\mathrm{NU}}$$

or

$$\mathbf{F} = F_1\mathbf{i} + F_2\mathbf{j} + F_3\mathbf{k},$$

where

$$F_1 = \mathbf{i}\,.\,\mathbf{F}, \qquad F_2 = \mathbf{j}\,.\,\mathbf{F}, \qquad F_3 = \mathbf{k}\,.\,\mathbf{F}.$$

We can therefore write

$$\mathbf{F} \underset{S}{\leftrightarrow} (F) \equiv \begin{pmatrix} F_1 \\ F_2 \\ F_3 \end{pmatrix}. \tag{8.6}$$

The law on particle motion on which all our discussion so far has been based is the group of three experimental observations by S:

$$F_1 = m\frac{d^2x_1}{dt^2}, \qquad F_2 = m\frac{d^2x_2}{dt^2}, \qquad F_3 = m\frac{d^2x_3}{dt^2},$$

which can be written together as the single matrix equation

$$\begin{pmatrix} F_1 \\ F_2 \\ F_3 \end{pmatrix} = m\begin{pmatrix} d^2x_1/dt^2 \\ d^2x_2/dt^2 \\ d^2x_3/dt^2 \end{pmatrix}, \tag{8.7}$$

or simply

$$(F) = m\left(\frac{d^2x}{dt^2}\right) = m(a_s). \tag{8.8}$$

(See Section 8.9.2, page 241.)

Using the relationships (8.5) and (8.6), the law expressed by (8.8) can be reformulated as the vector equation

$$\mathbf{F} = m\mathbf{a}. \tag{8.9}$$

Of the three equations (8.7), (8.8) and (8.9), the last has the simplest appearance. However, it must be emphasized that its meaning must be based upon the observations described by either of the first two and made in the S frame. The vector equation is derived from the matrix equations, not vice versa. That is why in this chapter, where we must be clear about which observer is making which measurements in which frame, we shall make considerable use of the matrix formulation.

We now embark upon the problem of finding how a law of mechanics, such as (8.8), appears to other observers. Here, too, we shall be repeating some of the work done earlier (for example in Chapter 5) in order to stress the precise correspondence between the physical observations and the mathematical descriptions.

8.2. Change of origin

Suppose S' is another observer with a frame of reference he regards as stationary. This we shall call the S' rest frame, or simply the S' frame. For it we shall also use an orthogonal Cartesian system, with origin at O′ and axes (defined by $\mathbf{i}'$, $\mathbf{j}'$, $\mathbf{k}'$) parallel with those of the S frame. If O′ is moving relative to O, S will describe this motion by measuring components of the vector $\overrightarrow{\mathrm{OO'}}$,

$$\overrightarrow{\mathrm{OO'}} \underset{S}{\leftrightarrow} \begin{pmatrix} X_1 \\ X_2 \\ X_3 \end{pmatrix} \equiv (X),$$

and observing how (X) varies with time (see Figure 8.3).

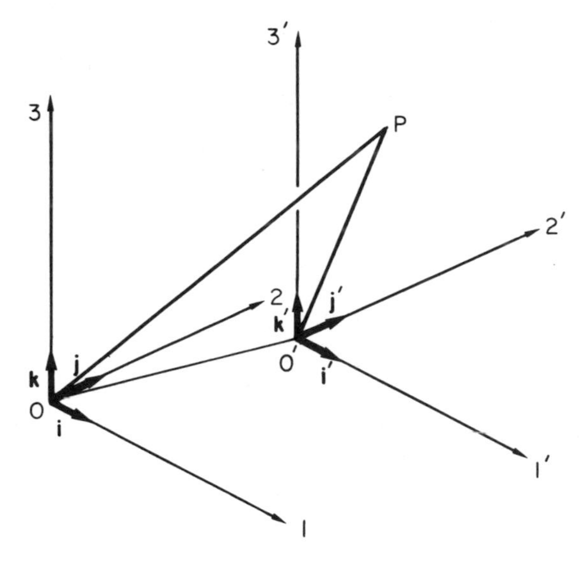

Figure 8.3

If P is the position of a moving point particle, S will measure the components of $\overrightarrow{\mathrm{OP}}$:

$$\overrightarrow{\mathrm{OP}} \underset{S}{\leftrightarrow} \begin{pmatrix} x_1 \\ x_2 \\ x_3 \end{pmatrix} \equiv (x).$$

S could also measure the components of $\overrightarrow{\mathrm{O'P}}$:

$$\overrightarrow{\mathrm{O'P}} \underset{S}{\leftrightarrow} \begin{pmatrix} y_1 \\ y_2 \\ y_3 \end{pmatrix} \equiv (y).$$

The fact that OO′, O′P, PO form the sides of a triangle can be expressed either as

$$\overrightarrow{OP} = \overrightarrow{OO'} + \overrightarrow{O'P}$$

or

$$(x) = (X) + (y). \tag{8.10}$$

This equation refers solely to measurements made by S (i.e. it concerns components or projections only in the S frame). If S' wishes to describe the motion of P he will do so in terms of the components of O′P in *his* frame:

$$\overrightarrow{O'P} \underset{S'}{\leftrightarrow} \begin{pmatrix} y_1' \\ y_2' \\ y_3' \end{pmatrix} \equiv (y').$$

Now since the axes of the S and S' frames are parallel, the corresponding components of O′P will be the same in each:

$$(y) = (y'). \tag{8.11}$$

Hence from equations (8.10) and (8.11),

$$(x) - (X) = (y'), \tag{8.12}$$

where the left-hand side describes measurements made by S and the right-hand side those by S'.

If S and S' now observe the time variation of these measurements, the relationship between them can be obtained by differentiating equation (8.12):

$$\left(\frac{dx}{dt}\right) - \left(\frac{dX}{dt}\right) = \left(\frac{dy'}{dt}\right). \tag{8.13}$$

Now (dx/dt) defines the velocity of P as measured by S, (v_s), and (dX/dt) defines the velocity of O′ as measured by S, which we shall denote by (U). (dy'/dt) defines the velocity of P as measured by S' in his frame. This we denote by $(v'_{s'})$. The subscript $_{s'}$ shows that S' has made the measurement of velocity relative to his rest frame, and the superscript $'$ shows that the components of this velocity are those in the S' frame. Then equation (8.13) may be written as

$$(v_s) - (U) = (v'_{s'}). \tag{8.14}$$

Differentiating equation (8.12) a second time:

$$\left(\frac{d^2x}{dt^2}\right) - \left(\frac{d^2X}{dt^2}\right) = \left(\frac{d^2y'}{dt^2}\right). \tag{8.15}$$

If (a_s), (A), $(a'_{s'})$ are used to denote the corresponding accelerations, equation (8.15) may be written as

$$(a_s) - (A) = (a'_{s'}). \tag{8.16}$$

We could equally well start with S', who would regard O as a moving point in his frame. Its motion would be measured by S' in terms of the components of O′O in the S' frame:

$$\overrightarrow{\mathrm{O'O}} \underset{S'}{\leftrightarrow} \begin{pmatrix} Y'_1 \\ Y'_2 \\ Y'_3 \end{pmatrix} \equiv (Y).$$

Arguments similar to these above enable us to relate positions of P in the two frames by

$$(y') - (Y') = (x), \tag{8.17}$$

the velocities by

$$\left(\frac{\mathrm{d}y'}{\mathrm{d}t}\right) - \left(\frac{\mathrm{d}Y'}{\mathrm{d}t}\right) = \left(\frac{\mathrm{d}x}{\mathrm{d}t}\right)$$

or

$$(v'_{s'}) - (V') = (v_s) \tag{8.18}$$

and the accelerations by

$$\left(\frac{\mathrm{d}^2y'}{\mathrm{d}t^2}\right) - \left(\frac{\mathrm{d}^2Y'}{\mathrm{d}t^2}\right) = \left(\frac{\mathrm{d}^2x}{\mathrm{d}t^2}\right)$$

or

$$(a'_{s'}) - (B') = (a_s). \tag{8.19}$$

In equations (8.18) and (8.19) (V') and (B') are the velocity and acceleration of O′ as measured by S' relative to his rest frame, expressed as components in that frame.

By comparing equation (8.18) with (8.14) and (8.19) with (8.16), we see that

$$(V') = -(U), \qquad (B') = -(A). \tag{8.20}$$

These results are the formal demonstration of the fact that S and S' agree about their relative motions—they are equal and opposite. Such a demonstration is hardly necessary in this simple case. It is obvious from inspection of Figure 8.3 that

$$(Y') = -(X),$$

from which equations (8.20) follow immediately. Moreover, the distinction between $(v'_{s'})$ and $(v_{s'})$ which we have made so far is not strictly necessary here. They both refer to the *same* velocity (i.e. the velocity of P as measured by S'). S' determines this vector, $\mathbf{v}_{s'}$, as a reconstruction in space of the three elements of $(\mathrm{d}y'/\mathrm{d}t)$, the reconstruction being described by the vector sum

$$\mathbf{v}_{s'} = \frac{\mathrm{d}y'_1}{\mathrm{d}t}\mathbf{i}' + \frac{\mathrm{d}y'_2}{\mathrm{d}t}\mathbf{j}' + \frac{\mathrm{d}y'_3}{\mathrm{d}t}\mathbf{k}'.$$

The superscript $'$ in $(v'_{s'})$ means that its elements are the projections of $\mathbf{v}_{s'}$ in the S' frame, which brings us back simply to $(\mathrm{d}y'_1/\mathrm{d}t, \mathrm{d}y'_2/\mathrm{d}t, \mathrm{d}y'_3/\mathrm{d}t)$. The unsuperscripted $(v_{s'})$ means that its elements are the projections of $\mathbf{v}_{s'}$ *on the axes of the S frame*. The first element, for example, is

$$\mathbf{v}_{s'} \,.\, \mathbf{i} = \frac{\mathrm{d}y'_1}{\mathrm{d}t}\mathbf{i}' \,.\, \mathbf{i} + \frac{\mathrm{d}y'_2}{\mathrm{d}t}\mathbf{j}' \,.\, \mathbf{i} + \frac{\mathrm{d}y'_3}{\mathrm{d}t}\mathbf{k}' \,.\, \mathbf{i},$$

but since $\mathbf{i}$ is parallel with $\mathbf{i}'$, and therefore perpendicular to $\mathbf{j}'$ and to $\mathbf{k}'$, this element reduces to

$$\mathbf{v}_{s'} \,.\, \mathbf{i} = \frac{\mathrm{d}y'_1}{\mathrm{d}t}.$$

Similar results are true for the other two elements. Hence

$$(v_{s'}) = (v'_{s'}),$$

although the two methods of expressing the same velocity $\mathbf{v}_{s'}$ are different in principle.

We shall see later that this difference in principle is a real difference in measurement when the two frames of reference no longer have parallel axes.

8.2.1. Uniform motion of origin

In the previous section (X) or (Y') might have any variation with time, which would allow S and S' to have any form of relative motion. If we now restrict this to uniform motion we may write

$$(X) = -(Y') = (b) + (U)t, \qquad (8.21)$$

where

$$(b) \equiv \begin{pmatrix} b_1 \\ b_2 \\ b_3 \end{pmatrix}$$

represents the displacement of O′ relative to O at time zero, and

$$(U) \equiv \begin{pmatrix} U_1 \\ U_2 \\ U_3 \end{pmatrix},$$

where U_1, U_2, U_3 are all constants, is the velocity of O′ relative to O. Then

$$(V') = -(U) = (\text{constant}), \qquad (B') = -(A) = 0,$$

and, in consequence,

$$(v_s) - (U) = (v'_{s'}), \qquad (a_s) = (a'_{s'}). \tag{8.22}$$

8.3. Equation of motion under uniform translation

Let us now return to the law of motion (8.8). (F) is the column matrix of projections of the vector F on the axes of the S rest frame. The projections of the same vector on the S' rest frame we denote by (F'), but since the corresponding axes are parallel, these will be the same:

$$\begin{pmatrix} F_1 \\ F_2 \\ F_3 \end{pmatrix} = \begin{pmatrix} F'_1 \\ F'_2 \\ F'_3 \end{pmatrix} \quad \text{or} \quad (F) = (F'). \tag{8.23}$$

Hence, from equations (8.8), (8.22) and (8.23),

$$(F') = m\frac{\mathrm{d}^2 y'}{\mathrm{d}t^2} = m(a'_{s'}). \tag{8.24}$$

Comparing (8.8) and (8.24) we see that the Newtonian law of motion is the same for S and S'.

S and S' therefore have a precisely similar view of the behaviour of the universe, or at least that part of it expressed in the law (8.8) above. Now S' is *any* observer moving with a uniform velocity relative to S, so that the law, instead of being reserved for a particular observer, is true for a wide category of observers, and its status is thereby raised considerably. Moreover, as far as observations based on this law are concerned, there is no way in which experimentally we can assert that S is stationary and S' moving with velocity U rather than that S' is stationary and S moving with velocity $-U$. The necessity for determining absolute rest disappears as far as the laws of mechanics are concerned; to all observers moving uniformly with respect to each other these laws have the same form.

We often gain a vivid impression of this fact when trains are moving slowly and steadily beside each other at night. It is quite easy to think that one's own train is stationary or even moving backwards, when it is the other train which is travelling forward more rapidly than one's own. It is then that we realize that we judge our own steady motion only relative to another's, and not by any observation of local phenomena (in this case what is happening in one's own carriage). The type of motion that we *can* determine from local observation is *acceleration* and *deceleration*—the pressure of the seat on our backs when a car or aircraft accelerates rapidly; being thrust forward against the seat belts when it brakes sharply.

Observations made by the class of observers who are in uniform motion relative to each other are linked by transformations such as (8.21), which are called *Galilean transformations.* A result, such as equation (8.8), which has the same form for all these observers, is said to be *invariant with respect to all Galilean transformations.*

8.4. Change of orientation

For this we shall suppose that S and S' have frames with a common origin $O \equiv O'$, but that the axes of S' defined by the orthogonal unit vectors $\mathbf{i}', \mathbf{j}', \mathbf{k}'$ are tilted with respect to $\mathbf{i}, \mathbf{j}, \mathbf{k}$, as shown in Figure 8.4.

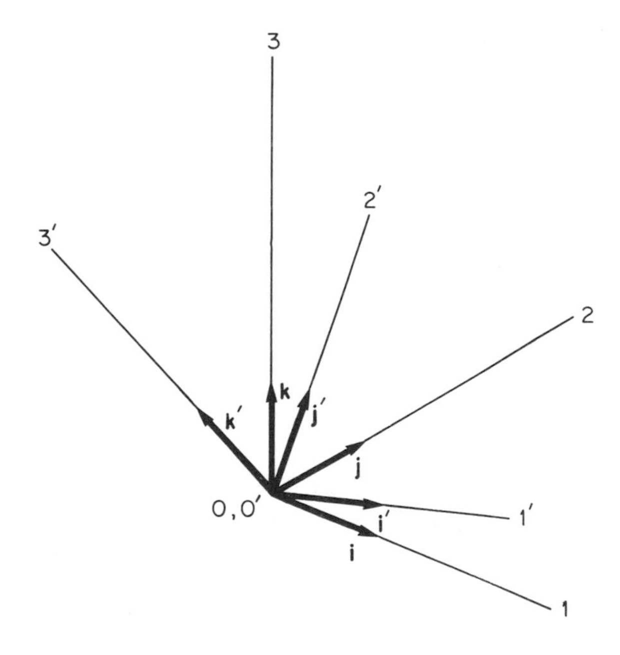

Figure 8.4

Using equation (8.1) we can describe any of these unit vectors in terms of the others; for example,

$$\mathbf{i} = (\mathbf{i}' \cdot \mathbf{i})\mathbf{i}' + (\mathbf{j}' \cdot \mathbf{i})\mathbf{j}' + (\mathbf{k}' \cdot \mathbf{i})\mathbf{k}'.$$

Thus if $\mathbf{x}$ is the position vector $\overrightarrow{\mathrm{OP}}$ which defines the position of a particle P relative to O (or O′), so that

$$\mathbf{x} \underset{S}{\leftrightarrow} (x) \equiv \begin{pmatrix} x_1 \\ x_2 \\ x_3 \end{pmatrix},$$

then

$$\begin{aligned} \mathbf{x} &= x_1\mathbf{i} + x_2\mathbf{j} + x_3\mathbf{k} \\ &= x_1[(\mathbf{i}' \cdot \mathbf{i})\mathbf{i}' + (\mathbf{j}' \cdot \mathbf{i})\mathbf{j}' + (\mathbf{k}' \cdot \mathbf{i})\mathbf{k}'] + \text{etc.} \\ &= [(\mathbf{i}' \cdot \mathbf{i})x_1 + (\mathbf{i}' \cdot \mathbf{j})x_2 + (\mathbf{i}' \cdot \mathbf{k})x_3]\mathbf{i}' + \text{etc.} \end{aligned} \tag{8.25}$$

But in the S' frame $\mathbf{x}$ will have components (x');

$$\mathbf{x} \underset{S'}{\leftrightarrow} (x') \equiv \begin{pmatrix} x_1' \\ x_2' \\ x_3' \end{pmatrix},$$

which implies

$$\mathbf{x} = x_1'\mathbf{i}' + x_2'\mathbf{j}' + x_3'\mathbf{k}'. \tag{8.26}$$

Hence, from equations (8.25) and (8.26),

$$\begin{aligned} x_1' &= (\mathbf{i}' \cdot \mathbf{i})x_1 + (\mathbf{i}' \cdot \mathbf{j})x_2 + (\mathbf{i}' \cdot \mathbf{k})x_3, \\ x_2' &= \text{etc.}, \\ x_3' &= \text{etc.} \end{aligned}$$

The scalar products on the right-hand side of these equations are cosines of the angles between pairs of axes. Thus $\mathbf{i}' \cdot \mathbf{j}$ is the cosine of the angle between the S' axis $1'$ and the S axis 2; we shall denote this by a_{12}. Hence

$$\begin{aligned} x_1' &= a_{11}x_1 + a_{12}x_2 + a_{13}x_3, \\ x_2' &= a_{21}x_1 + a_{22}x_2 + a_{23}x_3, \\ x_3' &= a_{31}x_1 + a_{32}x_2 + a_{33}x_3, \end{aligned} \tag{8.27}$$

or, in matrix form,

$$\begin{pmatrix} x_1' \\ x_2' \\ x_3' \end{pmatrix} = \begin{pmatrix} a_{11} & a_{12} & a_{13} \\ a_{21} & a_{22} & a_{23} \\ a_{31} & a_{32} & a_{33} \end{pmatrix} \begin{pmatrix} x_1 \\ x_2 \\ x_3 \end{pmatrix}, \tag{8.28}$$

or, more concisely,

$$(x') = (A)(x). \tag{8.29}$$

(See Section 8.9.3, page 241.)

In a similar way we can show that

$$\begin{pmatrix} x_1 \\ x_2 \\ x_3 \end{pmatrix} = \begin{pmatrix} a_{11} & a_{21} & a_{31} \\ a_{12} & a_{22} & a_{32} \\ a_{13} & a_{23} & a_{33} \end{pmatrix} \begin{pmatrix} x_1' \\ x_2' \\ x_3' \end{pmatrix}. \tag{8.30}$$

Thus

$$(x) = (\tilde{A})(x'), \tag{8.31}$$

where $(\tilde{A})$ is the transpose of (A).

From equations (8.29) and (8.31),

$$(x') = (A)(\tilde{A})(x') \quad \text{and} \quad (x) = (\tilde{A})(A)(x)$$

so that the products $(A)(\tilde{A})$ and $(\tilde{A})(A)$ leave unchanged any matrix into which they are multiplied. We may therefore write them as the unit matrix:

$$(A)(\tilde{A}) = (\tilde{A})(A) = (1) \equiv \begin{pmatrix} 1 & 0 & 0 \\ 0 & 1 & 0 \\ 0 & 0 & 1 \end{pmatrix}, \tag{8.32}$$

or, in terms of inverse matrices,

$$(\tilde{A}) = (A)^{-1}, \qquad (A) = (\tilde{A})^{-1}. \tag{8.33}$$

(See Section 8.9.4, page 242.)

The transformation relating measurements by S and S' is completely described by the 3×3 matrix (A). Such a matrix is called *orthogonal*, which we can take as defined either by the fact that it transforms between two sets of orthogonal axes or by the equivalent relationships (8.32) or (8.33).

It is important to realize that either of the equations (8.29) and (8.31) implies that S and S' *agree* that they are measuring the *same vector* $\mathbf{x}$. They will, of course, make different coordinate measurements, but provided these are related by (A) or $(\tilde{A})$ they describe a vector of the same length and direction.

If the two frames are fixed relative to each other, so that a_{11}, a_{12}, etc., are all constants, then by differentiating in detail equations (8.29) and (8.31) with respect to time, it is easy to show that

$$\left(\frac{\mathrm{d}x'}{\mathrm{d}t}\right) = (A)\left(\frac{\mathrm{d}x}{\mathrm{d}t}\right), \qquad \left(\frac{\mathrm{d}x}{\mathrm{d}t}\right) = (\tilde{A})\left(\frac{\mathrm{d}x'}{\mathrm{d}t}\right). \tag{8.34}$$

According to our previous definitions of velocity

$$\left(\frac{\mathrm{d}x'}{\mathrm{d}t}\right) = (v'_{s'}), \qquad \left(\frac{\mathrm{d}x}{\mathrm{d}t}\right) = (v_s), \tag{8.35}$$

and we may therefore write

$$(v'_{s'}) = (A)(v_s), \qquad (v_s) = (\tilde{A})(v'_{s'}). \tag{8.36}$$

Now (v_s) is the velocity of P as measured by S, expressed as components in his frame. When multiplied by (A) we obtain the components of that same velocity, but now in the S' frame. Thus, in accord with the notation of Section 8.2,

$$(A)(v_s) = (v'_s).$$

Similarly, $(v'_{s'})$ is the velocity of P as measured by S', expressed as components in his frame. Multiplication by $(\tilde{A})$ gives the components of that velocity in the S frame and is expressed by

$$(\tilde{A})(v'_{s'}) = (v_{s'}).$$

Equations (8.36) may then be written:

$$(v'_{s'}) = (v'_s), \qquad (v_s) = (v_{s'}). \tag{8.37}$$

These results express, in the S' and S frames, the fact that S and S' agree over the velocity vector of the particle:

$$\mathbf{v}_s = \mathbf{v}_{s'}.$$

Differentiating equations (8.34) a second time,

$$\left(\frac{\mathrm{d}^2x'}{\mathrm{d}t^2}\right) = (A)\left(\frac{\mathrm{d}^2x}{\mathrm{d}t^2}\right), \qquad \left(\frac{\mathrm{d}^2x}{\mathrm{d}t^2}\right) = (\tilde{A})\left(\frac{\mathrm{d}^2x'}{\mathrm{d}t^2}\right), \tag{8.38}$$

or

$$(a'_{s'}) = (A)(a_s), \qquad (a_s) = (\tilde{A})(a'_{s'}). \tag{8.39}$$

A similar argument to that above shows that

$$(a'_{s'}) = (a'_s), \qquad (a_s) = (a_{s'}), \tag{8.40}$$

which express the agreement of S and S' over the acceleration vector of the particle:

$$\mathbf{a}_s = \mathbf{a}_{s'}.$$

The components of a force $\mathbf{F}$ in the two frames are also related by the transformation matrix (A):

$$(F') = (A)(F), \qquad (F) = (\tilde{A})(F'). \tag{8.41}$$

Hence if the law

$$(F) = m\left(\frac{\mathrm{d}^2x}{\mathrm{d}t^2}\right) \tag{8.42}$$

is true for the observer S, multiplying this on the left by (A) and using equations (8.38) and (8.41) shows that

$$(F') = m\left(\frac{\mathrm{d}^2x'}{\mathrm{d}t^2}\right). \tag{8.43}$$

This result extends the Galilean transformations of Section 8.3 under which the Newtonian law remains invariant. All that is necessary is that the observers should move with constant relative velocity and that the angles between the axes of their rest frames should be constant. All inertial frames of reference are linked in this way, and in all of them the basic law of particle motion (8.8) is the same.

The simple results obtained so far in this chapter hardly warrant the rather elaborately detailed arguments that have been used. However, we shall shortly see how important the precision of these arguments is when they are no longer restricted to Galilean transformations.

8.4.1. Rotation: vector description

Suppose the S' frame is rotating relative to the S frame with angular speed w about an axis OW. We define the vector $\mathbf{w}$ as having magnitude w and the direction OW, as shown in Figure 8.5, considering this positive if the rotation is in the same sense as a right-handed screw motion. (See Section 8.9.5, page 243.) If P is a point fixed in the S' frame it will, because of the rotation of that frame, move to a neighbouring position Q relative to the S frame in a small interval of time δt. However, if, in addition, it is moving even in the S' frame, it will reach some other neighbouring point, R, say. Thus S observes the change in position as $\overrightarrow{\mathrm{PR}}$ and would say that the velocity of P is

$$\mathbf{v}_s = \lim_{\delta t \to 0} \frac{\overrightarrow{\mathrm{PR}}}{\delta t},$$

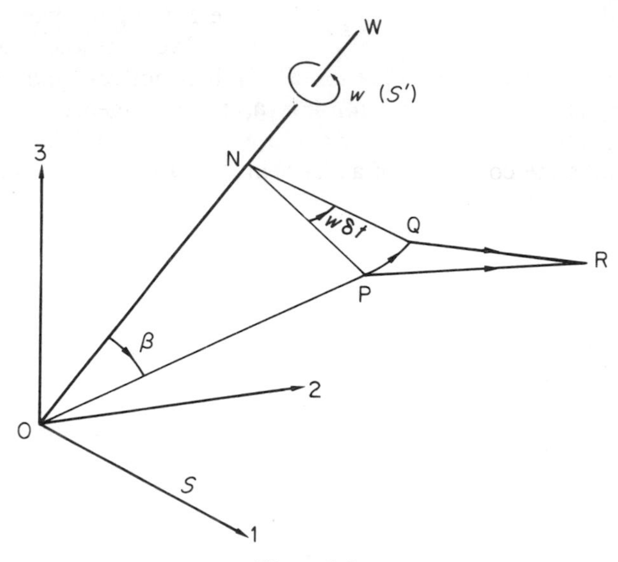

Figure 8.5

while S' sees the change only as $\overrightarrow{\mathrm{QR}}$ and would measure the velocity as

$$\mathbf{v}_{s'} = \lim_{\delta t \to 0} \frac{\overrightarrow{\mathrm{QR}}}{\delta t}.$$

If PN is the perpendicular from P to the rotation axis, PNQ is the angle turned through in a time δt (i.e. $w\,\delta t$). Hence

$$\mathrm{PQ} \simeq w\,\delta t\,\mathrm{PN} = w\,\delta t\,\mathrm{OP}\sin\beta,$$

where β is the angle between OP and the rotation axis OW. Now $w\,\mathrm{OP}\sin\beta$ is the magnitude of the vector product $\mathbf{w} \wedge \overrightarrow{\mathrm{OP}}$, and $\overrightarrow{\mathrm{PQ}}$ is in the direction of this product. Hence

$$\overrightarrow{\mathrm{PQ}} \simeq (\mathbf{w} \wedge \overrightarrow{\mathrm{OP}})\delta t.$$

Remembering that in the limit $\delta t \to 0$, the arc PQ and the straight line PQ become indistinguishable,

$$\overrightarrow{\mathrm{PR}} = \overrightarrow{\mathrm{QR}} + \overrightarrow{\mathrm{PQ}} \simeq \overrightarrow{QR} + (\mathbf{w} \wedge \overrightarrow{\mathrm{OP}})\delta t,$$

$$\frac{\overrightarrow{\mathrm{PR}}}{\delta t} \simeq \frac{\overrightarrow{\mathrm{QR}}}{\delta t} + \mathbf{w} \wedge \overrightarrow{\mathrm{OP}}. \tag{8.44}$$

$\overrightarrow{\mathrm{OP}}$ is the position vector $\mathbf{x}$, and the result (8.44) can therefore be written, in the limit $\delta t \to 0$,

$$\mathbf{v}_s = \mathbf{v}_{s'} + \mathbf{w} \wedge \mathbf{x}. \tag{8.45}$$

In this discussion we have thought of the S frame as fixed and the S' frame as rotating, and therefore of $\mathbf{v}_s$ as the 'true' velocity and $\mathbf{v}_{s'}$ as 'relative'. However, the derivation of equation (8.45) does not rest upon any special status being given to S, and we can interpret it more generally by calling $\mathbf{v}_s$ the 'S apparent velocity' and $\mathbf{v}_{s'}$ the 'S' apparent velocity'; $\mathbf{w}$ is the angular velocity of S' relative to S.

8.4.2. Rotation: matrix description

When the S' frame is twisting about O relative to S, in some general way, the elements a_{11}, a_{12}, of the matrix (A) will vary with time. By differentiating the detailed equations (8.29) and (8.31) it is easy to show that in this case

$$\left(\frac{\mathrm{d}x'}{\mathrm{d}t}\right) = (A)\left(\frac{\mathrm{d}x}{\mathrm{d}t}\right) + \left(\frac{\mathrm{d}A}{\mathrm{d}t}\right)(x), \tag{8.46}$$

$$\left(\frac{\mathrm{d}x}{\mathrm{d}t}\right) = (\tilde{A})\left(\frac{\mathrm{d}x'}{\mathrm{d}t}\right) + \left(\frac{\mathrm{d}\tilde{A}}{\mathrm{d}t}\right)(x'), \tag{8.47}$$

where $(\mathrm{d}A/\mathrm{d}t)$ is the 3×3 matrix, each of whose elements is the time derivative of the corresponding element of (A). Note that equations (8.46) and (8.47) show that we can differentiate matrix products according to the rules for ordinary algebraic products. Then

$$\begin{aligned}(A)\left(\frac{\mathrm{d}x}{\mathrm{d}t}\right) &= (A)(\tilde{A})\left(\frac{\mathrm{d}x'}{\mathrm{d}t}\right) + (A)\left(\frac{\mathrm{d}\tilde{A}}{\mathrm{d}t}\right)(x') \\ &= \left(\frac{\mathrm{d}x'}{\mathrm{d}t}\right) + (\Omega')(x'),\end{aligned} \tag{8.48}$$

where

$$(\Omega') = (A)\left(\frac{\mathrm{d}\tilde{A}}{\mathrm{d}t}\right). \tag{8.49}$$

Now

$$\left(\frac{\mathrm{d}x}{\mathrm{d}t}\right) = (v_s), \qquad \left(\frac{\mathrm{d}x'}{\mathrm{d}t}\right) = (v'_{s'}).$$

Thus equation (8.48) may be written

$$(A)(v_s) = (v'_{s'}) + (\Omega')(x'). \tag{8.50}$$

Since (v_s) describes the vector $\mathbf{v}_s$ by its components in the S frame, multiplying on the left by (A) will give the components of the *same* velocity $\mathbf{v}_s$ in the S' frame. Thus we can finally write equation (8.48) entirely in terms of quantities referred to the S' frame (denoted by a superscript prime),

$$(v'_s) = (v'_{s'}) + (\Omega')(x'). \tag{8.51}$$

We see from this that the apparent velocities of P, according to S and S', do *not* agree. To obtain agreement we must add the vector $(\Omega')(x')$ to the S' apparent velocity.

8.4.3. The angular velocity matrix

The similarity between the vector equation (8.45) and the matrix equation (8.51) must indicate some connexion between the angular velocity vector, $\mathbf{w}$, and the matrix (Ω'). In this section we shall see more precisely what this connexion is.

If we differentiate the product

$$(A)(\tilde{A}) = (1),$$

we obtain

$$(A)\left(\frac{\mathrm{d}\tilde{A}}{\mathrm{d}t}\right) + \left(\frac{\mathrm{d}A}{\mathrm{d}t}\right)(\tilde{A}) = (0). \tag{8.52}$$

The first term of this is (Ω'), by the definition (8.49), and the second is $(\tilde{\Omega}')$, since, by the rule for transposing a product,

$$(\tilde{\Omega}') = \widetilde{\left((A)\left(\frac{\mathrm{d}\tilde{A}}{\mathrm{d}t}\right)\right)} = \left(\frac{\mathrm{d}\tilde{\tilde{A}}}{\mathrm{d}t}\right)(\tilde{A}) = \left(\frac{\mathrm{d}A}{\mathrm{d}t}\right)(\tilde{A}).$$

(See Section 8.9.6, page 243.) Hence equation (8.52) is equivalent to

$$(\Omega') + (\tilde{\Omega}') = (0). \tag{8.53}$$

A matrix with this property is said to be *antisymmetric*. In terms of the individual elements the property (8.53) may be written as

$$\omega'_{ij} = -\omega'_{ji}. \tag{8.54}$$

Thus elements reflected across the leading diagonal of the matrix are the negative of each other. This implies, in particular (when $i = j$), that the elements of the diagonal itself are zero. Thus only three out of the nine elements of (Ω') are non-zero and independent, and it therefore has the form

$$(\Omega') = \begin{pmatrix} 0 & \omega'_{12} & \omega'_{13} \\ -\omega'_{12} & 0 & \omega'_{23} \\ -\omega'_{13} & -\omega'_{23} & 0 \end{pmatrix}.$$

The double suffix notation is cumbersome for so few elements, and the link with $\mathbf{w}$ is more easily established by writing instead

$$(\Omega') = \begin{pmatrix} 0 & -\omega'_3 & \omega'_2 \\ \omega'_3 & 0 & -\omega'_1 \\ -\omega'_2 & \omega'_1 & 0 \end{pmatrix}. \tag{8.55}$$

Then

$$(\Omega')(x') = \begin{pmatrix} 0 & -\omega_3' & \omega_2' \\ \omega_3' & 0 & -\omega_1' \\ -\omega_2' & \omega_1' & 0 \end{pmatrix} \begin{pmatrix} x_1' \\ x_2' \\ x_3' \end{pmatrix} = \begin{pmatrix} \omega_2' x_3' & -\omega_3' x_2' \\ \omega_3' x_1' & -\omega_1' x_3' \\ \omega_1' x_2' & -\omega_2' x_1' \end{pmatrix}.$$

From this we can see that if we define a vector $\boldsymbol{\omega}$ by the components ω_1', ω_2', ω_3' in the S' frame, then the components of the matrix product $(\Omega')(x')$ are the same as those of the vector product $\boldsymbol{\omega} \wedge \mathbf{x}$. Thus the vector equivalent of equation (8.51) is

$$\mathbf{v}_s = \mathbf{v}_{s'} + \boldsymbol{\omega} \wedge \mathbf{x}, \tag{8.56}$$

and we can identify $\boldsymbol{\omega}$ with the $\mathbf{w}$ of Section 8.4.1. To show that the matrix description of angular velocity is a consistent one we note first that, from equation (8.51),

$$(v_{s'}') = (v_s') - (\Omega')(x'). \tag{8.57}$$

Comparing this with equation (8.51) shows that we can equally well consider that S has an angular velocity relative to S' described in the S' frame by the matrix $-(\Omega')$. Both equations (8.51) and (8.57) use S' components. To transform to the S frame we multiply equation (8.51) by $(\tilde{A})$ and (A) as follows:

$$(\tilde{A})(v_s') = (A)(v_{s'}') + (\tilde{A})(\Omega')(A)(\tilde{A})(x').$$

(Since $(A)(\tilde{A})$ is a unit matrix, its insertion leaves the third term unchanged.) Thus

$$(v_s) = (v_{s'}) + (\Omega)(x) \tag{8.58}$$

where

$$(\Omega) = (\tilde{A})(\Omega')(A) = (\tilde{A})(A)\left(\frac{\mathrm{d}\tilde{A}}{\mathrm{d}t}\right)(A) = \left(\frac{\mathrm{d}\tilde{A}}{\mathrm{d}t}\right)(A) \tag{8.59}$$

represents the angular velocity of S' relative to S in terms of S components. (See Section 8.9.7, page 244.) Rewriting equation (8.58) as

$$(v_{s'}) = (v_s) - (\Omega)(x), \tag{8.60}$$

and comparing it with equation (8.58) shows that the angular velocity of S relative to S', when referred in the S frame, is $-(\Omega)$. Provided, then, that we transform the *angular velocity matrix* (Ω) between the frames according to the laws

$$(\Omega') = (A)(\Omega)(\tilde{A}), \qquad (\Omega) = (\tilde{A})(\Omega')(A) \tag{8.61}$$

this description gives the following self-consistent table:

Angular velocity of	S' relative to S	S relative to S'
as described in the S' frame	$(\Omega') = (A)\left(\frac{d\tilde{A}}{dt}\right)$	$-(\Omega')$
as described in the S frame	$(\Omega) = \left(\frac{d\tilde{A}}{dt}\right)(A)$	$-(\Omega)$

(See Section 8.9.8, page 245.)

Although the preceding analysis is necessary in order to show the meaning and consistency of the matrix description, it is usually unnecessary in practice to use the definitions (8.49) or (8.59) to calculate the angular velocity matrix. When an axis can be identified as the one about which the rotation is taking place, it is much simpler to describe the angular velocity using the angles $\varepsilon_1, \varepsilon_2, \varepsilon_3$, or $\varepsilon_1', \varepsilon_2', \varepsilon_3'$ which that axis makes with the axes of the two frames. If the angular speed of S' relative to S is ω, then the vector components in the two frames will be $\omega \cos \varepsilon_1$, $\omega \cos \varepsilon_2$, $\omega \cos \varepsilon_3$ and $\omega \cos \varepsilon_1'$, $\omega \cos \varepsilon_2'$, $\omega \cos \varepsilon_3'$. The corresponding matrices can then be written immediately as

$$(\Omega') = \omega \begin{pmatrix} 0 & -\cos \varepsilon_3' & \cos \varepsilon_2' \\ \cos \varepsilon_3' & 0 & -\cos \varepsilon_1' \\ -\cos \varepsilon_2' & \cos \varepsilon_1' & 0 \end{pmatrix},$$

$$(\Omega) = \omega \begin{pmatrix} 0 & -\cos \varepsilon_3 & \cos \varepsilon_2 \\ \cos \varepsilon_3 & 0 & -\cos \varepsilon_1 \\ -\cos \varepsilon_2 & \cos \varepsilon_1 & 0 \end{pmatrix}.$$

8.4.4. Acceleration under rotation

Writing the velocity, from equation (8.48), in the form

$$\left(\frac{dx}{dt}\right) = (\tilde{A})\left[\left(\frac{dx'}{dt}\right) + (\Omega')(x')\right],$$

and differentiating again, we obtain the acceleration,

$$\left(\frac{d^2x}{dt^2}\right) = (\tilde{A})\left[\left(\frac{d^2x'}{dt^2}\right) + (\Omega')\left(\frac{dx'}{dt}\right) + \left(\frac{d\Omega'}{dt}\right)(x')\right]$$

$$+ \left(\frac{d\tilde{A}}{dt}\right)\left[\left(\frac{dx'}{dt}\right) + (\Omega')(x')\right] \tag{8.62}$$

Hence

$$(A)\left(\frac{d^2x}{dt^2}\right) = \left(\frac{d^2x'}{dt^2}\right) + 2(\Omega')\left(\frac{dx'}{dt}\right) + (\Omega')^2(x') + \left(\frac{d\Omega'}{dt}\right)(x'). \qquad (8.63)$$

since

$$(\Omega') = (A)\left(\frac{d\tilde{A}}{dt}\right).$$

Following the previous notation,

$$\left(\frac{d^2x'}{dt^2}\right) = (a'_{s'}),$$

the S' apparent acceleration, referred to the S' frame, and

$$\left(\frac{d^2x}{dt^2}\right) = (a_s),$$

the S apparent acceleration, referred to the S frame. Multiplying on the left by (A) transforms the latter to the S' frame:

$$(A)\left(\frac{d^2x}{dt^2}\right) = (a'_s).$$

Thus equation (8.63) may be written as

$$(a'_s) = (a'_{s'}) + 2(\Omega')(v'_{s'}) + (\Omega')^2(x') + \left(\frac{d\Omega'}{dt}\right)(x'), \qquad (8.64)$$

where the whole equation now refers to the S′ frame. Multiplying by $(\tilde{A})$ and (A) as follows, we can refer it to the S frame:

$$(\tilde{A})(a'_s) = (\tilde{A})(a'_{s'}) + 2(\tilde{A})(\Omega')(A)(\tilde{A})(v'_{s'}) + (\tilde{A})(\Omega')(A)(\tilde{A})(\Omega')(A)(\tilde{A})(x') + (\tilde{A})\left(\frac{d\Omega'}{dt}\right)(A)(\tilde{A})(x')$$

or

$$(a_s) = (a_{s'}) + 2(\Omega)(v_{s'}) + (\Omega)^2(x) + \left(\frac{d\Omega}{dt}\right)(x). \qquad (8.65)$$

Either of equations (8.64) and (8.65) has the vector equivalent

$$\mathbf{a}_s = \mathbf{a}_{s'} + 2\boldsymbol{\omega} \wedge \mathbf{v}_{s'} + \boldsymbol{\omega} \wedge (\boldsymbol{\omega} \wedge \mathbf{x}) + \frac{d\boldsymbol{\omega}}{dt} \wedge \mathbf{x}. \qquad (8.66)$$

8.5. Equation of motion under rotation

We can now transform the simple equation of motion

$$(F) = m\left(\frac{d^2x}{dt^2}\right), \tag{8.67}$$

which is true for S, into measurements which S' can make. The force $\mathbf{F}$ will have different components or projections in the two rest frames, but since no time differentiation is involved we can repeat the arguments used in Section 8.4 regarding the vector $\mathbf{x}$ to show that they are related simply by the transformation matrix, (A):

$$(F') = (A)(F), \qquad (F) = (\tilde{A})(F'). \tag{8.68}$$

The acceleration transforms in the much more complicated manner of equation (8.63). Applying this and equation (8.68) to the law (8.67),

$$(\tilde{A})(F') = m(\tilde{A})\left[\left(\frac{d^2x'}{dt^2}\right) + 2(\Omega')\frac{dx'}{dt} + (\Omega')^2(x') + \left(\frac{d\Omega'}{dt}\right)(x')\right]. \tag{8.69}$$

Multiplying from the left by (A), and remembering that

$$(A)(\tilde{A}) = 1,$$

$$(F') = m\left[\left(\frac{d^2x'}{dt^2}\right) + 2(\Omega')\left(\frac{dx'}{dt}\right) + (\Omega')^2(x') + \left(\frac{d\Omega'}{dt}\right)(x')\right]. \tag{8.70}$$

Thus if each observer assumes that the force acting on the particle is mass $\times$ acceleration, S, as equation (8.67) shows, would simply get (F), while S', according to equation (8.70), would obtain

$$m\left(\frac{d^2x'}{dt^2}\right) = (F') - 2m(\Omega')\left(\frac{dx'}{dt}\right) - m(\Omega')^2(x') - m\left(\frac{d\Omega'}{dt}\right)(x'). \tag{8.71}$$

In addition to (F'), which is what S' would expect anyway from transforming the force (F) to his own frame of reference, there are three additional pseudo-forces arising from the rotation:

(1) The *Coriolis* force, $-2m(\Omega')(dx'/dt)$, which depends upon the angular velocity of the frame and the velocity of the particle. In vector form this is $-2m\boldsymbol{\omega} \wedge \mathbf{v}_{s'}$.

(2) The *centrifugal* force, $-m(\Omega')^2(x')$, which depends upon the angular velocity of the frame and the position of the particle. In vector form this is $-m\boldsymbol{\omega} \wedge (\boldsymbol{\omega} \wedge \mathbf{x})$.

(3) The *angular acceleration* force, $-m(d\Omega'/dt)(x')$, which depends upon the angular *acceleration* and the particle position. In vector form it is $-m(d\boldsymbol{\omega}/dt) \wedge \mathbf{x}$. In many applications the angular velocity is constant, and in consequence this term vanishes.

A summary of the preceding two sections is given in the following chart.

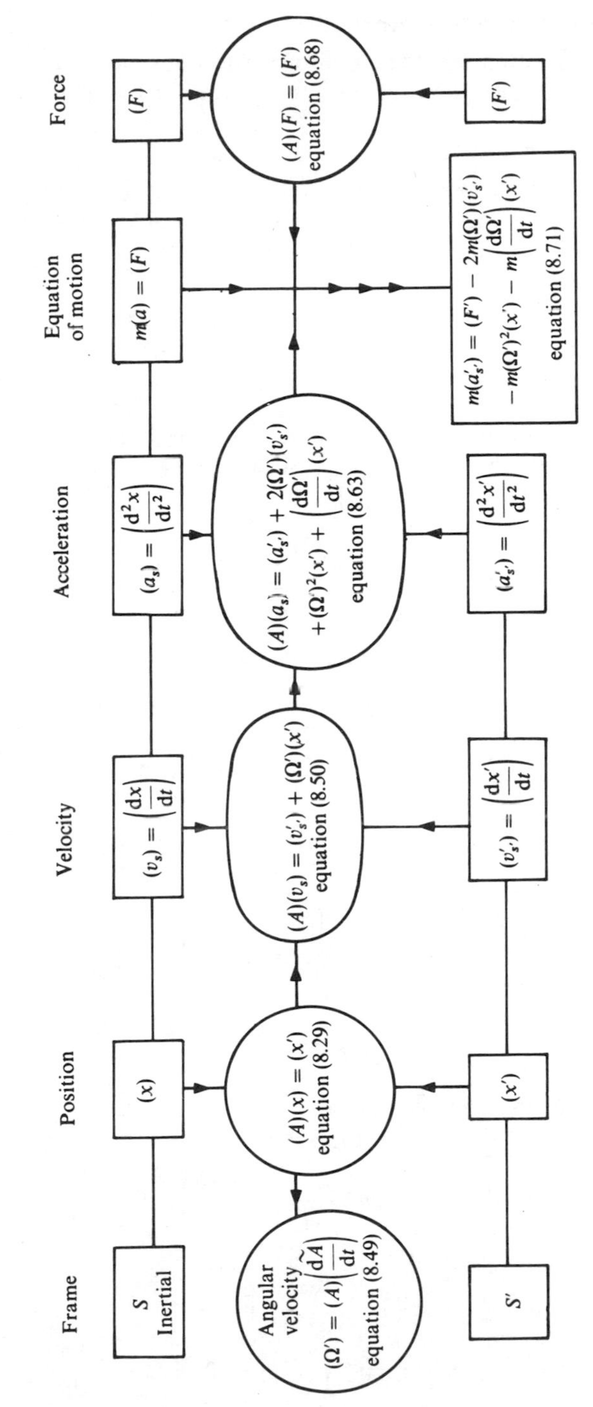

Chart summarizing Sections 8.4 and 8.5

8.6. Combined transformations

More complicated coordinate transformations in general can be split into successive transformations of one or other of two types just described.

8.6.1. Translation and rotation

Suppose, for example, that relative to the S rest frame the origin, O′, of the S' rest frame is moving with a constant velocity while its axes are also rotating. We transform first of all to an intermediate observer S'' whose rest frame also has for its origin O′, but whose axes remain parallel with those of S. If the position of a particle, P, is measured by S as (x) and by S'' as (y'') we know, by applying the results (8.10) and (8.21), that

$$(x) = (b) + t(U) + (y''). \tag{8.72}$$

We then transform from the rest frame of S'' to that of S'. Since these have the same origin, O′, this transformation is given by a matrix, (B), defined as above (for example $b_{12} = \mathbf{i}' . \mathbf{j}''$). Hence

$$(y'') = (\tilde{B})(y'), \tag{8.73}$$

where (y') is the position of P as measured by S'. Combining equations (8.72) and (8.73) we obtain the transformation between the S and S' frames,

$$(x) = (b) + t(U) + (\tilde{B})(y'). \tag{8.74}$$

From this we can obtain the transformation relating velocities,

$$\begin{aligned}\left(\frac{\mathrm{d}x}{\mathrm{d}t}\right) &= (U) + (\tilde{B})\left(\frac{\mathrm{d}y'}{\mathrm{d}t}\right) + \left(\frac{\mathrm{d}\tilde{B}}{\mathrm{d}t}\right)(y') \\ &= (U) + (\tilde{B})\left[\left(\frac{\mathrm{d}y'}{\mathrm{d}t}\right) + (B)\left(\frac{\mathrm{d}\tilde{B}}{\mathrm{d}t}\right)(y')\right] \\ &= (U) + (\tilde{B})\left[\left(\frac{\mathrm{d}y'}{\mathrm{d}t}\right) + (\Phi')(y')\right],\end{aligned} \tag{8.75}$$

where

$$(\Phi') = (B)\left(\frac{\mathrm{d}\tilde{B}}{\mathrm{d}t}\right)$$

describes the instantaneous angular velocity of S' relative to S''. Differentiating a second time we obtain the acceleration,

$$\begin{aligned}\left(\frac{\mathrm{d}^2x}{\mathrm{d}t^2}\right) &= (\tilde{B})\left[\left(\frac{\mathrm{d}^2y'}{\mathrm{d}t^2}\right) + (\Phi')\left(\frac{\mathrm{d}y'}{\mathrm{d}t}\right) + \left(\frac{\mathrm{d}\Phi'}{\mathrm{d}t}\right)(y')\right] \\ &\quad + \left(\frac{\mathrm{d}\tilde{B}}{\mathrm{d}t}\right)\left[\left(\frac{\mathrm{d}y'}{\mathrm{d}t}\right) + (\Phi')(y')\right] \\ &= (\tilde{B})\left[\left(\frac{\mathrm{d}^2y'}{\mathrm{d}t^2}\right) + 2(\Phi')\left(\frac{\mathrm{d}y'}{\mathrm{d}t}\right) + (\Phi')^2(y') + \left(\frac{\mathrm{d}\Phi'}{\mathrm{d}t}\right)(y')\right].\end{aligned} \tag{8.76}$$

The similarity of the transformations (8.63) and (8.76) shows that the uniform translation of the origin of S' makes no difference to the effects arising from the rotation.

8.6.2. Rotation and rotation

Suppose that the S, S' and S'' rest frames all have a common origin and that the first and second are related by the transformation (A):

$$(x') = (A)(x), \qquad (x) = (\tilde{A})(x'), \tag{8.77}$$

while the second and third are related by (B):

$$(x'') = (B)(x'), \qquad (x') = (\tilde{B})(x''). \tag{8.78}$$

Then if (C) describes the relation between the first and third:

$$(x'') = (C)(x), \qquad (x) = (\tilde{C})(x''), \tag{8.79}$$

we can see from equations (8.77) and (8.78) that

$$(C) = (B)(A), \qquad (\tilde{C}) = (\tilde{A})(\tilde{B}). \tag{8.80}$$

Now, from equation (8.79), the angular velocity of S'' relative to S is described, in the S'' frame, by the matrix

$$(\Psi'') = (C)\left(\frac{\mathrm{d}\tilde{C}}{\mathrm{d}t}\right).$$

Substituting from equation (8.80),

$$\begin{aligned}(\Psi'') &= [(B)(A)]\frac{\mathrm{d}}{\mathrm{d}t}[(\tilde{A})(\tilde{B})] \\ &= (B)(A)\left[(\tilde{A})\left(\frac{\mathrm{d}\tilde{B}}{\mathrm{d}t}\right) + \left(\frac{\mathrm{d}\tilde{A}}{\mathrm{d}t}\right)(\tilde{B})\right] \\ &= (B)\left(\frac{\mathrm{d}\tilde{B}}{\mathrm{d}t}\right) + (B)(A)\left(\frac{\mathrm{d}\tilde{A}}{\mathrm{d}t}\right)(\tilde{B}). \end{aligned} \tag{8.81}$$

$(B)(\mathrm{d}\tilde{B}/\mathrm{d}t)$ describes, in the S'' frame, the angular velocity of S'' relative to S'. Let us call this (Φ''). $(A)(\mathrm{d}\tilde{A}/\mathrm{d}t)$ describes, in the S' frame, the angular velocity of S' relative to S. This we have already called (Ω'). Multiplying before and after by (B) and $(\tilde{B})$ re-expresses this angular velocity in the S'' frame. This we call (Ω''). Hence equation (8.81) may be written as

$$(\Psi'') = (\Phi'') + (B)(\Omega')(\tilde{B}) = (\Phi'') + (\Omega''). \tag{8.82}$$

We see from this that superimposed angular velocities are expressed simply by adding the two angular velocity matrices. This is another aspect of the fact that for most purposes the three independent non-zero elements of the angular velocity matrix may be considered as components of a vector.

8.7. Motion at the earth's surface

Most of the elementary demonstrations of the laws of mechanics are given in laboratories fixed to the earth's surface. Now the earth is not only rotating about its own axis but is also orbiting the sun. In view of the complicated transformation that we have seen to arise from such motion we might well expect that the simple Newtonian laws of Chapter 1 would have a far from simple expression at the earth's surface.

The third law, which states that two vector forces are equal and opposite,

$$\mathbf{f}_{ij} = -\mathbf{f}_{ji},$$

presents no problem. In the S frame this is equivalent to

$$(f_{ij}) = (-f_{ji}). \tag{8.83}$$

We saw in Section 8.4 that, whatever the relative positions and motions of two frames of reference, the components of any vector (represented in space by a line of definite length in a definite direction) in the two frames are connected by the matrix (A). The elements of (A) are given by the instantaneous values of the cosines of the angles between the axes of the two frames. Thus

$$(f_{ij}) = (\tilde{A})(f'_{ij}) \quad \text{and} \quad (f_{ji}) = (\tilde{A})(f'_{ji}).$$

Hence, from equation (8.83),

$$(\tilde{A})(f'_{ij}) = -(\tilde{A})(f'_{ji}),$$

and on multiplying this from the left by (A)

$$(f'_{ij}) = -(f'_{ji}), \tag{8.84}$$

since

$$(A)(\tilde{A}) = (1).$$

Equations (8.83) and (8.84) have exactly the same form, which means that S and S' agree about this law, however S and S' may be moving relative to each other.

The first two laws can be summarized as

$$(F) = m\left(\frac{d^2x}{dt^2}\right) \tag{8.85}$$

for a stationary observer S. For the time being we shall assume that the sun is stationary, and that the S frame has its origin, O, at the sun's centre, with the axes defined by unit vectors **i**, **j** and **k** fixed in direction.

8.7.1. Orbital motion

Firstly, let us calculate the effect of the earth's orbital motion on the law (8.85). For simplicity we shall take the earth's orbit as circular, of radius a_0 and constant angular frequency ω_0, and use for S' a frame whose origin, O′, is at the earth's centre and whose axes remain parallel with those of S (see Figure 8.6a). The vector **k** is chosen to be perpendicular to the plane of the earth's orbit. The projection on this plane is shown in Figure 8.6b, from

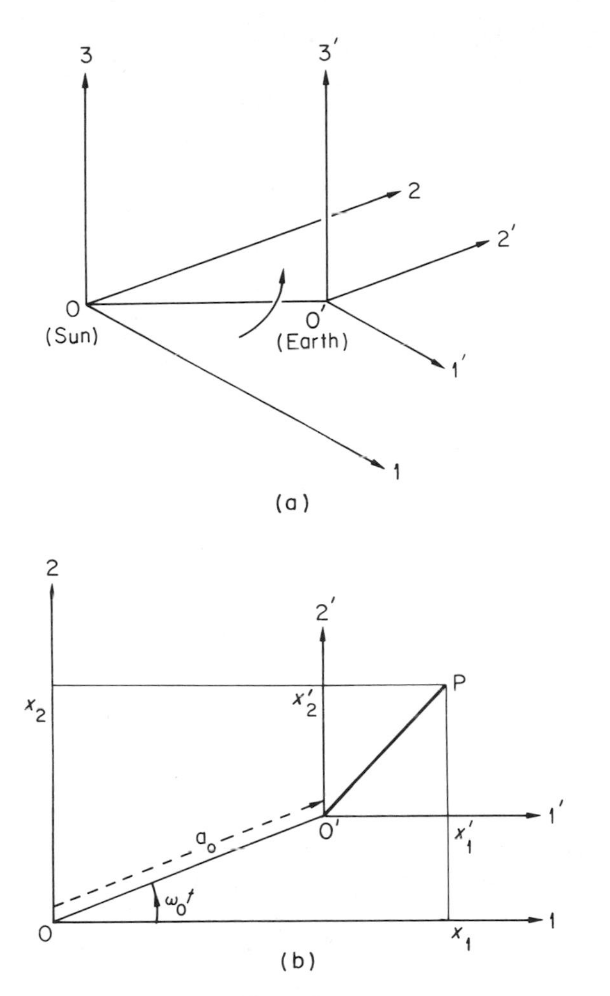

Figure 8.6

which we can see that the relationship between the position of a point P in the two frames is given by

$$x_1 = x'_1 + a_0 \cos \omega_0 t$$

$$x_2 = x'_2 + a_0 \sin \omega_0 t.$$

Time is taken to be zero when the axes O1 and O′1′ coincide. The third component is the same for both frames,

$$x_3 = x'_3,$$

and we may therefore write the transformation as

$$(x) = (x') + a_0(\sigma),$$

where

$$(\sigma) = \begin{pmatrix} \cos \omega_0 t \\ \sin \omega_0 t \\ 0 \end{pmatrix}. \tag{8.86}$$

Then

$$\left(\frac{dx}{dt}\right) = \left(\frac{dx'}{dt}\right) + a_0\left(\frac{d\sigma}{dt}\right),$$

$$\left(\frac{d^2x}{dt^2}\right) = \left(\frac{d^2x'}{dt^2}\right) + a_0\left(\frac{d^2\sigma}{dt^2}\right).$$

From equation (8.86),

$$\left(\frac{d^2\sigma}{dt^2}\right) = -\omega_0^2(\sigma).$$

Hence

$$\left(\frac{d^2x}{dt^2}\right) = \left(\frac{d^2x'}{dt^2}\right) - a_0\omega_0^2(\sigma). \tag{8.87}$$

Since corresponding pairs of axes in the two frames are parallel, the projections of any vector will be the same for both. Thus

$$(F) = (F'). \tag{8.88}$$

Substituting from equations (8.87) and (8.88) into equation (8.85), we obtain

$$(F') = m\left[\left(\frac{d^2x'}{dt^2}\right) - a_0\omega_0^2(\sigma)\right]$$

or

$$\left(\frac{d^2x'}{dt^2}\right) = \frac{1}{m}(F') + a_0\omega_0^2(\sigma). \tag{8.89}$$

Comparing equations (8.89) and (8.85), we see that for the laws of mechanics to retain their simple Newtonian form in an earthbound laboratory the extra term, $a_0\omega_0^2(\sigma)$, on the right-hand side of equation (8.89) must be negligible. Now the orbital period of the earth is

$$T_0 \simeq 365 \text{ days} \simeq 3{\cdot}16 \times 10^7 \text{ s},$$

and the (approximately) constant radius of its orbit is

$$a_0 \simeq 1{\cdot}5 \times 10^{11} \text{m}.$$

Hence

$$a_0\omega_0^2 = \frac{4\pi^2 a_0}{T_0^2} \simeq 6 \times 10^{-3} \text{ m s}^{-2},$$

and since the elements of the column matrix (σ) do not exceed unity in magnitude, 6×10^{-3} m s^{-2} is the maximum possible magnitude of any contribution from the extra term $a_0\omega_0^2(\sigma)$.

To gain some physical sense of this magnitude we may compare it with the acceleration at the earth's surface, arising from gravity,

$$g \simeq 9{\cdot}8 \text{ m s}^{-2}.$$

Thus, only when accuracy of an experiment becomes sensitive to forces one-thousandth that of gravity will the effect of orbital motion become noticeable. Otherwise the observer S' will have for his law

$$\left(\frac{d^2x'}{dt^2}\right) = \frac{1}{m}(F'),$$

the same as for the fixed observer, S.

8.7.2. Spin

Even if the earth's centre were at rest, or moving with constant velocity, the fact that it is spinning about its north–south axis would affect the equation of motion in a reference frame moving with the earth's surface. Suppose 1, 2 and 3 are the fixed axes, 1 and 2 lying in the equatorial plane and 3 along the rotation axis, as shown in Figure 8.7. Q is the point at the surface under consideration, at colatitude θ. OQ then defines the axis 3′; 1′ is perpendicular to this and in the same plane as 3 and 3′; 2′ completes the orthogonal set of axes rotating with the earth and with the same origin, O, as the fixed set.

The spin, or angular velocity, $\boldsymbol{\omega}$, is constant and in the direction 3. It will therefore have constant components $-\omega \sin\theta$, 0, $\omega\cos\theta$ along the axes

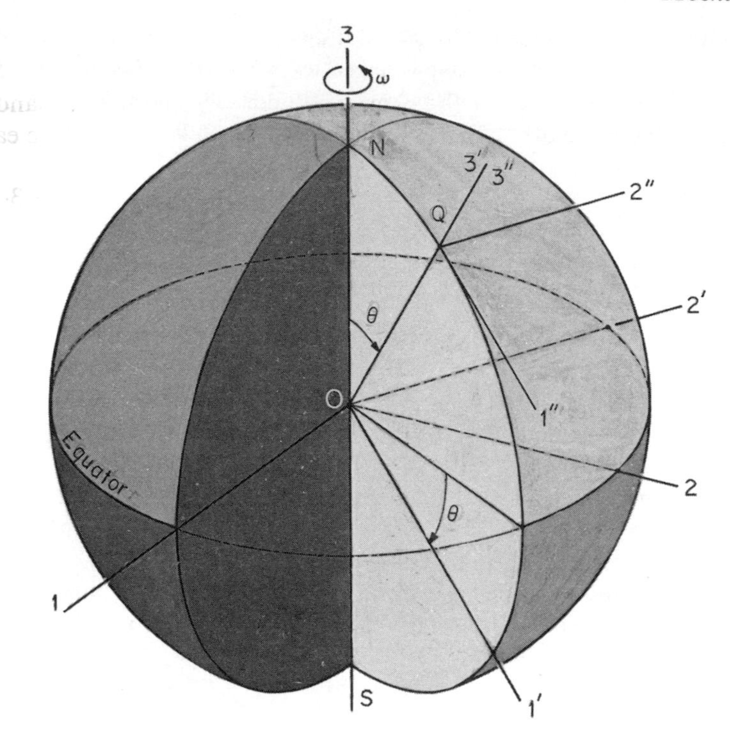

Figure 8.7

1′, 2′, 3′. Hence

$$(\Omega') = \omega\begin{pmatrix} 0 & -\cos\theta & 0 \\ \cos\theta & 0 & \sin\theta \\ 0 & -\sin\theta & 0 \end{pmatrix},$$

$$(\Omega')^2 = -\omega^2\begin{pmatrix} \cos^2\theta & 0 & \sin\theta\cos\theta \\ 0 & 1 & 0 \\ \sin\theta\cos\theta & 0 & \sin^2\theta \end{pmatrix}, \qquad (8.90)$$

$$\left(\frac{d\Omega'}{dt}\right) = (0).$$

From equation (8.71) the motion of a particle in the rotating frame will follow the law,

$$\left(\frac{d^2x'}{dt^2}\right) = \frac{1}{m}(F') - 2(\Omega')\left(\frac{dx'}{dt}\right) - (\Omega')^2(x'). \qquad (8.91)$$

Since we are concerned with motion near Q, we must now move to axes 1″, 2″, 3″ parallel with 1′, 2′, 3′ but with origin at Q. 1″ and 2″ point south and east along the earth's surface, 3″ normally and upwards. This involves a simple shift of origin along the axis 3′ by an amount a, the earth's radius:

$$(x') = (x'') + (a),$$

where

$$(a) = \begin{pmatrix} 0 \\ 0 \\ a \end{pmatrix}. \tag{8.92}$$

Thus

$$\left(\frac{\mathrm{d}x'}{\mathrm{d}t}\right) = \left(\frac{\mathrm{d}x''}{\mathrm{d}t}\right), \qquad \left(\frac{\mathrm{d}^2x'}{\mathrm{d}t^2}\right) = \left(\frac{\mathrm{d}^2x''}{\mathrm{d}t^2}\right).$$

Since the axes 1′, 2′, 3′ are parallel with axes 1″, 2″, 3″, the components of force and angular velocity will also be unchanged:

$$(F') = (F''), \qquad (\Omega') = (\Omega'').$$

Hence for the frame at the earth's surface the equation of motion (8.91) becomes

$$\left(\frac{\mathrm{d}^2x''}{\mathrm{d}t^2}\right) = \frac{1}{m}(F'') - 2(\Omega'')\left(\frac{\mathrm{d}x''}{\mathrm{d}t}\right) - (\Omega'')^2(x'') - (\Omega'')^2(a), \tag{8.93}$$

where (Ω''), $(\Omega'')^2$ are the same as (Ω'), $(\Omega')^2$ in equation (8.90).

The magnitude of any term in $(\Omega'')^2(a)$ will not exceed $a\omega^2$. The mean radius of the earth is

$$a = 6{\cdot}37 \times 10^6 \text{ m}$$

and the period of rotation is

$$\tau \simeq 1 \text{ day} = 8{\cdot}64 \times 10^4 \text{ s}.$$

Hence

$$a\omega^2 = \frac{4\pi^2 a}{\tau^2} \simeq 3{\cdot}37 \times 10^{-2} \text{ m s}^{-2}.$$

We can see from this that in the rotating frame there is a small constant centrifugal force well under one per cent of the gravitational force. The variable centrifugal force, $(\Omega'')^2(x'')$, will normally be quite negligible since any laboratory dimension, x'', will be extremely small compared with a. (See Section 8.9.9, page 247.)

Any term in the Coriolis force $2(\Omega'')(\mathrm{d}x''/\mathrm{d}t)$ will have a magnitude not exceeding $2\omega v^*$ where v^* is the laboratory speed. This will become comparable with gravity only at the very high value of

$$v^* = \frac{g}{\omega} = \frac{\tau g}{2\pi} \simeq 1{\cdot}35 \times 10^5 \text{ m s}^{-1}.$$

Thus for speeds of up to around 100 m s^{-1} the Coriolis force will also be well under one per cent of the gravitational force.

For most laboratory experiments, then, the spin as well as the orbital motion of the earth will give only very small deviations from the simple Newtonian laws of Chapter 1.

8.7.3. Motion on the horizontal plane

The centrifugal forces arising from the earth's spin are small and cause a slight change in the magnitude and direction of apparent gravity (see Example 27, page 247). Now we shall include in 'gravity' these small effects and take as the 'vertical' axis, Q3, at the earth's surface a line antiparallel with apparent gravity. It points back at the earth's axis at an angle $\lambda = \theta - \varepsilon$, where ε is less than $\frac{1}{10}^\circ$. The 'horizontal' plane is perpendicular to this and contains the axis Q1 pointing south and the axis Q2 (coincident with Q2″) pointing east (see Figure 8.8). (For convenience we again use unprimed axes 1, 2, 3 for this final frame. They are not to be confused with the axes 1, 2, 3, whose origin is O, which was the starting frame for Section 8.7.2.)

In this frame the gravitational force is

$$m(g) = m\begin{pmatrix} 0 \\ 0 \\ -g_a \end{pmatrix}.$$

Since this has absorbed the centrifugal forces, only the true force (F) and the Coriolis force $-2(\Omega)(\mathrm{d}x/\mathrm{d}t)$ remain and the equation of motion (8.93) reduces to

$$\frac{\mathrm{d}^2x}{\mathrm{d}t^2} = \frac{1}{m}(F) + (g) - 2(\Omega)\left(\frac{\mathrm{d}x}{\mathrm{d}t}\right). \tag{8.94}$$

The angular velocity matrix is now

$$(\Omega) = \begin{pmatrix} 0 & -\omega_3 & \omega_2 \\ \omega_3 & 0 & -\omega_1 \\ -\omega_2 & \omega_1 & 0 \end{pmatrix} = \omega\begin{pmatrix} 0 & -\cos\lambda & 0 \\ \cos\lambda & 0 & \sin\lambda \\ 0 & -\sin\lambda & 0 \end{pmatrix}. \tag{8.95}$$

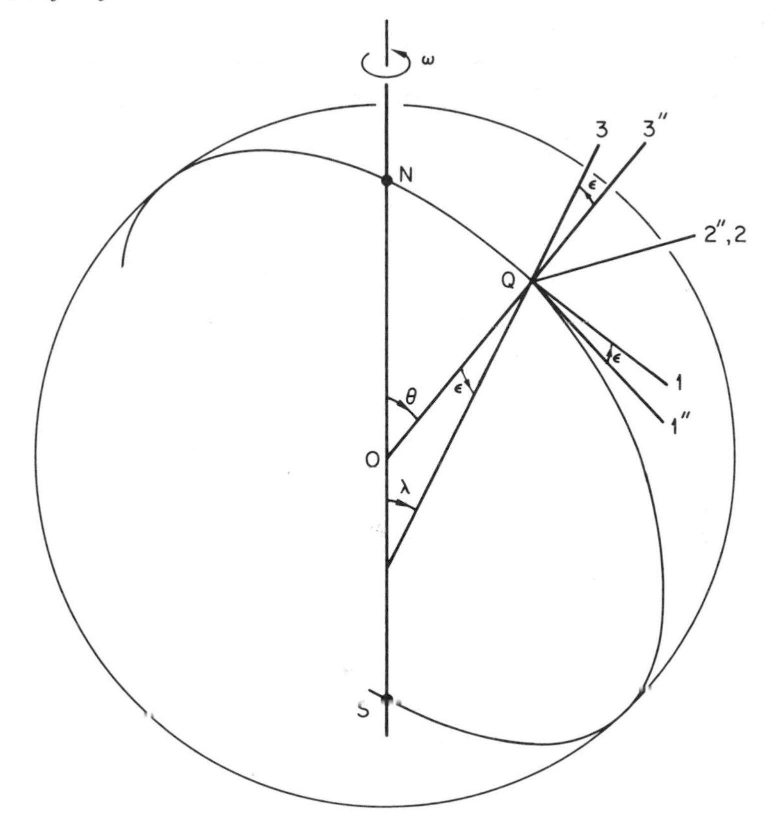

Figure 8.8

In the preceding section we saw that $(\Omega)(\mathrm{d}x/\mathrm{d}t)$ has elements which are small compared with g_a. However, when motion is confined to the horizontal plane the force (F) must necessarily have a vertical component that cancels g_a. In this case the effect of the Coriolis force must be judged by comparison with the other components of (F). If these are small the Coriolis force may be very important.

Now if motion is confined to the horizontal plane

$$x_3 = \dot{x}_3 = \ddot{x}_3 = 0,$$

and equation (8.94) is equivalent to

$$\ddot{x}_1 = \frac{1}{m}F_1 + 2\omega \cos \lambda \, \dot{x}_2,$$

$$\ddot{x}_2 = \frac{1}{m}F_2 - 2\omega \cos \lambda \, \dot{x}_1.$$

The Coriolis force components $-2\omega \cos \lambda\, \dot{x}_2$, $2\omega \cos \lambda\, \dot{x}_1$ have a resultant which is at an angle $-\pi/2$ with respect to the direction of the velocity, as we can see from Figures 8.9a and b. Thus motion on the earth's surface in the Northern Hemisphere is always diverted *to the right*. In the Southern Hemisphere it is *to the left*.

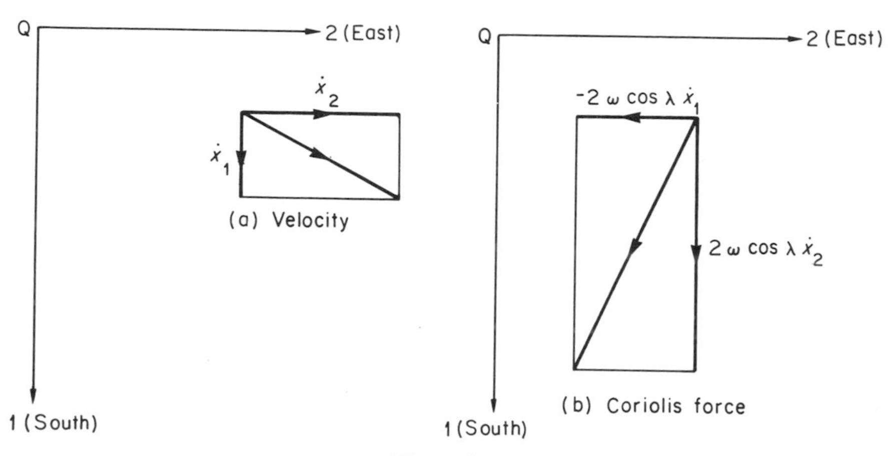

Figure 8.9

We can account for this effect by introducing a further rotation about the vertical axis 3. Note first that if a new frame with axes 1′, 2′, 3′ and the same origin Q rotates relative to the frame with axes 1, 2, 3 in a general manner described by the angular velocity matrix (Γ'), then the equation of motion (8.94) becomes in the new frame

$$\left(\frac{\mathrm{d}^2 x'}{\mathrm{d}t^2}\right) = \frac{1}{m}(F') + (g') - 2(\Omega' + \Gamma')\left(\frac{\mathrm{d}x'}{\mathrm{d}t}\right). \tag{8.96}$$

Now if the axes 3 and 3′ coincide, the element ω_3' will be unchanged:

$$\omega_3' = \omega_3 = \omega \cos \lambda,$$

while the only non-zero element of (Γ') will be γ_3'. Apparent gravity will also be unchanged:

$$(g) = (g') = \begin{pmatrix} 0 \\ 0 \\ -g_a \end{pmatrix}.$$

Motion in the horizontal plane implies

$$x_3' = \dot{x}_3' = \ddot{x}_3' = 0.$$

Then equation (8.96) is equivalent to

$$\ddot{x}'_1 = \frac{1}{m}F'_1 - 2(\omega \cos \lambda + \gamma'_3)\dot{x}'_2,$$

$$\ddot{x}'_2 = \frac{1}{m}F'_2 + 2(\omega \cos \lambda + \gamma'_3)\dot{x}'_1.$$

Hence if we choose

$$\gamma_3 = -\omega \cos \lambda$$

we simplify the equations of motion to

$$\ddot{x}'_1 = \frac{1}{m}F'_1,$$

$$\ddot{x}'_2 = \frac{1}{m}F'_2. \tag{8.97}$$

These are what we would expect for a stationary frame of reference. For any system of forces we therefore first find the motion they would produce in a stationary frame and then superimpose upon that a steady rotation, $-\omega \cos \lambda$, about the apparent vertical to give the motion that would be observed relative to a frame fixed in the earth's surface.

A simple case would be the motion of a particle attached to a horizontal spring. By choosing the axis $1'$ along the spring's axis, equation (8.97) would give simple harmonic motion,

$$x'_1 = A \cos (\omega_0 t + \alpha),$$

$$x'_2 = 0.$$

Assuming the frequency of this to be much greater than that of the earth's rotation,

$$\omega_0 \gg \omega,$$

then, to an observer on the earth, this linear oscillation would slowly turn about a vertical axis with angular velocity $-\omega \cos \lambda$. In practice, spring oscillations would be damped out too quickly for this rotation easily to be seen, since even at the North and South Poles,

$$\lambda = 0, \qquad \lambda = 180°,$$

the rotation is at a rate of only one turn in approximately twenty-four hours, while at colatitudes 30° North and South,

$$\lambda = 60°, \qquad \lambda = 120°,$$

one turn takes approximately forty-eight hours.

To show the effect we therefore need a free oscillation which persists for several hours. The well-known method of achieving this is Foucault's pendulum. This is simply a heavy bob (of several kilogrammes mass) suspended on a long thin wire (at least 20 or 30 metres in length). The motion of the bob is not strictly horizontal but lies on the surface of a sphere whose centre is the point of support of the pendulum. However, if the amplitude is small compared with the length of the pendulum the motion is confined near enough to the horizontal plane for the analysis above to give an accurate description.

Winds in the earth's atmosphere can also be considered approximately as motion in the horizontal plane since the thickness of the atmosphere is very small compared with the radius of the earth. We would normally expect winds to flow radially inwards towards a centre of low pressure and radially outwards from a centre of high pressure. The Coriolis force in the Northern Hemisphere causes them to veer to the right, so that in the first case the winds spiral anticlockwise towards the low pressure centre to give a cyclone, while in the second they spiral clockwise away from the high pressure centre to give an anticyclone. In the Southern Hemisphere the directions of rotation are reversed.

8.8. Charged particle in a magnetic field; Larmor rotation

A particle of mass m and charge q moving in an electromagnetic field defined by $\mathbf{E}$ and $\mathbf{B}$ has the equation of motion

$$m\mathbf{a} = q\mathbf{E} + q\mathbf{v} \wedge \mathbf{B} = q\mathbf{E} - q\mathbf{B} \wedge \mathbf{v}. \tag{8.98}$$

We saw in Section 8.4.3 that vector multiplication by $\boldsymbol{\omega}$ was equivalent to matrix multiplication by (Ω). In a similar way we can here define the anti-symmetric matrix

$$(\mathscr{B}) = \begin{pmatrix} 0 & -B_3 & B_2 \\ B_3 & 0 & -B_1 \\ -B_2 & B_1 & 0 \end{pmatrix},$$

and then write equation (8.98) in more precise terms, referred to the S frame, as

$$m\left(\frac{\mathrm{d}^2x}{\mathrm{d}t^2}\right) = q(E) - q(\mathscr{B})\left(\frac{\mathrm{d}x}{\mathrm{d}t}\right). \tag{8.99}$$

The similarity between magnetic effects and rotation may be exploited to simplify equation (8.99). If S' has a constant rotation relative to S, described in the S frame by (Ω), then S has a rotation relative to S', described in the S frame, also, by $-(\Omega)$. The relation inverse to (8.63) is

$$(\tilde{A})\left(\frac{\mathrm{d}^2x'}{\mathrm{d}t^2}\right) = \left(\frac{\mathrm{d}^2x}{\mathrm{d}t^2}\right) - 2(\Omega)\left(\frac{\mathrm{d}x}{\mathrm{d}t}\right) + (\Omega)^2(x) - \left(\frac{\mathrm{d}\Omega}{\mathrm{d}t}\right)(x), \tag{8.100}$$

while the electric field transforms as

$$(\tilde{A})(E') = (E). \qquad (8.101)$$

Substituting from equations (8.100) and (8.101) into equation (8.99), we obtain

$$m(\tilde{A})\left(\frac{\mathrm{d}^2x'}{\mathrm{d}t^2}\right) = q(\tilde{A})(E') - q\left[(\mathscr{B}) + \frac{2m}{q}(\Omega)\right]\left(\frac{\mathrm{d}x}{\mathrm{d}t}\right)$$
$$+ m(\Omega)^2(x) - m\left(\frac{\mathrm{d}\Omega}{\mathrm{d}t}\right)(x). \qquad (8.102)$$

If the magnetic induction is uniform (in the S frame) and constant with time, all the elements of $(\mathscr{B})$ will be constants. By choosing a rotation such that

$$(\Omega) = -\frac{q}{2m}(\mathscr{B}), \qquad (8.103)$$

the second term will vanish, and since (Ω) will be constant the last term vanishes also. If, in addition, the magnetic induction is very small we may ignore $(\mathscr{B})^2$ and therefore $(\Omega)^2$. On multiplying equation (8.102) by (A), we then finally have simply

$$m\left(\frac{\mathrm{d}^2x'}{\mathrm{d}t^2}\right) = q(E'), \qquad (8.104)$$

which is the equation of motion, *in the S' frame*, of a particle subject only to the electric field.

Thus if the magnetic induction has magnitude B, its effect is to superimpose upon the motion to be expected from the electric field a rotation opposite in direction to the induction and of angular speed

$$\omega_{\mathrm{L}} = \frac{qB}{2m}. \qquad (8.105)$$

This angular speed is known as the *Larmor* frequency.

The form (8.104) is particularly useful when the electric field is a central one. An example is the classical picture of a hydrogen atom whose one electron of charge $-e$ is attracted to the heavy central nucleus (a proton) of charge e, and is therefore acted on by a central force inversely proportional to the square of its distance from the nucleus. We know from Chapter 6 that its bound orbit is an ellipse. If now a weak magnetic field is switched on, the orbit will still be an ellipse in the frame S' and the normal to the plane of this ellipse will precess about the direction of the magnetic induction with the Larmor angular speed.

It must be remembered that this derivation of the Larmor rotation rests upon the validity of neglecting the $m(\Omega)^2(x)$ term in equation (8.102). More specifically, this implies that the term is small compared with $q(\tilde{A})(E')$ or $q(E)$. Hence if ω, x, E and B are magnitudes typical of the Larmor frequency, the distance of the particle from the centre of rotation, the electric field and the magnetic induction, then the condition for the Larmor rotation to arise is

$$m\omega^2 x \ll qE$$

or, using equation (8.105),

$$xqB^2 \ll mE. \tag{8.106}$$

(See Section 8.9.10, page 249.)

8.9. Comments and worked examples

8.9.1. See Section 8.1

A matrix is an ordered rectangular array of quantities. Thus

$$\begin{pmatrix} a_{11} & a_{12} & a_{13} \\ a_{21} & a_{22} & a_{23} \\ a_{31} & a_{32} & a_{33} \end{pmatrix}, \quad \begin{pmatrix} b_{11} & b_{12} & b_{13} & b_{14} \\ b_{21} & b_{22} & b_{23} & b_{24} \end{pmatrix}, \quad \begin{pmatrix} p_1 \\ p_2 \\ p_3 \\ p_4 \end{pmatrix},$$

are respectively a 3×3 (square), a 2×4 and a 4×1 (column) matrix, where the first figure refers to the number of rows and the second to the number of columns. Each of the individual quantities is called an *element* of the matrix, and usually has two indices giving its row and column position in the matrix.

The condensed notations (a_{ij}), (b_{ij}), (p_i) or, more simply, (a), (b), (p) can normally be used without confusion.

Matrices of similar dimensions may be added or subtracted, element by element,

$$\begin{pmatrix} a_{11} & a_{12} & a_{13} \\ a_{21} & a_{22} & a_{23} \\ a_{31} & a_{32} & a_{33} \end{pmatrix} + \begin{pmatrix} c_{11} & c_{12} & c_{13} \\ c_{21} & c_{22} & c_{23} \\ c_{31} & c_{32} & c_{33} \end{pmatrix} = \begin{pmatrix} a_{11} + c_{11} & a_{12} + c_{12} & a_{13} + c_{13} \\ a_{21} + c_{21} & a_{22} + c_{22} & a_{23} + c_{23} \\ a_{31} + c_{31} & a_{32} + c_{32} & c_{33} + c_{33} \end{pmatrix},$$

$$\begin{pmatrix} p_1 \\ p_2 \\ p_3 \\ p_4 \end{pmatrix} + \begin{pmatrix} q_1 \\ q_2 \\ q_3 \\ q_4 \end{pmatrix} = \begin{pmatrix} p_1 + q_1 \\ p_2 + q_2 \\ p_3 + q_3 \\ p_4 + q_4 \end{pmatrix},$$

or

$$(a) + (c) = (a + c), \qquad (p) + (q) = (p + q).$$

When the rows and columns of a matrix (b) are interchanged we obtain its *transpose*, denoted by $(\tilde{b})$. Thus

$$\widetilde{\begin{pmatrix} b_{11} & b_{12} & b_{13} & b_{14} \\ b_{21} & b_{22} & b_{23} & b_{24} \end{pmatrix}} = \begin{pmatrix} b_{11} & b_{21} \\ b_{12} & b_{22} \\ b_{13} & b_{23} \\ b_{14} & b_{24} \end{pmatrix} \quad \text{or} \quad \tilde{b}_{ij} = b_{ji}.$$

The difficulties in using matrices in mechanics come almost entirely from lack of familiarity rather than any intrinsic difficulty. Once the rules for multiplication and transposition are learnt (see Sections 8.9.3 and 8.9.6), equations and laws employing them need appear little more complicated than ordinary algebraic equations.

8.9.2. See Section 8.1

A factor common to all the elements of any matrix is a *scalar* factor and can always be written before or after the matrix. Thus

$$\begin{pmatrix} \dfrac{\alpha_{11}}{\gamma} & \dfrac{\alpha_{12}}{\gamma^2} \\ \dfrac{\alpha_{21}}{\gamma} & \dfrac{\alpha_{22}}{\gamma^2} \end{pmatrix} = \frac{1}{\gamma^2}\begin{pmatrix} \gamma\alpha_{11} & \alpha_{12} \\ \gamma\alpha_{21} & \alpha_{22} \end{pmatrix} = \begin{pmatrix} \gamma\alpha_{11} & \alpha_{12} \\ \gamma\alpha_{21} & \alpha_{22} \end{pmatrix}\frac{1}{\gamma^2}.$$

An operation such as differentiation can also be factored out if it is the same for each element of the matrix, although it is usually necessary to keep the correct order. Thus

$$\begin{pmatrix} \mathrm{d}^2x_1/\mathrm{d}t^2 \\ \mathrm{d}^2x_2/\mathrm{d}t^2 \\ \mathrm{d}^2x_3/\mathrm{d}t^2 \end{pmatrix} = \frac{\mathrm{d}^2}{\mathrm{d}t^2}\begin{pmatrix} x_1 \\ x_2 \\ x_3 \end{pmatrix}.$$

8.9.3. See Section 8.4

The multiplication rule is one of the few laws of matrix algebra that we shall require. It may be formulated as follows:

If $(a) = (b)(c)$, then

$$a_{ij} = \sum_k b_{ik}c_{kj}, \tag{8.107}$$

or, more fully,

$$\begin{pmatrix} b_{11} & b_{12} & \dots & b_{1m} \\ \vdots & & & \\ b_{i1} & b_{i2} & \dots & b_{im} \\ \vdots & & & \\ b_1 & \dots & \dots & b_m \end{pmatrix} \begin{pmatrix} c_{11} & c_{12} & \dots & c_{1j} & \dots & c_{1n} \\ c_{21} & \dots & \dots & c_{2j} & \dots & c_{2n} \\ \vdots & & & & & \\ c_{m1} & \dots & \dots & c_{mj} & \dots & c_{mn} \end{pmatrix}$$

$$= \begin{pmatrix} & j\text{th column} & \\ \dots & \vdots & \\ \dots & \vdots & \\ & \boxed{\begin{matrix} b_{i1}c_{1j} + b_{i2}c_{2j} \\ + \dots + b_{im}c_{mj} \end{matrix}} & \dots \\ & \uparrow & \\ & ij\text{th element, } a_{ij} & \end{pmatrix} i\text{th row}$$

Thus the ijth element of the product, $(b)(c)$, is obtained by multiplying together corresponding elements of the ith row of (b) and the jth column of (c). Hence (b) must have the same number of columns as (c) has rows. Note that only when (b) and (c) are both square and of the same dimensions will the products $(b)(c)$ and $(c)(b)$ be square *and* of the same dimensions, and that, even then, these products, in general, *will not be equal*. It is therefore essential to maintain the correct order in matrix products.

The extended product $(b)(c)(d)\dots$ follows from the above rules and requires the number of columns in each factor to be the same as the number of rows in the following one. It is easy to show that the factors can be grouped in any convenient way to give the final product provided, as always, that the order of the individual matrices is unchanged. Thus the product $(a)(b)(c)(d)$ may be calculated in any of the following ways:

$$(a)(b)(c)(d) = (a)[(b)\{(c)(d)\}] = [(a)(b)][(c)(d)]$$

$$= [(a)\{(b)(c)\}](d) = \text{etc.}$$

This is the *associative* law for matrix multiplication.

8.9.4. See Section 8.4

Equation (8.29),

$$(x') = (A)(x), \tag{8.108}$$

can be solved for (x) if we can find a matrix $(A)^{-1}$ such that

$$(A)^{-1}(A) = \begin{pmatrix} 1 & 0 & 0 \\ 0 & 1 & 0 \\ 0 & 0 & 1 \end{pmatrix} = (1).$$

The unit matrix (1), multiplying any matrix from the left or right, leaves it unchanged. Hence, on multiplying equation (8.108) from the left by $(A)^{-1}$ we obtain

$$(A)^{-1}(x') = (A)^{-1}(A)(x) = (1)(x) = (x).$$

A matrix does not necessarily have an inverse, and even when it does, the process of finding it is complicated. However, this is a problem that need not concern us in the case of the orthogonal matrix (A) of Section 8.4 since equations (8.32) and (8.33) show that its inverse is simply its transpose.

8.9.5. See Section 8.4.1

Calling quantities vectors does not make them so. For example a rotation of 90° about an axis can be described in terms of a magnitude and direction; the measurements of a beauty queen form an ordered set of three numbers. Superficially both might seem to qualify as vectors. However, a vector must satisfy the vector law of addition and transform between frames of reference in the way discussed in Section 8.4. It is easy to show that finite rotations cannot be compounded like vectors, and it would be difficult even to give any physical significance to the compounding of two sets of bodily measurements, or to their transformation between different frames.

It is therefore necessary to *show* that w compounds like a vector before we have the right to call it a vector. This is left as an exercise for the reader.

8.9.6. See Section 8.4.3

Suppose (a) is the product of two matrices,

$$(a) = (b)(c).$$

Then if $(\tilde{a})$ is the transpose of (a), by definition its ijth element will be

$$\tilde{a}_{ij} = a_{ji} = \sum_k b_{jk} c_{ki}.$$

Now if $(\tilde{b})$, $(\tilde{c})$ are the transposes of (b) and (c),

$$b_{jk} = \tilde{b}_{kj} \quad \text{and} \quad c_{ki} = \tilde{c}_{ik}.$$

Hence

$$\tilde{a}_{ij} = \sum_k \tilde{b}_{kj}\tilde{c}_{ik} = \sum_k \tilde{c}_{ik}\tilde{b}_{kj}.$$

Comparing this with the original definition of a matrix product (equation 8.107), we see that

$$(\tilde{a}) = (\tilde{c})(\tilde{b}).$$

This may easily be extended to give the general result: if

$$(a) = (b)(c)\ldots(p)(q),$$

then

$$(\tilde{a}) = (\tilde{q})(\tilde{p})\ldots(\tilde{c})(\tilde{b}).$$

To obtain the transpose of any product we simply take the transpose of each factor and reverse the order of multiplication.

8.9.7. See Section 8.4.3

We have seen that the vector equation (8.56)

$$\mathbf{v}_s = \mathbf{v}_{s'} + \boldsymbol{\omega} \wedge \mathbf{x} \tag{8.109}$$

may be expressed in the S frame (equation 8.58) as

$$(v_s) = (v_{s'}) + (\Omega)(x) \tag{8.110}$$

and in the S' frame (equation 8.51) as

$$(v'_s) = (v'_{s'}) + (\Omega')(x'). \tag{8.111}$$

These two matrix equations may be written explicitly in terms of the measurements that S and S' would make in their respective frames:

$$\left(\frac{\mathrm{d}x}{\mathrm{d}t}\right) = (\tilde{A})\left(\frac{\mathrm{d}x'}{\mathrm{d}t}\right) + (\Omega)(x), \tag{8.112}$$

and (equation 8.48)

$$(A)\left(\frac{\mathrm{d}x}{\mathrm{d}t}\right) = \left(\frac{\mathrm{d}x'}{\mathrm{d}t}\right) + (\Omega')(x'). \tag{8.113}$$

Although the matrix equations (8.112) and (8.113) may appear more complicated than the vector equivalent, there is often an advantage in their direct correspondence with physical measurements. Ultimately they must be interpreted in the same way. For example $(v'_{s'})$ is a vector obtained in the following three stages:

(a) The position vector is projected on to the axes of the frame S' to give its components x'_1, x'_2, x'_3.

(b) From the variation with time of these coordinates, $\mathrm{d}x'_1/\mathrm{d}t$, $\mathrm{d}x'_2/\mathrm{d}t$, $\mathrm{d}x'_3/\mathrm{d}t$ may be found.

(c) Using these three time derivatives as components in the S' frame a vector can be constructed which is defined as $(v'_{s'})$.

Such operational interpretations are unambiguous and immediately apparent from the matrix expression $(\mathrm{d}x'/\mathrm{d}t)$. The difficulty of expressing them in the vector form may be seen from the variety of notations used. Thus the following equations,

$$\frac{\mathrm{d}\mathbf{x}}{\mathrm{d}t} = \frac{\mathrm{d}^*\mathbf{x}}{\mathrm{d}t} + \boldsymbol{\omega} \wedge \mathbf{x},$$

$$\frac{\mathrm{D}\mathbf{x}}{\mathrm{D}t} = \frac{\mathrm{d}\mathbf{x}}{\mathrm{d}t} + \boldsymbol{\omega} \wedge \mathbf{x},$$

$$\frac{\mathrm{d}\mathbf{x}}{\mathrm{d}t} = \dot{\mathbf{x}} + \boldsymbol{\omega} \wedge \mathbf{x},$$

$$\dot{\mathbf{x}} = \frac{\delta\mathbf{x}}{\delta t} + \boldsymbol{\omega} \wedge \mathbf{x},$$

are all in use as ways of expressing equations (8.112) or (8.113).

8.9.8. See Section 8.4.3

Example 26

If the three angular velocity elements $\omega_1, \omega_2, \omega_3$ are grouped as a column matrix, or vector,

$$(\omega) = \begin{pmatrix} \omega_1 \\ \omega_2 \\ \omega_3 \end{pmatrix},$$

this will transform as

$$(\omega') = (A)(\omega). \tag{8.114}$$

If, however, they constitute the antisymmetric matrix

$$(\Omega) = \begin{pmatrix} 0 & -\omega_3 & \omega_2 \\ \omega_3 & 0 & -\omega_1 \\ -\omega_2 & \omega_1 & 0 \end{pmatrix},$$

this transforms as

$$(\Omega') = (A)(\Omega)(\tilde{A}). \tag{8.115}$$

Are the transformations (8.114) and (8.115) consistent?

Let us consider simply one element. According to equation (8.114),

$$\omega_1' = a_{11}\omega_1 + a_{12}\omega_2 + a_{13}\omega_3, \tag{8.116}$$

while equation (8.115) gives

$$\omega'_1 = (a_{22}a_{33} - a_{23}a_{32})\omega_1 + (a_{23}a_{31} - a_{21}a_{33})\omega_2$$
$$+ (a_{21}a_{32} - a_{22}a_{31})\omega_3. \tag{8.117}$$

We must therefore see whether or not the coefficients of ω_1, ω_2 and ω_3 in these two expressions are the same. To do this consider the vector product of the unit vectors defining the frame S':

$$\mathbf{j}' \wedge \mathbf{k}' = \mathbf{i}'.$$

Express this in terms of the unit vectors defining frame S:

$$(a_{21}\mathbf{i} + a_{22}\mathbf{j} + a_{23}\mathbf{k}) \wedge (a_{31}\mathbf{i} + a_{32}\mathbf{j} + a_{33}\mathbf{k}) = a_{11}\mathbf{i} + a_{12}\mathbf{j} + a_{13}\mathbf{k}.$$

Multiplying this out, and using the relationships

$$\mathbf{i} \wedge \mathbf{i} = 0, \qquad \mathbf{i} \wedge \mathbf{j} = \mathbf{k}, \quad \text{etc.,}$$

we have

$$(a_{22}a_{33} - a_{23}a_{32})\mathbf{i} + (a_{23}a_{31} - a_{21}a_{33})\mathbf{j} + (a_{21}a_{32} - a_{22}a_{31})\mathbf{k}$$
$$= a_{11}\mathbf{i} + a_{12}\mathbf{j} + a_{13}\mathbf{k}.$$

Hence

$$a_{22}a_{33} - a_{23}a_{32} = a_{11}, \qquad a_{23}a_{31} - a_{21}a_{33} = a_{12},$$
$$a_{21}a_{32} - a_{22}a_{31} = a_{13}. \tag{8.118}$$

We see from equations (8.118) that equations (8.116) and (8.117) are equivalent, and the value of ω'_1 (and by similar reasoning ω'_2 and ω'_3) is therefore the same whichever of the transformations (8.114) and (8.115) is used. The transformations are therefore consistent.

It must, however, be pointed out that we have considered only transformations between two *right-handed* frames. When this restriction is lifted the two transformations are *not* the same. For example if the two frames have the same axes 2 and 3, but axis 1′ is in the opposite sense from axis 1, then the transformation elements (the cosines of the angles between various pairs of elements) are

$$a_{11} = -1, \qquad a_{22} = a_{33} = 1,$$

$$a_{12} = a_{13} = a_{23} = a_{21} = a_{31} = a_{32} = 0,$$

or

$$(A) = \begin{pmatrix} -1 & 0 & 0 \\ 0 & 1 & 0 \\ 0 & 0 & 1 \end{pmatrix} = (\tilde{A}).$$

In this case the first of the relationships (8.118) is clearly wrong.

Thus when transformations between left- and right-handed frames are included we must distinguish between 'vectors', such as velocity, which are column matrices and obey the transformation law (8.114) and those, such as angular velocity, which are antisymmetric square matrices and obey the transformation law (8.115). The former are known as *polar vectors*, or simply *vectors*, and the latter as *axial vectors*, or *pseudo-vectors*.

8.9.9. See Section 8.7.2

Example 27

A particle released from rest at the earth's surface is observed to move with an initial acceleration whose magnitude and direction are independent of the particle's mass. This acceleration is normally called the gravitational acceleration and is used to describe the earth's gravity. Calculate the differences between 'true' gravity (for a stationary spherical earth) and apparent gravity (for a spinning spherical earth).

Let us use the axes 1″, 2″, 3″ of Section 8.7.2 with origin at Q, the point on the earth's surface under consideration. For a spinning earth the motion of the particle is described by equations (8.93) and (8.90), where ω is the angular speed of the earth and θ is the colatitude of Q.

If the particle is released from Q at rest then

$$(x'')_{\text{initial}} = \left(\frac{\mathrm{d}x''}{\mathrm{d}t}\right)_{\text{initial}} = (0).$$

The only true force is the earth's gravitational attraction, mg, directed towards its centre. Hence

$$(F'') = m\begin{pmatrix} 0 \\ 0 \\ -g \end{pmatrix} = m(g).$$

Equation (8.93) then gives for the initial acceleration

$$\begin{aligned}
\left(\frac{\mathrm{d}^2x''}{\mathrm{d}t^2}\right)_{\text{initial}} &= (g) - (\Omega'')^2(a) \\
&= \begin{pmatrix} 0 \\ 0 \\ -g \end{pmatrix} + \omega^2\begin{pmatrix} \cos^2\theta & 0 & \sin\theta\cos\theta \\ 0 & 1 & 0 \\ \sin\theta\cos\theta & 0 & \sin^2\theta \end{pmatrix}\begin{pmatrix} 0 \\ 0 \\ a \end{pmatrix} \\
&= \begin{pmatrix} a\omega^2\sin\theta\cos\theta \\ 0 \\ a\omega^2\sin^2\theta - g \end{pmatrix}.
\end{aligned}$$

Hence if (g_t) is the true, and (g_a) the apparent, gravitational acceleration, then

$$(g_t) = \begin{pmatrix} 0 \\ 0 \\ -g \end{pmatrix}, \qquad (g_a) = \begin{pmatrix} a\omega^2 \sin\theta\cos\theta \\ 0 \\ a\omega^2 \sin^2\theta - g \end{pmatrix}.$$

The true and apparent gravitational forces are then $m(g_t)$ and $m(g_a)$. The differences are illustrated in Figure 8.10.

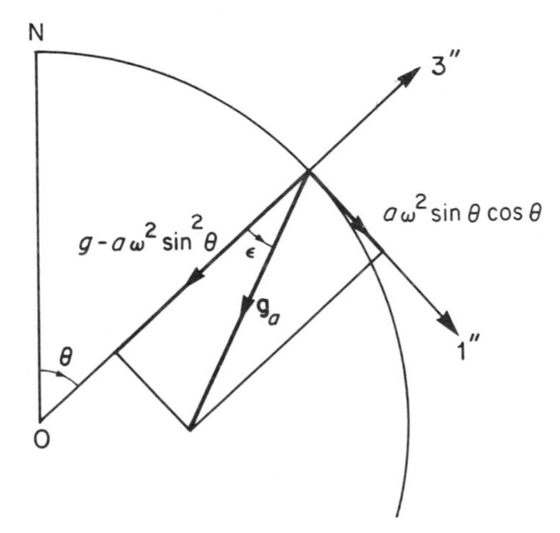

Figure 8.10

The magnitude of apparent gravity is

$$g_a = |g_a| = [(a\omega^2 \sin^2\theta - g)^2 + a^2\omega^4 \sin^2\theta\cos^2\theta]^{1/2}$$
$$= g\left(1 - \frac{a\omega^2}{g}\sin^2\theta + \ldots\right),$$

and its direction is displaced through a small angle ε towards the South Pole (assuming Q to be in the Northern Hemisphere), where

$$\varepsilon \simeq \tan\varepsilon = \frac{a\omega^2 \sin\theta\cos\theta}{g - a\omega^2\sin^2\theta} = \frac{a\omega^2}{2g}\sin 2\theta + \ldots.$$

The greatest change in magnitude is at the equator ($\theta = \pi/2$),

$$g_a - g = -a\omega^2 = -0{\cdot}034 \text{ m s}^{-2}.$$

The greatest change in direction is at latitude 45° ($\theta = \pi/4$),

$$\varepsilon \simeq \frac{a\omega^2}{2g} = 1{\cdot}7 \text{ mrad} \simeq 6'.$$

8.9.10. See Section 8.8

Example 28

Show that a particle of charge q and mass m moving perpendicularly to a uniform magnetic field of induction B will travel in a circular path with angular velocity

$$\omega_c = \frac{qB}{m}$$

According to equation (8.105), it could remain at rest in the frame S' since there is no electric field, and would therefore travel in a circular path in the frame S with angular velocity

$$\omega_L = \frac{qB}{2m}.$$

How do you account for this apparent contradiction?

Suppose that δs is a small element of the path, traversed at a speed v (see Figure 8.11). The force acting on the particle will be qvB perpendicular

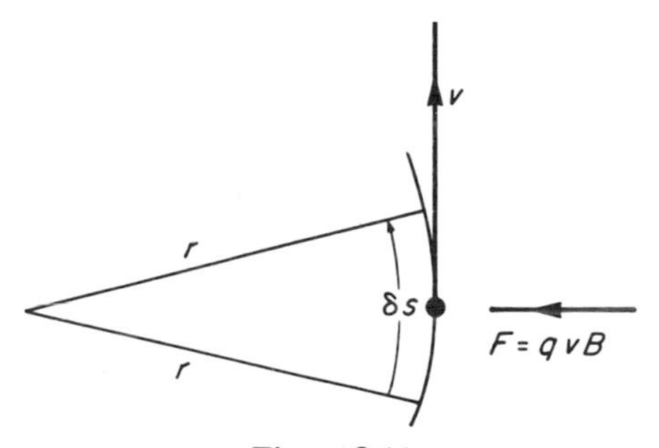

Figure 8.11

to v and the acceleration will therefore be qvB/m. Hence the radius of curvature, r, of δs must be such that

$$\frac{v^2}{r} = \frac{qvB}{m}.$$

Since the force is perpendicular to the velocity it will do no work on the particle and the kinetic energy will be unchanged. Its speed along δs will therefore be constant, and the next element of the path will have the same radius of curvature. Proceeding in this way along the whole path we see that it will be a circle of radius

$$r = \frac{mv}{qB},$$

and the angular speed with which it is traversed is

$$\omega_c = \frac{v}{r} = \frac{qB}{m}.$$

ω_c is known as the *cyclotron frequency.*

The condition (8.106) which is necessary for equation (8.105) to be true is clearly broken since E is zero. To derive the cyclotron frequency using the rotating frame method we must use the complete equation (8.102) which becomes, for zero field and a constant rotation,

$$m(\tilde{A})\left(\frac{d^2x'}{dt^2}\right) = -q\left[(\mathscr{B}) + \frac{2m}{q}(\Omega)\right]\left(\frac{dx}{dt}\right) + m(\Omega)^2(x).$$

Converting this all to the S' frame by the transformations,

$$(A)(x) = (x'), \qquad (A)\left(\frac{dx}{dt}\right) = \left(\frac{dx'}{dt}\right) + (\Omega')(x'),$$

$$(A)(\mathscr{B})(\tilde{A}) = (\mathscr{B}'), \qquad (A)(\Omega)(\tilde{A}) = (\Omega'),$$

we have

$$m\left(\frac{d^2x'}{dt^2}\right) = -q\left[(\mathscr{B}') + \frac{2m}{q}(\Omega')\right]\left[\left(\frac{dx'}{dt}\right) + (\Omega')(x')\right] + m(\Omega')^2(x'). \tag{8.119}$$

If the particle is to be at rest in the S' frame,

$$\left(\frac{dx'}{dt}\right) = 0, \qquad \left(\frac{d^2x'}{dt^2}\right) = 0$$

and equation (8.119) reduces to

$$0 = -q\left[(\mathscr{B}') + \frac{m}{q}(\Omega')\right](\Omega')(x').$$

Hence the rotation which will bring this about is

$$(\Omega') = -\frac{q}{m}(\mathscr{B}'),$$

which *is* in accord with the cyclotron frequency, ω_c.

8.10. Problems

8.1. Show that the Coriolis force acting on a particle, which is otherwise moving freely on a horizontal plane at the earth's surface, will cause it to move in a circle. Also find its radius.

8.2. Show that the force required to counteract the Coriolis force on a mass m, and thus enable it to move with constant horizontal velocity $\mathbf{v}$ relative to the earth, has magnitude

$|2m\omega v \cos\lambda|$ where λ is the colatitude of the position at the earth's surface and ω is the angular speed of the earth. What is the direction of this force?

The mainland of the United Kingdom stretches from latitude 50° N to latitude 58·5° N approximately. If a uniform westerly wind of 20 km hour^{-1} is blowing across the whole country, what is the pressure difference between its northern and southern extremes?

(The radius of the earth is approximately 6400 km; the density of air is 1·3 kg m^{-3}.)

8.3. Relative to the rotating earth a body is projected normally outwards from the surface at latitude 45° N with speed ωa where ω is the angular speed of the earth and a is its radius. Calculate the maximum height above the earth's surface reached by the body using (a) a frame of reference rotating with the earth, (b) the methods of Chapter 6. Account for the differences, if any, between your results.
(*Hint:* use the fact that $a\omega^2 \ll g$ to calculate the results as successive approximations.)

8.4. A particle is released from a point P at a height h above the earth's surface at colatitude λ. Determine its subsequent motion relative to the earth, showing that, throughout it, the angular momentum of the particle about the earth's centre is conserved. Show that if P is 100 m above the surface at latitude 45° N the particle will reach the surface approximately 15 mm east of a plumbline hanging from P.

8.5. Describe how the orientation of an orthogonal Cartesian frame is changed by the successive transformations:

$$(A) = \begin{pmatrix} \cos\phi & \sin\phi & 0 \\ -\sin\phi & \cos\phi & 0 \\ 0 & 0 & 1 \end{pmatrix}, \qquad (B) = \begin{pmatrix} \cos\theta & 0 & -\sin\theta \\ 0 & 1 & 0 \\ \sin\theta & 0 & \cos\theta \end{pmatrix},$$

$$(C) = \begin{pmatrix} \cos\psi & \sin\psi & 0 \\ -\sin\psi & \cos\psi & 0 \\ 0 & 0 & 1 \end{pmatrix}.$$

Hence show that by suitable choice of ϕ, θ, ψ, any reorientation of the axes may be achieved by the transformation

$$(D) = (C)(B)(A).$$

If ϕ and θ are held constant while ψ is allowed to vary, show that the matrix product $(D)(\mathrm{d}\tilde{D}/\mathrm{d}t)$ represents an angular rotation at the rate $\dot{\psi}$ about an axis whose polar angles are θ, ϕ in the original frame. (ϕ, θ, ψ are known as the *Euler angles.* They provide one way of expressing the most general change of direction of a set of axes.)

8.6. If $(\mathscr{X})$ and $(\mathscr{Y})$ are 3×3 antisymmetric matrices and (z) is a 3×1 matrix, state what type of matrix the product $(\mathscr{X})(\mathscr{Y})(z)$ is, and find its first element. Hence prove the vector relationship:

$$\mathbf{x} \wedge (\mathbf{y} \wedge \mathbf{z}) = (\mathbf{x} \cdot \mathbf{z})\mathbf{y} - (\mathbf{x} \cdot \mathbf{y})\mathbf{z}.$$

8.7. If $(\mathscr{X})$ is a 3×3 antisymmetric matrix and (y) is a 3×1 matrix, show that the product $(\mathscr{X})(y)$ transforms between right-handed Cartesian frames as a vector. Hence show that, for most purposes, the vector product $\mathbf{x} \wedge \mathbf{y}$ may be considered as a vector. Prove the distributive law for vector products:

$$(\mathbf{a} + \mathbf{b} + \ldots) \wedge (\mathbf{x} + \mathbf{y} + \ldots) = \mathbf{a} \wedge \mathbf{x} + \mathbf{b} \wedge \mathbf{x} + \ldots + \mathbf{a} \wedge \mathbf{y} + \mathbf{b} \wedge \mathbf{y} + \ldots.$$

Chapter 9

Motion of Rigid Bodies

Nearly all the equations we have used so far to describe the motion of an object have contained only one quantity intrinsic to that object—its mass. In the case of an uncharged particle this is its *only* property, and it is sufficient to describe the response of the particle to any influence the rest of the universe might have upon it. In the case of a collection of particles the total mass performs the same role in determining the motion of the centre-of-mass.

Mass is a scalar quantity—a single number with no implied sense of direction. It can link either two other scalar quantities, for example

$$\mathscr{V}_g = m\phi_g$$

(potential energy = mass × potential),

or two vector quantities, for example

$$\mathbf{F} = m\frac{\mathrm{d}\mathbf{v}}{\mathrm{d}t}$$

(force = mass × acceleration).

In the second type of equation the two vectors are always parallel and the mass does not change with their direction. Thus a force of fixed magnitude F, applied in turn along the axes 1 and 2, would give

$$F = m\frac{\mathrm{d}v_1}{\mathrm{d}t}, \qquad F = m\frac{\mathrm{d}v_2}{\mathrm{d}t}.$$

We could say that the property of the object (in this case its reluctance to move) exerts its influence equally in all directions, or is *isotropic*.

The rotation of a rigid body has characteristics which are very different. Suppose we try to rotate a long thin cylinder by applying a torque, G, firstly about its long axis and secondly about a perpendicular axis through its centre, as shown in Figure 9.1. If ω_1 is the angular velocity about the first axis, we know from Section 5.2.1 that the angular acceleration depends upon the torque according to an equation

$$G = I_1\frac{\mathrm{d}\omega_1}{\mathrm{d}t},$$

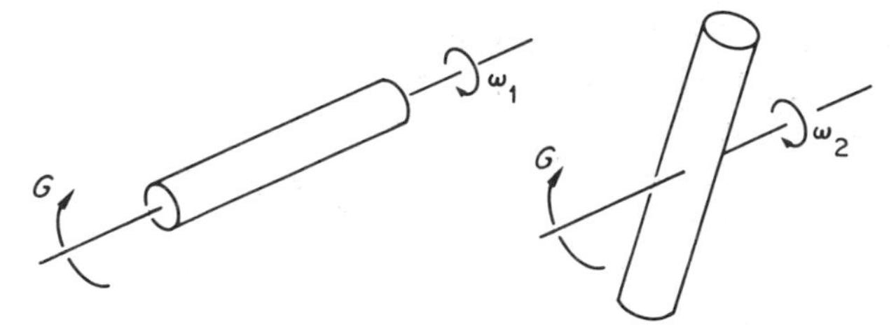

Figure 9.1

where I_1 (the moment of inertia) is the property of the cylinder that measures its reluctance to rotate about the long axis. If the same magnitude of torque is applied to change ω_2 about the second axis we would expect a relationship of the form:

$$G = I_2 \frac{d\omega_2}{dt}.$$

Common experience tells us that it is much easier to set the cylinder spinning in the first case than in the second. Hence

$$I_2 > I_1.$$

This alone shows that we cannot find a rotational equation of general application,

$$\mathbf{G} = I \frac{d\boldsymbol{\omega}}{dt} \quad \text{or} \quad (G) = I\left(\frac{d\omega}{dt}\right),$$

to correspond to the equation for linear motion,

$$\mathbf{F} = m \frac{d\mathbf{v}}{dt} \quad \text{or} \quad (F) = m\left(\frac{dv}{dt}\right).$$

The quantity that describes the inertial properties of a body in respect of rotation will have to be something more complicated than a scalar.

9.1. Rotation about a fixed axis

In Section 5.2 we saw that for a collection of particles, the torque and angular momentum referred to any fixed point O are related by (equation 5.12)

$$\mathbf{G}_{\text{ext}} = \frac{d\mathbf{J}}{dt} \tag{9.1}$$

or

$$\sum_i \mathbf{X}_i \wedge \mathbf{F}_{i\,\text{ext}} = \frac{d}{dt}\left(\sum_i m_i \mathbf{X}_i \wedge \mathbf{V}_i\right),$$

where $\mathbf{X}_i$, $\mathbf{V}_i$ are the position and velocity of the mass m_i, and $\mathbf{F}_{i\,\text{ext}}$ is the external force acting on it.

When a rigid body is rotating about a fixed axis we may take a fixed frame of reference with O1 as the rotation axis (see Figure 9.2). For the observer S using this frame the angular velocity will therefore be

$$\boldsymbol{\omega} \underset{S}{\leftrightarrow} \begin{pmatrix} \omega \\ 0 \\ 0 \end{pmatrix},$$

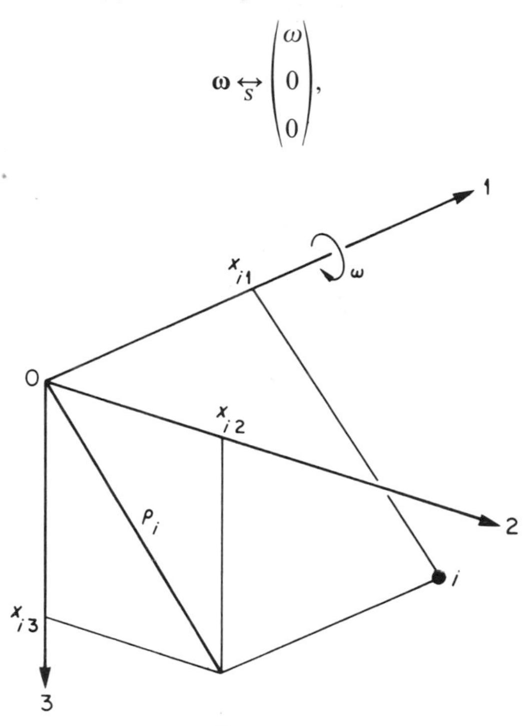

Figure 9.2

and since the body is rigid this will be the same for all parts of the body. Then the velocity of any particle in the body will be

$$\mathbf{V}_i = \boldsymbol{\omega} \wedge \mathbf{X}_i \underset{S}{\leftrightarrow} \begin{pmatrix} \omega_2 X_{i3} - \omega_3 X_{i2} \\ \omega_3 X_{i1} - \omega_1 X_{i3} \\ \omega_1 X_{i2} - \omega_2 X_{i1} \end{pmatrix} = \omega \begin{pmatrix} 0 \\ -X_{i3} \\ X_{i2} \end{pmatrix}.$$

Hence

$$\mathbf{X}_i \wedge \mathbf{V}_i \underset{S}{\leftrightarrow} \omega \begin{pmatrix} X_{i2}^2 + X_{i3}^2 \\ -X_{i1} X_{i2} \\ -X_{i1} X_{i3} \end{pmatrix}$$

and the angular momentum of the particle i will be

$$\mathbf{J}_i \underset{S}{\leftrightarrow} \omega m_i \begin{pmatrix} X_{i2}^2 + X_{i3}^2 \\ -X_{i1}X_{i2} \\ -X_{i1}X_{i3} \end{pmatrix}.$$

The total angular momentum of the body is therefore

$$\mathbf{J} = \sum_i \mathbf{J}_i \underset{S}{\leftrightarrow} \omega \begin{pmatrix} \sum_i m_i(X_{i2}^2 + X_{i3}^2) \\ -\sum_i m_i X_{i1} X_{i2} \\ -\sum_i m_i X_{i1} X_{i3} \end{pmatrix}. \tag{9.2}$$

As the body rotates X_{i1} remains constant while X_{i2} and X_{i3} vary. Thus the quantities

$$I_{21} = -\sum_i m_i X_{i1} X_{i2}, \qquad I_{31} = -\sum_i m_i X_{i1} X_{i3}, \tag{9.3}$$

which are called *products of inertia*, are variables. Since

$$X_{i2}^2 + X_{i3}^2 = \rho_i^2,$$

where ρ_i is the (*fixed*) distance of the ith particle from the rotation axis,

$$I_{11} = \sum_i m_i \rho_i^2, \tag{9.4}$$

which we already know as the *moment of inertia*, is a constant. Using these inertial quantities

$$\mathbf{J} \underset{S}{\leftrightarrow} \omega \begin{pmatrix} I_{11} \\ I_{21} \\ I_{31} \end{pmatrix}. \tag{9.5}$$

Note that $\mathbf{J}$ and $\boldsymbol{\omega}$ are *not* parallel, unlike the corresponding quantities $\mathbf{P}$ and $\mathbf{V}$ for linear motion.

Since $\boldsymbol{\omega}$ is fixed in direction the motion can be completely determined by using the first components of equations (9.1) and (9.5),

$$G_1 = \frac{\mathrm{d}J_1}{\mathrm{d}t} = I_{11}\frac{\mathrm{d}\omega}{\mathrm{d}t}.$$

The speed of any particle is

$$V_i = \omega \rho_i,$$

from which the kinetic energy of the rotating body is

$$\mathscr{T} = \sum_i \frac{m_i}{2} V_i^2 = \frac{\omega^2}{2} \sum_i m_i \rho_i^2 = \tfrac{1}{2} I_{11} \omega^2.$$

Thus for any fixed axis the moment of inertia has a role in respect of rotation that is analogous to that of mass for linear motion. It is clear, however, from the expression (9.4) that, in general, the moment of inertia will change with the axis.

9.2. The skew rotator

After the point particle one of the next simplest objects, and one whose motion demonstrates the basic ideas of the previous sections, is that shown in Figure 9.3. It consists of two point particles, P and Q, each of mass m

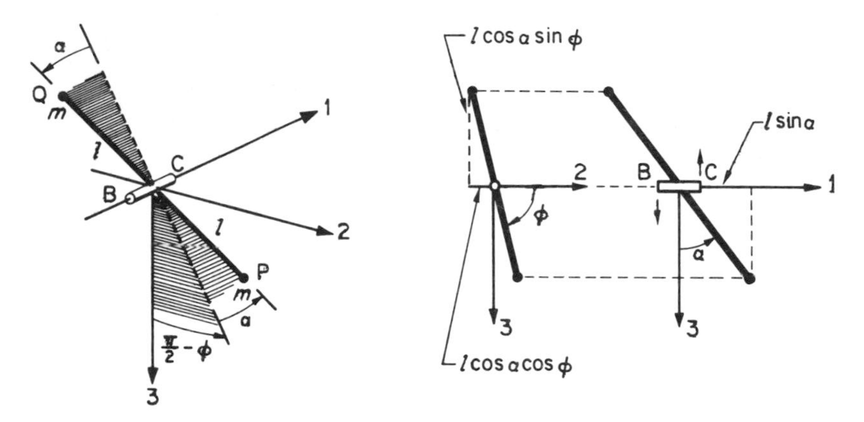

Figure 9.3

fixed on opposite sides by light rods of length l to a bearing BC. These rods make an angle α with the plane perpendicular to BC through its centre. When BC is supported on a fixed axle, the position of the object is determined by the angle, ϕ, that the plane through P, Q and BC makes with the fixed plane through BC containing the axis 2. 1 is the axis of rotation and 3 the third of the right-handed set of Cartesian axes.

This is a model of some practical importance since it gives a fair approximation to the properties of a skew fly-wheel. The angular velocity is

$$\boldsymbol{\omega} \leftrightarrow \omega \begin{pmatrix} 1 \\ 0 \\ 0 \end{pmatrix} = \frac{\mathrm{d}\phi}{\mathrm{d}t} \begin{pmatrix} 1 \\ 0 \\ 0 \end{pmatrix}.$$

The position of the particle P is

$$\mathbf{X}_1 \leftrightarrow \begin{pmatrix} l \sin \alpha \\ l \cos \alpha \cos \phi \\ l \cos \alpha \sin \phi \end{pmatrix}$$

and of Q is

$$\mathbf{X}_2 \leftrightarrow -\begin{pmatrix} l \sin \alpha \\ l \cos \alpha \cos \phi \\ l \cos \alpha \sin \phi \end{pmatrix}.$$

Hence the moment and products of inertia are

$$I_{11} = \sum_i m_i(X_{i2}^2 + X_{i3}^2) = 2ml^2 \cos^2 \alpha,$$

$$I_{21} = -\sum_i m_i X_{i1} X_{i2} = -ml^2 \sin 2\alpha \cos \phi,$$

$$I_{31} = -\sum_i m_i X_{i1} X_{i3} = -ml^2 \sin 2\alpha \sin \phi,$$

and the angular momentum is therefore

$$\mathbf{J} \leftrightarrow \omega \begin{pmatrix} I_{11} \\ I_{21} \\ I_{31} \end{pmatrix} = ml^2 \frac{\mathrm{d}\phi}{\mathrm{d}t} \begin{pmatrix} 2 \cos^2 \alpha \\ -\sin 2\alpha \cos \phi \\ -\sin 2\alpha \sin \phi \end{pmatrix}.$$

Thus if **G** is the torque about O,

$$\mathbf{G} = \frac{\mathrm{d}\mathbf{J}}{\mathrm{d}t},$$

or

$$G_1 = \frac{\mathrm{d}}{\mathrm{d}t}\left(2ml^2 \cos^2 \alpha \frac{\mathrm{d}\phi}{\mathrm{d}t}\right) = 2ml^2 \cos^2 \alpha \frac{\mathrm{d}^2\phi}{\mathrm{d}t^2}, \tag{9.6}$$

$$G_2 = -\frac{\mathrm{d}}{\mathrm{d}t}\left(ml^2 \sin 2\alpha \cos \phi \frac{\mathrm{d}\phi}{\mathrm{d}t}\right) = -ml^2 \sin 2\alpha \frac{\mathrm{d}^2(\sin \phi)}{\mathrm{d}t^2}, \tag{9.7}$$

$$G_3 = -\frac{\mathrm{d}}{\mathrm{d}t}\left(ml^2 \sin 2\alpha \sin \phi \frac{\mathrm{d}\phi}{\mathrm{d}t}\right) = ml^2 \sin 2\alpha \frac{\mathrm{d}^2(\cos \phi)}{\mathrm{d}t^2}. \tag{9.8}$$

If there are no external forces acting other than those at the bearings the torque can arise only from these. A frictionless bearing implies

$$G_1 = 0$$

and therefore, from equation (9.6),

$$\frac{d^2\phi}{dt^2} = 0, \qquad \phi = \omega t + \phi_0,$$

where $\omega = d\phi/dt$ is the constant angular velocity with which the object rotates and ϕ_0 is the initial value of ϕ. Hence

$$G_2 = ml^2 \sin 2\alpha \sin \phi\, \omega^2,$$

$$G_3 = -ml^2 \sin 2\alpha \cos \phi\, \omega^2.$$

Thus there will always be a torque of magnitude $ml^2 \sin 2\alpha\, \omega^2$ perpendicular to the plane through the axis and the two masses. This is qualitatively what we would expect. As they rotate the masses tend to fly outwards and reduce the angle α. This has to be resisted by the torque at the bearing BC as shown in Figure 9.3. Only when the masses are symmetrically placed,

$$\alpha = 0, \pi, \qquad \sin 2\alpha = 0,$$

does this torque vanish.

9.3. Rotation about a fixed point

To deal generally with this we apply the ideas introduced in Chapter 8 relating to measurements in two frames S and S' with a common origin O. The frame S is fixed (or is an inertial frame), while S' moves with the body so that any particle of the body has a constant position (x') in the S' framc. The velocity measured by S' will therefore be zero, and will have zero components in *any* frame. Hence in the S frame

$$(v_{s'}) = (0).$$

In the S frame the particle's velocity is therefore (see equation 8.58)

$$(v_s) = \left(\frac{dx}{dt}\right) = (\Omega)(x), \tag{9.9}$$

which is equivalent to

$$\mathbf{v}_s = \boldsymbol{\omega} \wedge \mathbf{x}.$$

Since we can rewrite the latter as

$$\mathbf{v}_s = -\mathbf{x} \wedge \boldsymbol{\omega},$$

we can also rewrite equation (9.9) as

$$(v_s) = \left(\frac{dx}{dt}\right) = -(\mathscr{X})(\omega), \tag{9.10}$$

where $(\mathscr{E})$ is the antisymmetric matrix

$$(\mathscr{X}) = \begin{pmatrix} 0 & -x_3 & x_2 \\ x_3 & 0 & -x_1 \\ -x_2 & x_1 & 0 \end{pmatrix}$$

and (ω) is the column matrix

$$(\omega) = \begin{pmatrix} \omega_1 \\ \omega_2 \\ \omega_3 \end{pmatrix}.$$

9.3.1. Angular momentum

Thus if $\mathbf{X}_i$ is the position vector of a mass m_i, its velocity in the S frame is

$$(V_i) = -(\mathscr{X}_i)(\omega), \tag{9.11}$$

(dropping the suffix $_S$) and its angular momentum, in vector form

$$\mathbf{J}_i = m_i\mathbf{X}_i \wedge \mathbf{V}_i,$$

will be, in matrix form,

$$\begin{aligned}(J_i) = m(\mathscr{X}_i)(V_i) &= -m_i(\mathscr{X}_i)^2(\omega) \\ &= (\mathscr{I}_i)(\omega),\end{aligned} \tag{9.12}$$

where

$$\begin{aligned}(\mathscr{I}_i) &= -m_i \begin{pmatrix} 0 & -X_{i3} & X_{i2} \\ X_{i3} & 0 & -X_{i1} \\ -X_{i2} & X_{i1} & 0 \end{pmatrix} \begin{pmatrix} 0 & -X_{i3} & X_{i2} \\ X_{i3} & 0 & -X_{i1} \\ -X_{i2} & X_{i1} & 0 \end{pmatrix} \\ &= m_i \begin{pmatrix} X_{i2}^2 + X_{i3}^2 & -X_{i1}X_{i2} & -X_{i1}X_{i3} \\ -X_{i2}X_{i1} & X_{i3}^2 + X_{i1}^2 & -X_{i2}X_{i3} \\ -X_{i3}X_{i1} & -X_{i3}X_{i2} & X_{i1}^2 + X_{i2}^2 \end{pmatrix}.\end{aligned} \tag{9.13}$$

Hence the total angular momentum is

$$(J) = \sum_i (J_i) = \sum_i (\mathscr{I}_i)(\omega) = (\mathscr{I})(\omega) \tag{9.14}$$

where $(\mathscr{I})$ is a symmetric matrix known as the *inertia matrix* (or tensor). The three positive diagonal elements are the *moments of inertia*:

$$I_{11} = \sum_i m_i(X_{i2}^2 + X_{i3}^2) = \sum_i m_i\rho_{i1}^2, \quad \text{etc.} \tag{9.15}$$

where ρ_{i1} is the distance of the ith particle from axis 1. The off-diagonal elements are the *products of inertia*:

$$I_{12} = I_{21} = -\sum_i m_i X_{i1} X_{i2}, \quad \text{etc.} \tag{9.16}$$

9.3.2. Kinetic energy

Since the transpose of the column matrix

$$(V_i) = \begin{pmatrix} V_{i1} \\ V_{i2} \\ V_{i3} \end{pmatrix}$$

is the row matrix

$$(\tilde{V}_i) = (V_{i1} \quad V_{i2} \quad V_{i3}),$$

their product is

$$(\tilde{V}_i)(V_i) = V_{i1}^2 + V_{i2}^2 + V_{i3}^2.$$

Hence the kinetic energy of a mass m_i moving with this velocity is

$$\mathscr{T}_i = \frac{m_i}{2}(V_{i1}^2 + V_{i2}^2 + V_{i3}^2) = \frac{m_i}{2}(\tilde{V}_i)(V_i).$$

Using equation (9.11), and remembering that $(\mathscr{X}_i)$ is antisymmetric,

$$\begin{aligned} \mathscr{T}_i &= \frac{m_i}{2}[\widetilde{(\mathscr{X}_i)(\omega)}][(\mathscr{X}_i)(\omega)] \\ &= \frac{m_i}{2}(\tilde{\omega})(\tilde{\mathscr{X}}_i)(\mathscr{X}_i)(\omega) \\ &= -\frac{m_i}{2}(\tilde{\omega})(\mathscr{X}_i)^2(\omega) \\ &= \tfrac{1}{2}(\tilde{\omega})(\mathscr{I}_i)(\omega). \end{aligned}$$

Summing overall the particles of the body we may express its total kinetic energy as

$$\mathscr{T} = \tfrac{1}{2}(\tilde{\omega})(\mathscr{I})(\omega). \tag{9.17}$$

9.3.3. Equations of motion

Expressed in the S frame the results (9.14) and (9.17) do not separate properties of the body from those of the motion, since the elements of the inertia matrix may vary with time. However, suppose we now find the components of the angular momentum $\mathbf{J}_i$ in the S' frame:

$$(J_i') = (A)(J_i) = -m_i(A)(\mathscr{X}_i)^2(\omega)$$
$$= -m_i(A)(\mathscr{X}_i)(\tilde{A})(A)(\mathscr{X}_i)(\tilde{A})(A)(\omega). \tag{9.18}$$

Now

$$(A)(\mathscr{X}_i)(\tilde{A}) = (\mathscr{X}_i'),$$
$$(A)(\omega) = (\omega')$$

and therefore

$$(J_i') = (\mathscr{I}_i')(\omega'),$$

where

$$(\mathscr{I}_i') = m_i\begin{pmatrix} X_{i2}'^2 + X_{i3}'^2 & -X_{i1}'X_{i2}' & -X_{i1}'X_{i3}' \\ & \text{etc.} & \end{pmatrix}.$$

The elements of $(\mathscr{I}_i')$ are now all constants, since X_{i1}', X_{i2}', X_{i3}' are the coordinates of the ith particle in a frame fixed in the body. Hence, summing over all particles,

$$(J') = (\mathscr{I}')(\omega'), \tag{9.19}$$

where the inertia matrix $(\mathscr{I}')$ is a constant, determined only by the body and the position of the axes within it. Its diagonal elements are the moments of inertia

$$I_{11}' = \sum_i m_i(X_{i2}'^2 + X_{i3}'^2) = \sum_i m_i\rho_{i1}'^2, \quad \text{etc.,} \tag{9.20}$$

and the off-diagonal elements are the products of inertia

$$I_{12}' = I_{21}' = -\sum_i m_i X_{i1}'X_{i2}', \quad \text{etc.} \tag{9.21}$$

(See Section 9.6.1, page 272.) Equation (9.17) may be expressed as

$$\mathscr{T} = \tfrac{1}{2}[\widetilde{(\tilde{A})(\omega')}](\mathscr{I})[(\tilde{A})(\omega')]$$
$$= \tfrac{1}{2}(\tilde{\omega}')(A)(\mathscr{I})(\tilde{A})(\omega')$$
$$= \tfrac{1}{2}(\tilde{\omega}')(\mathscr{I}')(\omega'). \tag{9.22}$$

The relationships (9.19) and (9.22) are similar in form to those describing linear motion,

$$(P) = M(V)$$

and

$$\mathscr{T} = \tfrac{1}{2}M(\tilde{V})(V),$$

and show that the rotational equivalent of inertial mass is the inertia matrix. However, $(\mathscr{I}')$ is not a scalar. In general, nine elements are required to describe it and one of its properties is to effect a *change of direction* between **J** and $\boldsymbol{\omega}$.

The equation of (rotational) motion,

$$\mathbf{G} = \frac{\mathrm{d}\mathbf{J}}{\mathrm{d}t},$$

which, in the inertial S frame, means

$$(G) = \left(\frac{\mathrm{d}J}{\mathrm{d}t}\right),$$

can, from equations (9.18) and (9.19), be written as

$$\begin{aligned}(G) &= \frac{\mathrm{d}}{\mathrm{d}t}[(\tilde{A})(J')] = \frac{\mathrm{d}}{\mathrm{d}t}[(\tilde{A})(\mathscr{I}')(\omega')] \\ &= (\tilde{A})(\mathscr{I}')\left(\frac{\mathrm{d}\omega'}{\mathrm{d}t}\right) + \left(\frac{\mathrm{d}\tilde{A}}{\mathrm{d}t}\right)(\mathscr{I}')(\omega').\end{aligned}$$

Therefore, remembering that

$$(A)\left(\frac{\mathrm{d}\tilde{A}}{\mathrm{d}t}\right) = (\Omega'),$$

the equation of motion in the S' frame is

$$(G') = (A)(G) = (\mathscr{I}')\left(\frac{\mathrm{d}\omega'}{\mathrm{d}t}\right) + (\Omega')(\mathscr{I}')(\omega'). \tag{9.23}$$

In practice, the usefulness of this equation depends very much upon the choice of axes in the body. They can be chosen to simplify the form of $(\mathscr{I}')$ and reduce the complexity considerably.

9.4. Inertia matrix

There are two general properties of the inertia matrix which can greatly simplify its calculation.

9.4.1. Parallel axes

Suppose the centre-of-mass, C, is at (X'_1, X'_2, X'_3) in the S' frame and that the ith particle is at (x_{i1}, x_{i2}, x_{i3}) in a frame whose axes are parallel with those of S' and whose origin is at C. Hence

$$X'_{i1} = X'_1 + x_{i1}, \quad \text{etc.}$$

A typical element of the inertia matrix is

$$\begin{aligned}\mathscr{I}'_{12} &= -\sum_i m_i X'_{i1} X'_{i2}\\ &= -\sum_i m_i (X'_1 + x_{i1})(X'_2 + x_{i2})\\ &= -\left(\sum_i m_i\right) X'_1 X'_2 - \left(\sum_i m_i x_{i1}\right) X'_2 - \left(\sum_i m_i x_{i2}\right) X'_1 - \sum_i m_i x_{i1} x_{i2}.\end{aligned}$$

Now

$$\sum_i m_i = M,$$

the total mass, and since C is the centre-of-mass,

$$\sum_i m_i x_{i1} = \sum_i m_i x_{i2} = 0.$$

Thus

$$\mathscr{I}'_{12} = -MX'_1 X'_2 - \sum_i m_i x_{i1} x_{i2}.$$

Similar results apply to all the elements of $(\mathscr{I}')$. Hence

$$(\mathscr{I}') = (\mathscr{I}'_{\mathrm{cm}}) + (\mathscr{I}'_{\mathrm{rel}}) \tag{9.24}$$

where $(\mathscr{I}'_{\mathrm{cm}})$ is the inertia matrix of a single particle of mass M at the centre-of-mass and $(\mathscr{I}'_{\mathrm{rel}})$ is the inertia matrix of the body calculated for parallel axes through the centre-of-mass. (See Section 9.6.2, page 274.)

9.4.2. *Principal axes*

Clearly all the elements of $(\mathscr{I}')$ are real. Also, since

$$\mathscr{I}'_{12} = -\sum_i m_i X'_{i1} X'_{i2} = \mathscr{I}'_{21}, \quad \text{etc.,}$$

elements reflected across the leading diagonal of the matrix are equal—the matrix is symmetric. There is a fundamental theorem in matrix algebra which states that a real symmetric matrix can always be transformed by suitable choice of perpendicular axes, with the same origin, to the *diagonal* form:

$$(\mathscr{I}^0) = \begin{pmatrix} I_1 & 0 & 0 \\ 0 & I_2 & 0 \\ 0 & 0 & I_3 \end{pmatrix}. \tag{9.25}$$

For these axes, known as *principal axes*, the products of inertia all vanish and only moments of inertia (the *principal* moments of inertia) remain. The angular momentum and the energy then take on the simple forms:

$$(J^0) = (\mathscr{I}^0)(\omega^0) = \begin{pmatrix} I_1 & 0 & 0 \\ 0 & I_2 & 0 \\ 0 & 0 & I_3 \end{pmatrix} \begin{pmatrix} \omega_1^0 \\ \omega_2^0 \\ \omega_3^0 \end{pmatrix} = \begin{pmatrix} I_1\omega_1^0 \\ I_2\omega_2^0 \\ I_3\omega_3^0 \end{pmatrix}, \tag{9.26}$$

$$\begin{aligned} \mathscr{T} &= \tfrac{1}{2}(\tilde{\omega}^0)(\mathscr{I}^0)(\omega^0) \\ &= \tfrac{1}{2}(\omega_1^0 \quad \omega_2^0 \quad \omega_3^0) \begin{pmatrix} I_1\omega_1^0 \\ I_2\omega_2^0 \\ I_3\omega_3^0 \end{pmatrix} = \tfrac{1}{2}(I_1\omega_1^{0^2} + I_2\omega_2^{0^2} + I_3\omega_3^{0^2}). \end{aligned} \tag{9.27}$$

(See Sections 9.6.3, page 276, and 9.6.4, page 277.)

9.4.3. Symmetrical bodies

The principal axes theorem enables us to start with any Cartesian frame, to calculate the inertia matrix for this frame and then to use this to find the principal axes and the principal moments of inertia. However, in practice, it is often possible, and much simpler, to base a search for the principal axes upon the symmetry characteristics of the body under consideration.

Consider, for example, the cylinder shown in Figure 9.4. This has one plane of symmetry which intersects it in the rectangle ABCD, and another

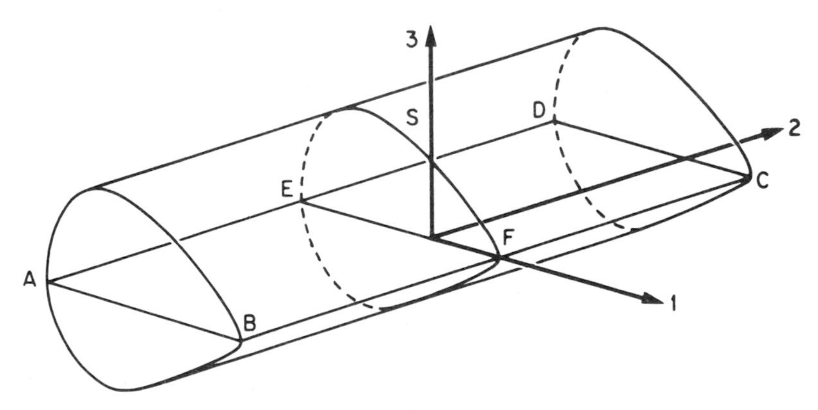

Figure 9.4

which intersects it in the plane curve S. Suppose these two planes intersect in the line EF. Then, without any detailed calculation, we can assert that a set of principal axes may be formed from EF (axis 1), any line perpendicular to EF in the plane ABCD (axis 2), the third principal axis forming an orthogonal set with these two (axis 3).

To prove this, consider first any slice of the body parallel with the plane 31. This will have the shape shown in Figure 9.5a. Every element in this with coordinates (X_1, X_2, X_3) will have a corresponding element with coordinates $(X_1, X_2, -X_3)$. Their contributions to the products of inertia, I_{13} and I_{31}, will therefore cancel. The contributions of every such slice will therefore be zero and

$$I_{13} = I_{31} = 0.$$

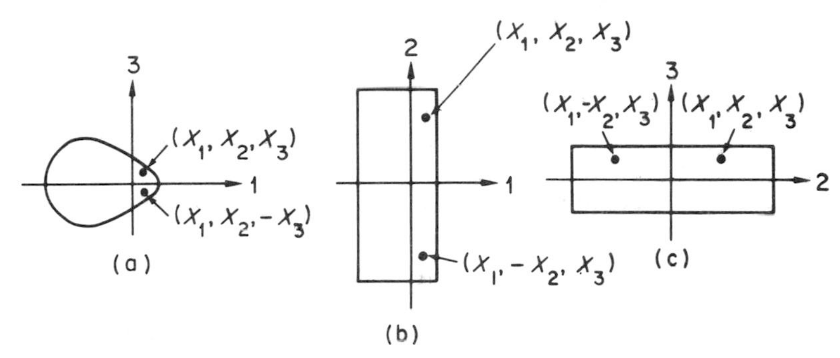

Figure 9.5

Any slice parallel with the plane 12 will have a shape like that of Figure 9.5b. Any element of this with coordinates (X_1, X_2, X_3) will match one with coordinates $(X_1, -X_2, X_3)$. Similar reasoning to the above then proves that

$$I_{12} = I_{21} = 0.$$

Finally, consider slices of the body parallel with the plane 23. These will have shapes like that of Figure 9.5c. Any element with coordinates (X_1, X_2, X_3) matches one with coordinates $(X_1, -X_2, X_3)$. Hence

$$I_{23} = I_{32} = 0.$$

Thus all the products of inertia vanish and the axes 1, 2, 3 are principal axes.

9.4.4. Routh's rule

When a body of uniform density has three perpendicular planes of symmetry, the three lines in which they intersect will be a set of principal axes. It is then often possible to simplify calculation of the moments of inertia by a process known as *Routh's rule*. Suppose, for example, we have an elliptical cylinder. There are three obvious planes of symmetry, giving an orthogonal set of principal axes through the centre. Any particle in the cylinder will have coordinates satisfying

$$\frac{X_1^2}{a_1^2} + \frac{X_2^2}{a_2^2} \leqslant 1, \qquad X_3^2 \leqslant a_3^2, \tag{9.28}$$

where $2a_1, 2a_2$ are the major and minor diameters of the elliptic cross-section of the cylinder and $2a_3$ is its length. The first principal moment of inertia is

$$I_1 = \rho \iiint (X_2^2 + X_3^2)\,\mathrm{d}X_1\,\mathrm{d}X_2\,\mathrm{d}X_3 \tag{9.29}$$

where ρ is the density. By using the variables

$$\eta_1 = \frac{X_1}{a_1}, \qquad \eta_2 = \frac{X_2}{a_2}, \qquad \eta_3 = \frac{X_3}{a_3}$$

we may rewrite this moment of inertia as

$$I_1 = \rho a_1 a_2 a_3 \iiint (a_2^2\eta_2^2 + a_3^2\eta_3^2)\,\mathrm{d}\eta_1\,\mathrm{d}\eta_2\,\mathrm{d}\eta_3, \tag{9.30}$$

where, from equations (9.28), the integration is over values of η_1, η_2, η_3 satisfying

$$\eta_1^2 + \eta_2^2 \leqslant 1, \qquad \eta_3^2 \leqslant 1. \tag{9.31}$$

There will be similar expressions for I_2 and I_3. We can see that they may be expressed in the form

$$I_1 = P_2 + P_3, \qquad I_2 = P_3 + P_1, \qquad I_3 = P_1 + P_2, \tag{9.32}$$

where

$$P_1 = \rho a_1^3 a_2 a_3 \iiint \eta_1^2\,\mathrm{d}\eta_1\,\mathrm{d}\eta_2\,\mathrm{d}\eta_3, \quad \text{etc.} \tag{9.33}$$

Now if we consider η_1, η_2, η_3 as the coordinates of a point, equations (9.31) show that this must be a point in or on a *circular cylinder* with diameter 2 and length 2. Let us call this the *unit cylinder*, and denote its volume by V_u. Then P_1 may be written as

$$P_1 = a_1^2\left(\frac{1}{V_u}\iiint \eta_1^2\,\mathrm{d}\eta_1\,\mathrm{d}\eta_2\,\mathrm{d}\eta_3\right) \times (\rho a_1 a_2 a_3 V_u). \tag{9.34}$$

The first factor in brackets is the mean squared distance of all the points in the unit cylinder from the plane 23. We shall call this the *extension*, λ_1, of the unit cylinder about the first plane of symmetry. (This is analogous to the square of the standard deviation, σ^2, of a statistical distribution, which measures its spread about the mean.) The second term is

$$\begin{aligned}\rho a_1 a_2 a_3 V_u &= \rho \iiint a_1 a_2 a_3\,\mathrm{d}\eta_1\,\mathrm{d}\eta_2\,\mathrm{d}\eta_3 \\ &= \rho \iiint \mathrm{d}X_1\,\mathrm{d}X_2\,\mathrm{d}X_3 \\ &= \rho V = M,\end{aligned} \tag{9.35}$$

where V and M are the volume and mass of the elliptical cylinder. (Note that since

$$V_u = \pi \times 1^2 \times 2 = 2\pi,$$

equation (9.35) provides a simple way of calculating the volume and mass of an elliptical cylinder.)

Then

$$P_1 = M\lambda_1 a_1^2,$$

with similar results for P_2 and P_3. From equation (9.32) the principal moments of inertia are therefore

$$I_1 = M(\lambda_2 a_2^2 + \lambda_3 a_3^2), \qquad I_2 = M(\lambda_3 a_3^2 + \lambda_1 a_1^2),$$
$$I_3 = M(\lambda_1 a_1^2 + \lambda_2 a_2^2). \tag{9.36}$$

A straightforward calculation based on the unit cylinder shows that

$$\lambda_1 = \lambda_2 = \tfrac{1}{4}, \qquad \lambda_3 = \tfrac{1}{3}. \tag{9.37}$$

An exactly similar argument can be used to show that any body with three perpendicular planes of symmetry and three 'diameters' $2a_1$, $2a_2$ and $2a_3$ will have principal moments of inertia of the form (9.36), where the λ's may be calculated from the corresponding 'unit body' with 'diameters' 2, 2 and 2. Examples are the ellipsoid and the rectangular box (see Section 9.6.4, page 277).

9.5. Principal axes motion

Using equation (9.23) referred to principal axes fixed in the body,

$$(G^0) = (\mathscr{I}^0)\left(\frac{d\omega^0}{dt}\right) + (\Omega^0)(\mathscr{I}^0)(\omega^0) \tag{9.38}$$

or

$$\begin{pmatrix} G_1^0 \\ G_2^0 \\ G_3^0 \end{pmatrix} = \begin{pmatrix} I_1 & 0 & 0 \\ 0 & I_2 & 0 \\ 0 & 0 & I_3 \end{pmatrix} \begin{pmatrix} d\omega_1^0/dt \\ d\omega_2^0/dt \\ d\omega_3^0/dt \end{pmatrix} + \begin{pmatrix} 0 & -\omega_3^0 & \omega_2^0 \\ \omega_3^0 & 0 & -\omega_1^0 \\ -\omega_2^0 & \omega_1^0 & 0 \end{pmatrix} \begin{pmatrix} I_1\omega_1^0 \\ I_2\omega_2^0 \\ I_3\omega_3^0 \end{pmatrix}.$$

In separated form,

$$G_1 = I_1 \frac{d\omega_1^0}{dt} - (I_2 - I_3)\omega_2^0\omega_3^0, \quad \text{etc.,} \tag{9.39}$$

which are known as *Euler's equations.*

9.5.1. *Motion of a symmetrical body*

Symmetrical here means that two of the moments of inertia are equal, for example

$$I_1 = I_2 = I.$$

If all sections through a body perpendicular to axis 3 are circular discs centred on this axis, this symmetry property is true for any pair of axes 1 and 2 completing an orthogonal frame. Figure 9.6 shows that for every particle at a

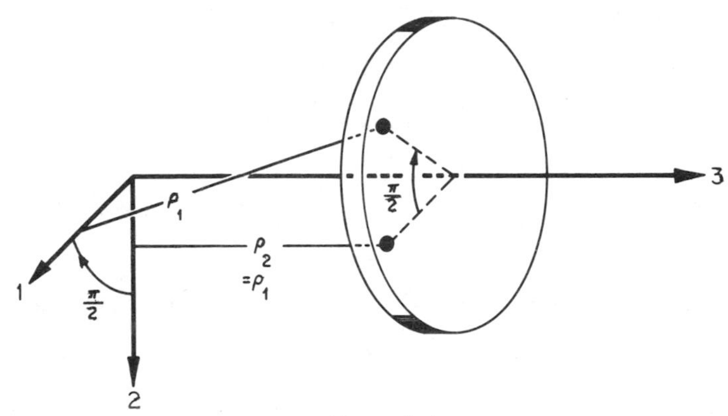

Figure 9.6

distance ρ, from axis 1 there is another (advanced by $\pi/2$) at the same distance from axis 2. The moments of inertia about these two axes are therefore equal. The top and gyroscope come into this category. Then equations (9.39) reduce to

$$\left.\begin{aligned} G_1^0 &= I\frac{\mathrm{d}\omega_1^0}{\mathrm{d}t} - (I - I_3)\omega_2^0\omega_3^0, \\ G_2^0 &= I\frac{\mathrm{d}\omega_2^0}{\mathrm{d}t} - (I_3 - I)\omega_3^0\omega_1^0, \\ G_3^0 &= I_3\frac{\mathrm{d}\omega_3^0}{\mathrm{d}t}. \end{aligned}\right\} \quad (9.40)$$

Thus if G_3^0 is known ω_3^0 can be determined from the third equation alone. If G_1^0 and G_2^0 are also known, they and ω_3^0 determine ω_1^0 and ω_2^0 from the first and second. The simplest case is no torque at all:

$$G_1^0 = G_2^0 = G_3^0 = 0.$$

Then

$$I\frac{\mathrm{d}\omega_1^0}{\mathrm{d}t} = (I - I_3)\omega_3^0\omega_2^0,$$

$$I\frac{\mathrm{d}\omega_2^0}{\mathrm{d}t} = -(I - I_3)\omega_3^0\omega_1^0,$$

$$\omega_3^0 = \text{constant}.$$

The solution to these equations is

$$\omega_1^0 = \omega^0 \sin(qt + \beta), \qquad \omega_2^0 = \omega^0 \cos(qt + \beta),$$

where

$$q = \frac{I - I_3}{I}\omega_3^0$$

and ω^0 and β are adjustable constants.

The direction of the angular velocity (i.e. the axis of rotation) itself rotates about the axis 3, or *precesses*, with angular velocity q.

If a body has the rotational symmetry of a gyroscope or top, it is no longer necessary to use axes fixed in the body to obtain the equations (9.40). The reason for doing so arose from the expression for angular momentum in the inertial frame S (equation 9.14):

$$(J) = (\mathscr{I})(\omega). \tag{9.41}$$

When this is substituted into the equation of motion we have

$$(G) = \left(\frac{\mathrm{d}J}{\mathrm{d}t}\right) = (\mathscr{I})\left(\frac{\mathrm{d}\omega}{\mathrm{d}t}\right) + \left(\frac{\mathrm{d}\mathscr{I}}{\mathrm{d}t}\right)(\omega). \tag{9.42}$$

In general $(\mathrm{d}\mathscr{I}/\mathrm{d}t)$ will be a non-zero matrix which is very difficult to calculate.

However, for a rotating frame, S', not necessarily fixed in the body, equation (9.42) will be replaced by

$$(G') = \left(\frac{\mathrm{d}J'}{\mathrm{d}t}\right) + (\Sigma'_{s'})(J')$$

$$= \frac{\mathrm{d}}{\mathrm{d}t}[(\mathscr{I}')(\omega')] + (\Sigma'_{s'})(\mathscr{I}')(\omega'). \tag{9.43}$$

We must be careful to distinguish between $(\Sigma'_{s'})$, which describes in the S' frame the angular velocity of S' relative to S, and (ω'), which describes, also in the S' frame, the angular velocity of the body relative to S. Now if S' can be chosen so that $(\mathscr{I}')$ is constant, equation (9.43) reduces to

$$(G') = (\mathscr{I}')\left(\frac{\mathrm{d}\omega'}{\mathrm{d}t}\right) + (\Sigma'_{s'})(\mathscr{I}')(\omega'). \tag{9.44}$$

An obvious choice for S' is a frame fixed in the body, as in Section 9.3.3. In this case not only is $(\mathscr{I}')$ constant, but the *same* angular velocity (that of the body relative to S) is described by both $(\Sigma'_{s'})$ and (ω'). $(\Sigma'_{s'})$ is just the antisymmetric form of (ω'), and equation (9.44) simplifies to equation (9.23).

However, when a body has rotational symmetry $(\mathscr{I}')$ will be constant for *any* orthogonal frame, provided one axis coincides with the axis of symmetry of the body. For suppose this is the axis $3'$. Then any element in the body remains at the same distance from this axis and I'_3 will be constant, as the body rotates relative to the frame. Individual elements will change their distances from the axes 1′ and 2′ as the body rotates, but equal ones move in to take their place so that on summation over the whole body I'_1 and I'_2 will also remain constant (and equal to each other), whatever the orientation of the body. These three moments of inertia are the same as the three principal moments of inertia for axes fixed in the body:

$$I'_1 = I'_2 = I, \qquad I'_3 = I_3.$$

The symmetry of the body ensures that the products of inertia are not only constant, but zero. Then equation (9.44) becomes, in detailed form,

$$(G') = \begin{pmatrix} I \ \mathrm{d}\omega'_1/\mathrm{d}t \\ I \ \mathrm{d}\omega'_2/\mathrm{d}t \\ I_3 \, \mathrm{d}\omega'_3/\mathrm{d}t \end{pmatrix} + \begin{pmatrix} 0 & -\sigma'_3 & \sigma'_2 \\ \sigma'_3 & 0 & -\sigma'_1 \\ -\sigma'_2 & \sigma'_1 & 0 \end{pmatrix} \begin{pmatrix} I\omega'_1 \\ I\omega'_2 \\ I_3\omega'_3 \end{pmatrix}. \qquad (9.45)$$

σ'_1, σ'_2, σ'_3 are the components in the S' frame of the angular velocity of S' relative to S. Hence

$$\left.\begin{aligned} G'_1 &= I\frac{\mathrm{d}\omega'_1}{\mathrm{d}t} - I\omega'_2\sigma'_3 + I_3\omega'_3\sigma'_2, \\ G'_2 &= I\frac{\mathrm{d}\omega'_2}{\mathrm{d}t} + I\omega'_1\sigma'_3 - I_3\omega'_3\sigma'_1, \\ G'_3 &= I_3\frac{\mathrm{d}\omega'_3}{\mathrm{d}t} - I(\omega'_1\sigma'_2 - \omega'_2\sigma'_1). \end{aligned}\right\} \qquad (9.46)$$

To see how these equations may be used let us consider the case of a steadily precessing top (Figure 9.7a). The top spins with a constant angular

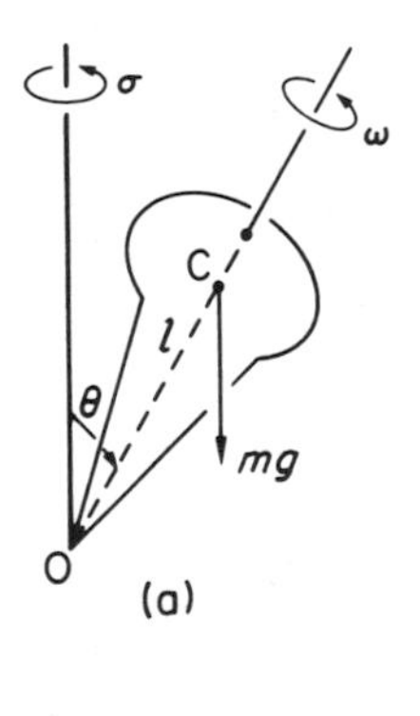

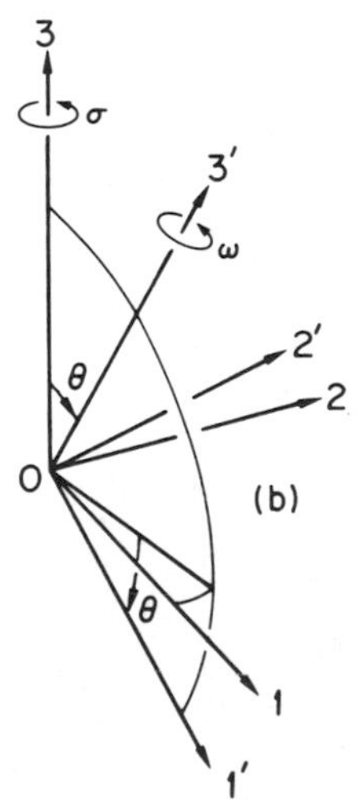

Figure 9.7

speed ω relative to S' about its axis, and the axis itself remains at a fixed angle to the vertical and rotates about it at a constant angular speed σ. (See Section 9.6.5, page 281.) In Figure 9.7b, 1, 2, 3 are the axes of the inertial frame S and $1', 2', 3'$ of the rotating frame S'. $3'$ is the axis of the top and $1'$ lies in the vertical plane through $3'$. Both frames have as origin the point of support of the top, O. Then the components in the S' frame of the angular velocities are

$$\sigma_1' = -\sigma \sin\theta, \qquad \sigma_2' = 0, \qquad \sigma_3' = \sigma\cos\theta$$

$$\omega_1' = -\sigma \sin\theta, \qquad \omega_2' = 0, \qquad \omega_3' = \omega + \sigma\cos\theta.$$

For these to be constant, equations (9.46) require

$$G_1' = G_3' = 0,$$

$$G_2' = I_3\sigma\omega\sin\theta + (I_3 - I)\sigma^2\sin\theta\cos\theta. \tag{9.47}$$

If the support point at O is smooth the force exerted by the support on the top will pass through O and therefore have no moment about O. The only external force which can have a moment is the gravitational force, mg, acting vertically downwards through C, the centre-of-mass. If OC has length l, the torque on the top is

$$G_1' = G_3' = 0, \qquad G_2' = mgl\sin\theta. \tag{9.48}$$

From equations (9.47) and (9.48) we see that, provided the top is not vertical ($\theta \neq 0$), steady motion can occur when

$$(I - I_3)\sigma^2\cos\theta - I_3\sigma\omega + mgl = 0. \tag{9.49}$$

(See Section 9.6.6, page 282.) Thus for a given top (which determines m, l, I and I_3) and a given inclination and spin (θ and ω) there are two possible rates of precession, σ, given by

$$\sigma = \frac{I_3\omega}{2(I - I_3)\cos\theta}\left[1 \pm \sqrt{1 - \frac{4mgl(I - I_3)\cos\theta}{I_3^2\omega^2}}\right]. \tag{9.50}$$

We normally think of a top as spinning rapidly and therefore possessing large ω. For this case the solutions (9.50) have the approximate values

$$\sigma_1 = \frac{I_3\omega}{(I - I_3)\cos\theta}, \qquad \sigma_2 = \frac{mgl}{I_3\omega}.$$

The first of these describes a rapid precession at a rate of the same order as ω. Note that it is independent of g and is therefore a precession that can occur even in the absence of external forces. Depending upon the relative values of I and I_3, σ may or may not have the same sign as ω. The second solution is the more familiar one of a slow precession about the (gravitational) vertical. It is inversely proportional to ω but is independent of θ, and always has the same sign as ω.

The derivation of equation (9.49) and its solution (9.50) does not, however, require that ω should be large. The only physical requirement is that σ should be real and ω must therefore satisfy

$$\omega^2 = \geqslant \frac{4mgl(I - I_3)\cos\theta}{I_3^2}.$$

A top usually has the property

$$I > I_3.$$

Then if the top is above its support point ($0 < \theta < \pi/2$) there is a minimum spin for which precession is possible. If it hangs below ($\pi/2 < \theta < \pi$) any spin can give steady precession.

It must not be thought that steady precession is necessarily observed under all the conditions outlined above. The support itself is never, in practice, free of friction and this will give rise to torques additional to that arising from gravity. Even if there were no friction the steady precession might be unstable; we have not shown whether a small perturbation gives rise simply to a small oscillation about the steady motion or destroys it completely.

9.6. Comments and worked examples

9.6.1. See Section 9.3.3

Example 29

A thin circular disc of uniform material has mass m, radius R and centre C. A Cartesian reference frame fixed in the disc, with origin at C, has axes $1'$ and $2'$ in the plane of the disc and axis $3'$ along the axis of the disc. With reference to this frame, show by consideration of symmetry and shape that (a) all the products of inertia are zero, (b) $I'_{11} = I'_{22} = \frac{1}{2}I'_{33}$. Hence calculate the inertia matrix.

(a) Consider an element i of the disc with mass δm_i, as shown in Figure 9.8a. Its contribution to the product of inertia I'_{23} is

$$\delta I'_{23} = -\delta m'_i X'_{i2} X'_{i3},$$

which is zero since all the X_{i3}'s are zero. Hence

$$I'_{23} = I'_{32} = 0.$$

The contribution of the element to I'_{12} is not zero. However, for every element with coordinates X'_{i1}, X'_{i2}, 0 there will be an element with coordinates X'_{i1}, $-X'_{i2}$, 0, as we can see in Figure 9.8a. The contribution of these two will cancel. Consequently, on summing the contributions from the whole disc,

$$I'_{12} = -\sum_i \delta m_i X'_{i1} X'_{i2} = I'_{21} = 0.$$

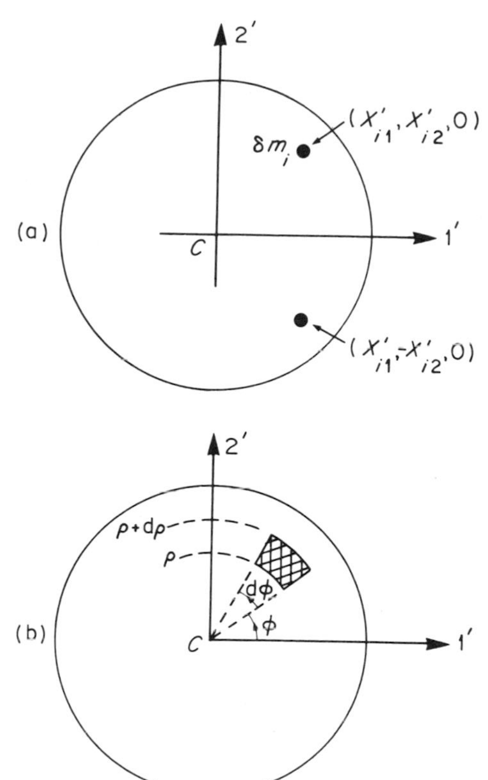

Figure 9.8

(b) The contribution of the element i to the three moments of inertia is

$$\delta I'_{11} = \delta m_i (X'^2_{i2} + X'^2_{i3}) = \delta m_i X'^2_{i2}, \tag{9.51}$$

$$\delta I'_{22} = \delta m_i (X'^2_{i3} + X'^2_{i1}) = \delta m_i X'^2_{i1}, \tag{9.52}$$

$$\delta I'_{33} = \delta m_i (X'^2_{i1} + X'^2_{i2}), \tag{9.53}$$

since X'_{i3} is zero for all elements. Hence

$$\delta I'_{11} + \delta I'_{22} = \delta I'_{33}$$

and, on summing over the whole disc, from symmetry

$$I'_{11} = I'_{22}.$$

Thus

$$I'_{11} = I'_{22} = \tfrac{1}{2} I'_{33}.$$

Now

$$X'^2_{i1} + X'^2_{i2} = \rho^2,$$

where ρ is the distance of the element i from the centre of the disc. Hence, if we take as this element the shaded area in Figure 9.8b, its mass will be

$$\delta m_i = \frac{\rho \, d\rho \, d\phi}{\pi R^2} m,$$

and equation (9.53) may be written

$$\delta I'_{33} = \frac{m}{\pi R^2} \rho^3 \, d\rho \, d\phi.$$

$$I'_{33} = \sum \delta I'_{33} = \frac{m}{\pi R^2} \int_0^{2\pi} \int_0^R \rho^3 \, d\rho \, d\phi$$

$$= \frac{2m}{R^2} \int_0^R \rho^3 \, d\rho$$

$$= \frac{mR^2}{2}.$$

Hence

$$I'_{11} = I'_{22} = \tfrac{1}{2} I'_{33} = \frac{mR^2}{2}$$

and the inertia matrix is

$$(\mathscr{I}') = \frac{mR^2}{4} \begin{pmatrix} 1 & 0 & 0 \\ 0 & 1 & 0 \\ 0 & 0 & 2 \end{pmatrix}. \tag{9.54}$$

9.6.2. See Section 9.4.1

Example 30

What is the inertia matrix, $(\mathscr{I}')$, of a circular disc referred to axes parallel with those of Example 29 but displaced a distance X along the axis of symmetry of the disc? Use your result to show that the inertia matrix of a homogeneous circular cylinder of mass M, length $2l$ and radius R is

$$(\mathscr{I}^0) = \frac{M}{12} \begin{pmatrix} 3R^2 + 4l^2 & 0 & 0 \\ 0 & 3R^2 + 4l^2 & 0 \\ 0 & 0 & 6R^2 \end{pmatrix}.$$

(The axis 3^0 is the axis of the cylinder. Axes 1^0 and 2^0 are perpendicular to this in the mid-plane of the cylinder.)

The centre of the disc is clearly its centre-of-mass. Hence we may use the parallel axes theorem (equation 9.24),

$$(\mathscr{I}') = (\mathscr{I}'_{cm}) + (\mathscr{I}'_{rel}), \tag{9.55}$$

where $(\mathscr{I}'_{\text{rel}})$ is as calculated in Example 29 (equation 9.54):

$$(\mathscr{I}'_{\text{rel}}) = \frac{mR^2}{4}\begin{pmatrix} 1 & 0 & 0 \\ 0 & 1 & 0 \\ 0 & 0 & 2 \end{pmatrix}. \tag{9.56}$$

In the given frame the position of the centre-of-mass is 0, 0, X. Hence

$$(\mathscr{I}'_{\text{cm}}) = mX^2\begin{pmatrix} 1 & 0 & 0 \\ 0 & 1 & 0 \\ 0 & 0 & 0 \end{pmatrix},$$

and therefore

$$(\mathscr{I}') = \frac{m}{4}\begin{pmatrix} R^2 + 4X^2 & 0 & 0 \\ 0 & R^2 + 4X^2 & 0 \\ 0 & 0 & 2R^2 \end{pmatrix}. \tag{9.57}$$

To find the inertia matrix of the cylinder let us first slice it into a set of parallel thin circular discs. A typical disc will be at a distance X from the mid-plane and have a thickness $\mathrm{d}X$, as shown in Figure 9.9. Its mass is

$$m = \frac{\mathrm{d}X}{2l}M$$

and its inertia matrix, referred to the axes 1^0, 2^0, 3^0, is, according to the result (9.57),

$$(\mathrm{d}\mathscr{I}^0) = \frac{M}{8l}\begin{pmatrix} (R^2 + 4X^2)\,\mathrm{d}X & 0 & 0 \\ 0 & (R^2 + 4X^2)\,\mathrm{d}X & 0 \\ 0 & 0 & 2R^2\,\mathrm{d}X \end{pmatrix}.$$

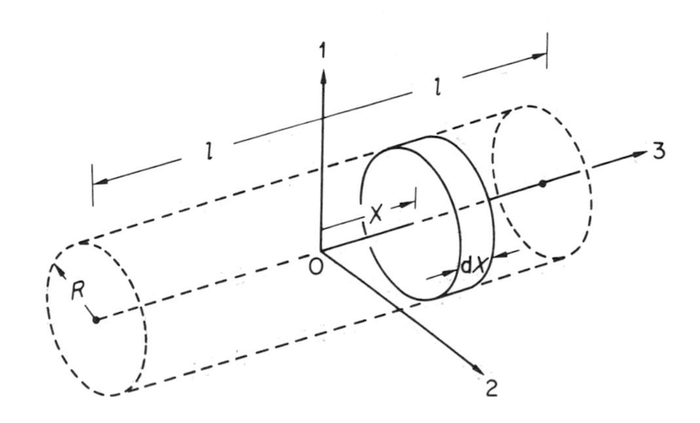

Figure 9.9

To find the total inertia matrix we sum over all the discs, which is equivalent to an integration over X from $-l$ to $+l$. Now

$$\int_{-l}^{l} R^2\,\mathrm{d}X = 2R^2 l, \qquad \int_{-l}^{l} 4X^2\,\mathrm{d}X = \frac{8l^3}{3}.$$

Then

$$(\mathcal{I}^0) = \frac{M}{12}\begin{pmatrix} 3R^2 + 4l^2 & 0 & 0 \\ 0 & 3R^2 + 4l^2 & 0 \\ 0 & 0 & 6R^2 \end{pmatrix}. \tag{9.58}$$

9.6.3. See Section 9.4.2

Example 31

The turbine rotor in a ship of mass 10^7 kg may be considered to be a uniform cylinder of mass 3×10^3 kg and of length 2 m and radius 0·5 m. Under cruising conditions the turbine rotates at 6000 revolutions minute^{-1} and the ship travels at 24 km hour^{-1}. Calculate the kinetic energy of the ship and of the rotor.

Let us take as axes in the rotor those described in Example 30. The result (9.58) shows them to be principal and we may therefore use the result (9.27) to give the kinetic energy, $\mathcal{T}_r$ in the spinning rotor:

$$\mathcal{T}_r = \tfrac{1}{2}(I_1\omega_1^{0^2} + I_2\omega_2^{0^2} + I_2\omega_3^{0^2}).$$

When the ship is travelling in a straight line the only angular velocity component is ω_3^0. Thus

$$\mathcal{T}_r = \tfrac{1}{2}I_3\omega_3^{0^2},$$

where, from equation (9.58),

$$I_3 = \tfrac{1}{2}MR^2.$$

(This is the same result as for a disc; the extension of the cylinder along its axis makes no difference to the moment of inertia about that axis.) Substituting the values

$$M = 3 \times 10^3\ \mathrm{kg},\ R = 0{\cdot}5\ \mathrm{m},\ \omega_3^0 = 6 \times 10^3\ \mathrm{rev\ min^{-1}} = 200\pi\ \mathrm{radians\ s^{-1}},$$

$$\mathcal{T}_r = \tfrac{1}{4} \times 3 \times 10^3 \times 0{\cdot}5^2 \times (200\pi)^2\ \mathrm{J} = 7{\cdot}4 \times 10^7\ \mathrm{J}.$$

For the kinetic energy of the ship

$$\mathcal{T}_s = \tfrac{1}{2}M_s V_s^2,$$

where the mass and speed have the values

$$M_s = 10^7\ \mathrm{kg},\ V_s = 24\ \mathrm{km\ hr^{-1}} = \tfrac{20}{3}\ \mathrm{m\ s^{-1}},$$

we find

$$\mathscr{T}_s = \tfrac{1}{2} \times 10^7 \times (\tfrac{20}{3})^2 \text{ J} = 22{\cdot}2 \times 10^7 \text{ J}.$$

Notice that the kinetic energy of the rotor is one-third that of the ship, although by comparison with the ship its mass is insignificant. The result illustrates how surprisingly large is the energy possessed by rapidly rotating machinery, and shows how great the destruction can be when such machinery breaks.

9.6.4. See Section 9.4.2

Example 32

(a) A body rotates about a fixed point, $\mathbf{G}^{(1)}$ being the torque acting on the body when it is rotating with angular velocity $\boldsymbol{\omega}$ and acceleration $d\boldsymbol{\omega}/dt$. If parts of the body are removed, show that the torque $\mathbf{G}^{(3)}$ that must act to give the incomplete body the same angular velocity and acceleration is given by

$$\mathbf{G}^{(3)} = \mathbf{G}^{(1)} - \mathbf{G}^{(2)},$$

where $\mathbf{G}^{(2)}$ is the torque required to give only those parts that are removed the same angular velocity and acceleration.

(b) Two turbine blades each of mass 0·5 kg are missing from the rotor of Example 31, at the positions A and B shown in Figure 9.10. Determine the torque acting on the rotor if it accelerates from rest at a constant rate of 6000 revolutions minute^{-2}. Show that, except briefly when the rotor starts to turn, the torque exerted by the bearings arises almost entirely from centrifugal forces. If the bearings are 2 m apart, show that when the rotor speed is 6000 revolutions minute^{-1} a vibrating force of approximately 5000 kg wt is set up at each bearing.

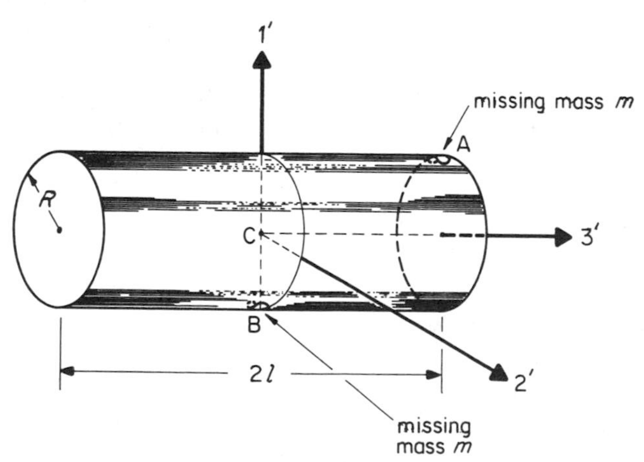

Figure 9.10

(a) Let us use axes fixed in the body. Then if $(\mathscr{I}^{(1)'})$ is the inertia matrix of the complete body, equation (9.23) shows that

$$(G^{(1)'}) = (\mathscr{I}^{(1)'})\left(\frac{\mathrm{d}\omega'}{\mathrm{d}t}\right) + (\Omega')(\mathscr{I}^{(1)'})(\omega'). \tag{9.59}$$

Now it is clear from the equations (9.20) and (9.21) defining the inertia matrix that, since each of its elements is a sum over all the particles of the body, we may write

$$(\mathscr{I}^{(1)'}) = (\mathscr{I}^{(2)'}) + (\mathscr{I}^{(3)'}),$$

where $(\mathscr{I}^{(2)'})$ is the inertia matrix of the parts that are to be removed and $(\mathscr{I}^{(3)'})$ is that of the remaining body. Hence

$$\begin{aligned}(G^{(1)'}) &= (\mathscr{I}^{(2)'})\left(\frac{\mathrm{d}\omega'}{\mathrm{d}t}\right) + (\Omega')(\mathscr{I}^{(2)'})(\omega') + (\mathscr{I}^{(3)'})\left(\frac{\mathrm{d}\omega'}{\mathrm{d}t}\right) + (\Omega')(\mathscr{I}^{(3)'})(\omega') \\ &= (G^{(2)'}) + (G^{(3)'}),\end{aligned}$$

according to the definitions of $\mathbf{G}^{(2)}$ and $\mathbf{G}^{(3)}$. Then

$$(G^{(3)'}) = (G^{(1)'}) - (G^{(2)'}) \tag{9.60}$$

or, in vector terms,

$$\mathbf{G}^{(3)} = \mathbf{G}^{(1)} - \mathbf{G}^{(2)}.$$

(b) We shall take the origin of the axes at C, the centre of the rotor, axis $3'$ along the rotor axis, axis $1'$ such that the positions of the missing blades lie in the plane $3'1'$, and axis $2'$ to complete the orthogonal frame, as shown in Figure 9.10.

It is clear from the symmetry of the complete rotor that these axes are a set of principal axes for the body. Hence its inertia matrix is of the form

$$(\mathscr{I}^{(1)'}) = \begin{pmatrix} I_1 & 0 & 0 \\ 0 & I_2 & 0 \\ 0 & 0 & I_3 \end{pmatrix}. \tag{9.61}$$

Since the rotation is solely about the axis $3'$,

$$(\omega') = \begin{pmatrix} 0 \\ 0 \\ \omega \end{pmatrix}, \quad (\Omega') = \begin{pmatrix} 0 & -\omega & 0 \\ \omega & 0 & 0 \\ 0 & 0 & 0 \end{pmatrix}, \quad \frac{\mathrm{d}\omega'}{\mathrm{d}t} = \begin{pmatrix} 0 \\ 0 \\ \mathrm{d}\omega/\mathrm{d}t \end{pmatrix}. \tag{9.62}$$

From equations (9.59), (9.61) and (9.62),

$$(G^{(1)'}) = \begin{pmatrix} 0 \\ 0 \\ I_3(\mathrm{d}\omega/\mathrm{d}t) \end{pmatrix} + \begin{pmatrix} 0 & -\omega & 0 \\ \omega & 0 & 0 \\ 0 & 0 & 0 \end{pmatrix} \begin{pmatrix} 0 \\ 0 \\ I_3\omega \end{pmatrix}$$

$$= \begin{pmatrix} 0 \\ 0 \\ I_3(\mathrm{d}\omega/\mathrm{d}t) \end{pmatrix}. \tag{9.63}$$

The turbine blades are missing from the positions $(R, 0, l)$ and $(-R, 0, 0)$. Their (joint) inertia matrix is therefore

$$(\mathscr{I}^{(2)'}) = m \begin{pmatrix} l^2 & 0 & -Rl \\ 0 & l^2 + 2R^2 & 0 \\ -Rl & 0 & 2R^2 \end{pmatrix}, \tag{9.64}$$

where m is the mass of each blade. Hence, from equations (9.62) and (9.64),

$$(G^{(2)'}) = (\mathscr{I}^{(2)'})\left(\frac{\mathrm{d}\omega'}{\mathrm{d}t}\right) + (\Omega')(\mathscr{I}^{(2)'})(\omega')$$

$$= \begin{pmatrix} -mRl\,\mathrm{d}\omega/\mathrm{d}t \\ 0 \\ 2mR^2\,\mathrm{d}\omega/\mathrm{d}t \end{pmatrix} + \begin{pmatrix} 0 & -\omega & 0 \\ \omega & 0 & 0 \\ 0 & 0 & 0 \end{pmatrix} \begin{pmatrix} -mRl\omega \\ 0 \\ 2mR^2\omega \end{pmatrix}$$

$$= \begin{pmatrix} -mRl\,\mathrm{d}\omega/\mathrm{d}t \\ -mRl\omega^2 \\ 2mR^2\,\mathrm{d}\omega/\mathrm{d}t \end{pmatrix}. \tag{9.65}$$

Then, from equations (9.60), (9.63) and (9.65), the torque for the incomplete rotor will be

$$(G^{(3)'}) = \begin{pmatrix} G_1^{(3)'} \\ G_2^{(3)'} \\ G_3^{(3)'} \end{pmatrix} = \begin{pmatrix} mRl\,\mathrm{d}\omega/\mathrm{d}t \\ mRl\omega^2 \\ (I_3 - 2mR^2)\,\mathrm{d}\omega/\mathrm{d}t \end{pmatrix}. \tag{9.66}$$

The driving torque (exerted by the steam on the rotor) is $G_3^{(3)'}$, and we can see from the third elements of equation (9.66) how this determines the angular acceleration $\mathrm{d}\omega/\mathrm{d}t$. Initially, the only other torque component is

$$G_1^{(3)'} = mRl\frac{\mathrm{d}\omega}{\mathrm{d}t}, \tag{9.67}$$

which arises from the angular acceleration and is constant. However, the component

$$G_2^{(3)'} = mRl\omega^2, \tag{9.68}$$

which, from the presence of the factor ω^2, arises only from centrifugal forces, will overtake $G_1^{(3)'}$ when

$$\omega^2 = \frac{d\omega}{dt}. \tag{9.69}$$

From the given acceleration

$$\frac{d\omega}{dt} = \frac{6000 \times 2\pi}{60 \times 60} = \frac{10\pi}{3} \text{ radian s}^{-2}. \tag{9.70}$$

Then equation (9.69) shows that $G_1^{(3)'}$ and $G_2^{(3)'}$ are equal when

$$\omega = \left(\frac{10\pi}{3}\right)^{1/2} \text{ radian s}^{-1}. \tag{9.71}$$

Since the rotor has the constant acceleration (9.70) it will reach the speed (9.71) in the short time

$$t = \frac{\omega}{d\omega/dt} = \left(\frac{3}{10\pi}\right)^{1/2} = 0{\cdot}31 \text{ s}.$$

Thenceforth, the component $G_2^{(3)'}$ exceeds $G_1^{(3)'}$ and rapidly predominates because of its ω^2 dependence. It will take 60 s to reach a speed of 6000 revolutions minute^{-1}, or

$$\omega = \frac{6000 \times 2\pi}{60} = 200\pi \text{ radian s}^{-1}. \tag{9.72}$$

Using the values (9.70) and (9.72) in equations (9.67) and (9.68), we can see that at this speed $G_1^{(3)'}$ is quite negligible compared with $G_2^{(3)'}$. For the given values

$$m = 0{\cdot}5 \text{ kg}, \qquad l = 1 \text{ m}, \qquad R = 0{\cdot}5 \text{ m}$$

we find the value

$$G_2^{(3)'} = 0{\cdot}5 \times 1{\cdot}0 \times 0{\cdot}5 \times (200\pi)^2 = 9{\cdot}9 \times 10^4 \text{ Nm}. \tag{9.73}$$

Now the centre-of-mass of the complete rotor and of the two blades considered together are both fixed points on the axis 3′. The centre-of-mass of the incomplete rotor is therefore also a fixed point on this axis. Hence, the resultant external force on the rotor must be zero, and the forces exerted on the rotor by the two bearings will be equal and opposite (gravity not being considered here), as shown in Figure 9.11.

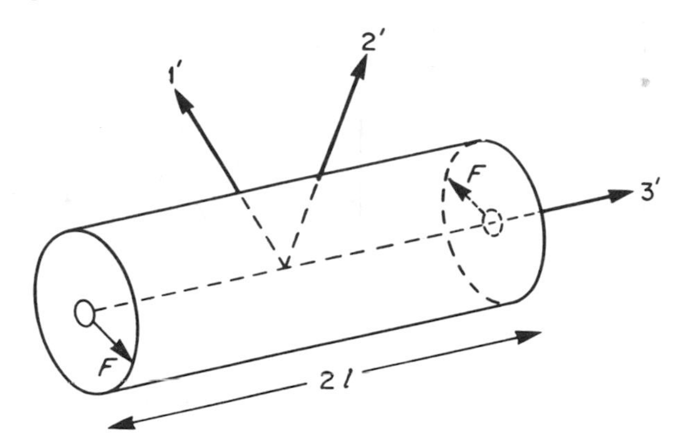

Figure 9.11

For the force F at each bearing to give the torque (9.73) we must have

$$2Fl = G_2^{(3)'} = 9{\cdot}9 \times 10^4 \text{ Nm},$$

from which

$$F = 4{\cdot}9 \times 10^4 \text{ N}. \tag{9.74}$$

Now 1 kg wt is the gravitational force on 1 kg (i.e. 9·81 N). The force (9.74) is therefore approximately 5000 kg wt, and since its direction rotates with the rotor, this is effectively a vibrating force at each bearing.

We see from this example how important it is that high-speed rotating machinery should be accurately balanced. The products of inertia that arise from quite small imperfections can give rise to severe stresses at the bearings (see problem 9.8).

9.6.5. See Section 9.5.1

It may at first sight seem strange that the gravitational force, acting downwards through the centre-of-mass, does not simply cause the top to fall down. However, we must remember that a spinning body possesses angular momentum, and therefore any change in the direction of the spin will, in general, result in a change in angular momentum. This is only possible if the forces acting on the body possess a torque which will account for this change.

Suppose that a spinning top did in fact 'fall' from θ_1 to θ_2 while its axis remained in the same vertical plane, as shown in Figure 9.12. Its momentum would change from $\mathbf{J}_1$ to $\mathbf{J}_2$, and, whether or not these have the same magnitude, this would require a change of angular momentum

$$\mathbf{J}^* = \mathbf{J}_2 - \mathbf{J}_1,$$

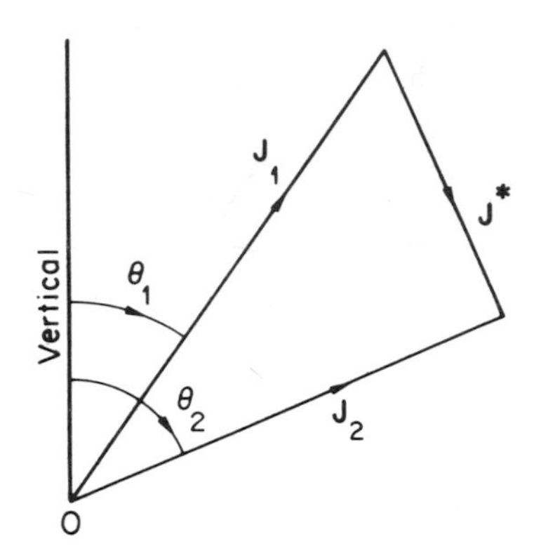

Figure 9.12

which lies in the same vertical plane (the plane of the diagram). To achieve a torque about O in this direction of **J*** would require a force acting on the top which had a component *perpendicularly outwards* from the plane of the diagram. Clearly the gravitational force has no such component. To cause the required change the force must be perpendicular (or at least have a component perpendicular) to the direction in which we wish the axis of the top to move.

9.6.6. See Section 9.5.1

It must be remembered that ω is the spin of the top about its axis *relative to the frame* S'. Since S' has a component of angular velocity $\sigma \cos \theta$ in the same direction, the total component of angular velocity of the top, *relative to the inertial frame* S, is

$$\omega'_3 = \omega + \sigma \cos \theta.$$

In terms of ω'_3 equation (9.49) becomes

$$I\sigma^2 \cos \theta - I_3 \sigma \omega'_3 + mgl = 0, \tag{9.75}$$

an equation which is often used, in place of equation (9.49), to discuss the precession of a top.

9.7. Problems

9.1. In a Cartesian frame of reference a rod of mass m and length $2a$ has its ends at the points $(a, x_2, 0)$, $(-a, x_2, 0)$. Show that its inertia matrix is

$$(\mathscr{I}) = \frac{m}{3}\begin{pmatrix} 3x_2^2 & 0 & 0 \\ 0 & a^2 & 0 \\ 0 & 0 & a^2 + 3x_2^2 \end{pmatrix}.$$

By treating a flat rectangular sheet as a set of parallel rods and a rectangular solid box as a set of parallel sheets, show that for principal axes at its centre the box has the inertia matrix

$$(\mathscr{I}^0) = \frac{M}{3}\begin{pmatrix} a_2^2 + a_3^2 & 0 & 0 \\ 0 & a_3^2 + a_1^2 & 0 \\ 0 & 0 & a_1^2 + a_2^2 \end{pmatrix},$$

where $2a_1, 2a_2, 2a_3$ are the lengths of its sides and M is its mass.

9.2. What is the 'unit body' (see Section 9.4.4) corresponding to the following uniform solids:

(a) A rectangular box of sides $2a_1, 2a_2, 2a_3$.
(b) An ellipsoid with principal diameters $2a_1, 2a_2, 2a_3$.
(c) A solid formed from two right circular cones each of height a_1 and base radius a_2, joined base to base.

Use these unit bodies to calculate the principal moments of inertia for axes through the centre-of-mass of each solid.

9.3. A rectangular solid box of sides $2a_1$, $2a_2$, $2a_3$ is allowed to swing freely under gravity about an axis along one of the edges of length $2a_3$. What is the length of the equivalent simple pendulum?

The box is held with sides $2a_1$ horizontal so that the centre-of-mass is above the axis of rotation and is allowed to fall. This procedure is repeated but with the centre-of-mass initially below the axis of rotation. Show that the ratio of the maximum angular velocities attained in the two cases is $[(a_1^2 + a_2^2)^{1/2} + a_2]^{1/2}/[(a_1^2 + a_2^2)^{1/2} - a_2]^{1/2}$.

9.4. A body is rotating with angular speed, ω, about a fixed axis which is parallel with a principal axis through its centre-of-mass. Show that there will be a torque about any point, O, on the axis equal to that arising from the 'centrifugal force', $MR\omega^2$, acting through the centre-of-mass perpendicular to the axis, where M is the mass of the body and R is the distance of the centre-of-mass from the axis of rotation.

A gate 4 m wide and 2 m high, of mass 100 kg, is pivoted about a vertical post at one end. Find the maximum rate at which it should be opened if the torque on the post is not to exceed twice its value when the gate is stationary and determine this maximum torque.

9.5. Show that when a body rotates about a fixed axis the torque **G** that it exerts about any point O on the axis has direction perpendicular to that axis and magnitude

$$G = (I_{12}^2 + I_{13}^2)^{1/2}\omega^2,$$

where ω is the instantaneous angular speed and I_{12} and I_{13} are the products of inertia about any two axes through O, fixed in the body, which are perpendicular to each other and to the axis.

9.6. A fly-wheel consists of a circular disc of radius a and thickness $2b$, with mass M. It is put out of balance by adding a mass m at a point on one of its circular edges. Show that in a suitable fixed frame of reference with origin at the centre of the disc two products of inertia are

$$I_{31} = -mab\cos\phi, \qquad I_{32} = -mab\sin\phi,$$

where ϕ is the angular distance of m from the axis 1, and that one moment of inertia is

$$I_{33} = \left(m + \frac{M}{2}\right)a^2.$$

For the case

$$M = 10\,\text{kg}, \qquad m = 0{\cdot}1\,\text{kg}, \qquad a = 20\,\text{cm}, \qquad b = 1\,\text{cm},$$

show that a constant couple applied about the axis, of magnitude $10\,\text{kg}\,\text{m}^2\,\text{s}^{-2}$, will accelerate the fly-wheel to 3000 rev min^{-1} in $4{\cdot}04\pi$ s. If it is then allowed to run freely on smooth bearings, determine the magnitude of the couple the fly-wheel exerts on them.

9.7. A rigid body is moving with respect to an inertial frame S, a particular point O′ in the body having the (variable) position (X) in the S frame. S' is a frame fixed in the body with the origin at O′. (A) is the transformation linking S' and the frame with origin at O′ and axes parallel with those of S. (x_i') is the position in the S' frame of any mass m_i forming the body.

Show that the kinetic energy of the mass m_i in the inertial frame S may be written in the form

$$\mathscr{T}_i = \frac{1}{2}m_i\left(\frac{\widetilde{\mathrm{d}X}}{\mathrm{d}t}\right)\left(\frac{\mathrm{d}X}{\mathrm{d}t}\right) + \frac{1}{2}m_i\left(\frac{\widetilde{\mathrm{d}X}}{\mathrm{d}t}\right)\left(\frac{\widetilde{\mathrm{d}A}}{\mathrm{d}t}\right)(x_i') + \frac{1}{2}\dot{m}_i(\tilde{x}_i')\left(\frac{\mathrm{d}A}{\mathrm{d}t}\right)\left(\frac{\mathrm{d}X}{\mathrm{d}t}\right)$$
$$+ \frac{1}{2}m_i(\tilde{\omega}')(\tilde{\mathscr{X}}_i')(\mathscr{X}_i')(\omega').$$

Hence show that if O′ is the centre-of-mass of the rigid body, its total kinetic energy in the inertial frame may be expressed as

$$\mathscr{T} = \tfrac{1}{2}M(\tilde{V})(V) + \tfrac{1}{2}(\tilde{\omega}')(\mathscr{I}')(\omega'),$$

and explain the various symbols used.

9.8. A mass $2m$ is added to the rotor of Example 32, page 277, at the point

$$X_i' = \tfrac{1}{2}R, \qquad X_2' = 0, \qquad X_3' = \tfrac{1}{2}l$$

to balance the loss of a single blade of mass m at A. Show that while this regains the *static* balance of the original whole rotor, it is not equivalent to it *dynamically*, and calculate the torque set up at the bearings when the rotor is turning at 6000 rev min^{-1} for the values given in Example 32.

9.9. A top consists of a heavy circular disc of mass m and radius a through the centre of which a light shaft is fixed. It extends on one side a distance $a\sqrt{10}$ and on the other a distance $2a\sqrt{10}$. Find the magnitudes and directions of precession that are possible when the top is spinning with angular speed ω and supported at either end of the shaft.

9.10. A fly-wheel is rotating at high speed on bearings fixed at the earth's surface. Determine the torque on the bearings arising from the rotation of the earth, showing how it depends both on the orientation of the bearings and on their position on the earth's surface. Substitute typical values of the physical quantities involved to obtain an order of magnitude for this effect.

Chapter 10

Generalized Mechanics; Lagrange's and Hamilton's Equations

10.1. Static and dynamic equilibrium; variational principles

Physical systems show a general tendency to settle down to conditions of minimum energy. Roughly speaking, this is because if there is energy to spare some process will be available which will transfer this to the rest of the universe. The simplest example is the condition of minimum potential energy for a system in static equilibrium. By expressing equilibrium as

$$\delta \mathscr{V} = 0 \tag{10.1}$$

rather than by saying that the resultant forces on all parts of the system are zero, we obtain a formulation of a physical law that is simple and independent, or largely independent, of the coordinate system. More precisely, equation (10.1) says that when a system is in static equilibrium any perturbation, or small change in the positions of the various parts of the system, will give a change in potential energy which is, *to first order*, zero. This is one of the simplest examples of a *variational principle*. By itself equation (10.1) does not tell us whether the equilibrium is stable or not. (See Section 10.17.1, page 319.) Often this question can be settled without further mathematical analysis. If not, the second-order term $\delta^2 \mathscr{V}$ must be calculated: if it is positive the equilibrium is stable, if negative it is unstable.

Is it possible to show that the equations of motion of a system are equivalent to a condition of *dynamic equilibrium*, expressed by a variational principle analogous to equation (10.1)? We may expect this, too, to involve energy, although here we must remember that there are kinetic energies as well as the work or potential energies arising from forces.

10.2. One particle in one dimension

Let us start with the simplest case: that of a single particle moving in one linear dimension, x, starting at $x_{(1)}$ at time $t_{(1)}$ and finishing at $x_{(2)}$ at time $t_{(2)}$ under the influence of a force F_x. At any intermediate time t, its actual position is $x(t)$ and its kinetic energy will be

$$\mathscr{T} = \frac{m}{2}\left(\frac{dx}{dt}\right)^2. \tag{10.2}$$

If some law embracing the whole path is to involve the kinetic energy we cannot use its value at a particular point of the path, since it will usually vary along the path. The simplest expression which does involve $\mathscr{T}$ and treats all parts of the path with equal importance is

$$Q = \int_{t_{(1)}}^{t_{(2)}} \mathscr{T}\,\mathrm{d}t = \frac{m}{2}\int_{t_{(1)}}^{t_{(2)}} \left(\frac{\mathrm{d}x}{\mathrm{d}t}\right)^2 \mathrm{d}t.$$

$(Q/(t_{(2)} - t_{(1)})$ is the time average of the kinetic energy.

The analogue of an infinitesimal displacement for the dynamical case is an infinitesimally perturbed path. Any such neighbouring path must still have the same end-points:

$$x(t_{(1)}) = x_{(1)}, \qquad x(t_{(2)}) = x_{(2)},$$

and in one sense the path must also be the same when only one dimension is involved. However, if by a perturbed path we mean motion which at a time t differs from the actual position $x(t)$ by an infinitesimal amount δx, as shown in Figure 10.1, then the corresponding change in $\mathscr{T}$ is given by

$$\begin{aligned}\mathscr{T} + \delta\mathscr{T} &= \frac{m}{2}\left[\frac{\mathrm{d}}{\mathrm{d}t}(x + \delta x)\right]^2 \\ &= \frac{m}{2}\left[\left(\frac{\mathrm{d}x}{\mathrm{d}t}\right)^2 + 2\frac{\mathrm{d}x}{\mathrm{d}t}\frac{\mathrm{d}(\delta x)}{\mathrm{d}t} + \left(\frac{\mathrm{d}(\delta x)^2}{\mathrm{d}t}\right)\right].\end{aligned}$$

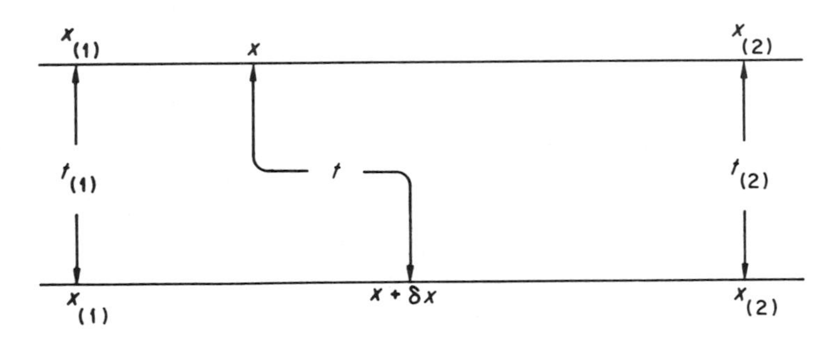

Figure 10.1

Using equation (10.2) and neglecting the last, second-order term,

$$\delta\mathscr{T} = m\frac{\mathrm{d}x}{\mathrm{d}t}\frac{\mathrm{d}(\delta x)}{\mathrm{d}t}.$$

The corresponding change in Q may be found by integrating by parts:

$$\delta Q = \int_{t(1)}^{t(2)} \delta\mathcal{T}\,\mathrm{d}t = m\int_{t(1)}^{t(2)} \frac{\mathrm{d}x}{\mathrm{d}t}\frac{\mathrm{d}(\delta x)}{\mathrm{d}t}\,\mathrm{d}t$$

$$= \left[\frac{\mathrm{d}x}{\mathrm{d}t}\delta x\right]_{(1)}^{(2)} - m\int_{t(1)}^{t(2)} \frac{\mathrm{d}^2x}{\mathrm{d}t^2}\delta x\,\mathrm{d}t.$$

The first term is zero because the perturbation δx is zero at both end-points. Also, since

$$F_x = m\frac{\mathrm{d}^2x}{\mathrm{d}t^2},$$

the second term may be written as

$$-\int_{t(1)}^{t(2)} F_x\,\delta x\,\mathrm{d}t = -\int_{t(1)}^{t(2)} \delta\mathcal{W}\,\mathrm{d}t,$$

where $\delta\mathcal{W}$ is the work which would be done by the true force, F_x, acting on the particle at any time if the particle were moved to the perturbed position. Thus

$$\int_{t(1)}^{t(2)} (\delta\mathcal{T} + \delta\mathcal{W})\,\mathrm{d}t = 0. \qquad (10.3)$$

10.3. Many particles in three dimensions

Suppose now that we have N particles, of masses $m_1, m_2, \ldots, m_N$, observed in a three-dimensional Cartesian frame. The total kinetic energy of the system will be

$$\mathcal{T} = \frac{m_1}{2}\left[\left(\frac{\mathrm{d}x_{11}}{\mathrm{d}t}\right)^2 + \left(\frac{\mathrm{d}x_{12}}{\mathrm{d}t}\right)^2 + \left(\frac{\mathrm{d}x_{13}}{\mathrm{d}t}\right)^2\right]$$

$$+ \frac{m_2}{2}\left[\left(\frac{\mathrm{d}x_{21}}{\mathrm{d}t}\right)^2 + \left(\frac{\mathrm{d}x_{22}}{\mathrm{d}t}\right)^2 + \left(\frac{\mathrm{d}x_{23}}{\mathrm{d}t}\right)^2\right] + \ldots,$$

where (x_{11}, x_{12}, x_{13}), $(x_{21}, x_{22}, x_{23}), \ldots$ are the positions of particles $1, 2, \ldots$. To avoid the double summation we can suppose that the position of the system as a whole is determined by $3N$ coordinates $x_1, x_2, \ldots, x_{3N}$, and we can then write

$$\mathcal{T} = \sum_{i=1}^{3N} \frac{m_i}{2}\left(\frac{\mathrm{d}x_i}{\mathrm{d}t}\right)^2,$$

remembering that in this summation m_1, m_2, m_3 are all equal to the mass of the first particle and x_1, x_2, x_3 are its coordinates; m_3, m_4, m_5 are all equal to the mass of the second particle and x_4, x_5, x_6 are its coordinates; and so on.

We can then follow exactly the argument of the preceding section, starting with

$$Q = \int_{t(1)}^{t(2)} \left[\sum_{i=1}^{3N} \frac{m_i}{2} \left(\frac{dx_i}{dt} \right)^2 \right] dt.$$

The perturbed path may now be visibly distinct from the true path except at the end-points, although the perturbation of each coordinate can still separately be described in the one-dimensional diagram of Figure 10.2. In each equation we have a summation over all particles, which finally yields

$$\int_{t(1)}^{t(2)} (\delta \mathscr{T} + \delta \mathscr{W})\, dt = 0, \tag{10.4}$$

where

$$\delta \mathscr{W} = \sum_{i=1}^{3N} F_{xi}\, \delta x_i$$

is the total work which would be done by all the forces if all the particles were moved to their perturbed positions.

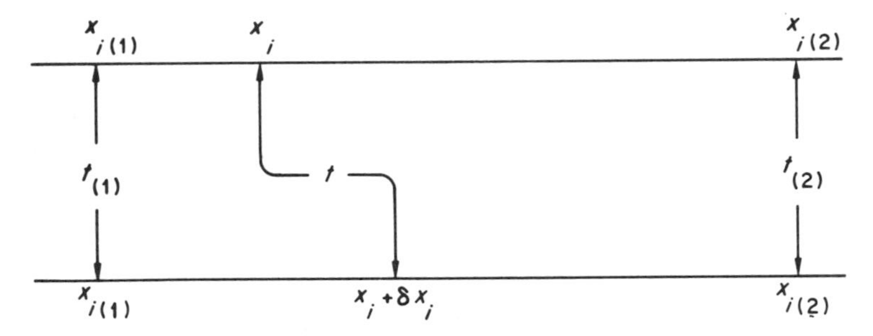

Figure 10.2

10.4. Generalized coordinates and forces

So far we have achieved little in obtaining the condition (10.4) except perhaps a more elegant way of expressing the $3N$ equations of motion for the N particles. To determine the motion, $x_i(t)$, for all the coordinates would, in effect, require regaining the equations of motion from condition (10.4).

However, a return to Cartesian coordinates is not the only way of using this condition. There may exist an alternative set of coordinates, $q_1, q_2, \ldots, q_n$, which will also serve to describe the motion of the system. The q's need not necessarily be distances. They may be, for example, angles or functions of distances and angles that are convenient for a particular problem. Thus the position of a single particle in two dimensions can be specified either by its Cartesian coordinates (x_1, x_2) or its polar coordinates (r, θ):

$$x_1 = r\cos\theta, \qquad x_2 = r\sin\theta.$$

Time may be explicitly involved. For example a bead may be free to slide along a rod which is kept at a constant angle β to the vertical and is kept rotating at a constant angular velocity ω about the vertical, as shown in Figure 10.3.

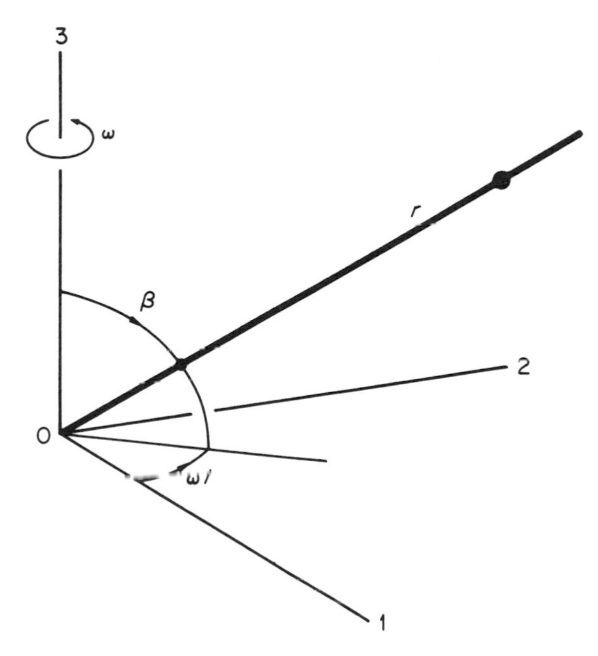

Figure 10.3

In this case

$$x_1 = r\sin\beta\cos\omega t, \qquad x_2 = r\sin\beta\sin\omega t, \qquad x_3 = r\cos\beta,$$

where r is the distance of the bead from the axis of rotation. Only r and t can vary here, and only their *two* values are required to determine the *three* coordinates, x_1, x_2, x_3.

In general, such relationships may be written

$$x_1 = x_1(q_1, q_2, \ldots, q_n; t), \quad x_2 = x_2(q_1, q_2, \ldots, q_n; t),$$
$$\ldots, \quad x_{3N} = x_{3N}(q_1, q_2, \ldots, q_n; t), \tag{10.5}$$

where, as we have just seen, $3N$ and n are not necessarily equal. The q's constitute a set of *generalized coordinates* for the system.

From equations (10.5) infinitesimal changes in the Cartesian coordinates can be expressed in terms of infinitesimal changes in the generalized coordinates:

$$\delta x_1 = \frac{\partial x_1}{\partial q_1}\delta q_1 + \frac{\partial x_1}{\partial q_2}\delta q_2 + \ldots + \frac{\partial x_1}{\partial q_n}\delta q_n + \frac{\partial x_1}{\partial t}\delta t,$$

$$\delta x_2 = \frac{\partial x_2}{\partial q_2}\delta q_2 + \ldots, \quad \text{etc.} \tag{10.6}$$

The last terms in all these equations are zero, since the displacements we are considering are those from the true position *at a definite time.* Hence the work which would be done by the forces is

$$\begin{aligned}\delta \mathscr{W} &= F_{x1}\,\delta x_1 + F_{x2}\,\delta x_2 + \ldots \\ &= \left(F_{x1}\frac{\partial x_1}{\partial q_1} + F_{x2}\frac{\partial x_2}{\partial q_1} + \ldots + F_{x3N}\frac{\partial x_{3N}}{\partial q_1}\right)\delta q_1 \\ &\quad + \left(F_{x1}\frac{\partial x_1}{\partial q_2} + F_{x2}\frac{\partial x_2}{\partial q_2} + \ldots + F_{x3N}\frac{\partial x_{3N}}{\partial q_2}\right)\delta q_2 \\ &\quad + \ldots + \left(F_{x1}\frac{\partial x_1}{\partial q_n} + F_{x2}\frac{\partial x_2}{\partial q_n} + \ldots + F_{x3N}\frac{\partial x_{3N}}{\partial q_n}\right)\delta q_n.\end{aligned}$$

The terms in brackets can all be expressed as function of the q's and possibly of t:

$$F_{x1}\frac{\partial x_1}{\partial q_1} + \ldots + F_{x3N}\frac{\partial x_{3N}}{\partial q_1} = F_1(q_1, q_2, \ldots, q_n; t) \equiv F_1(q_j; t), \quad \text{etc.,}$$

and therefore

$$\delta \mathscr{W} = F_1\,\delta q_1 + F_2\,\delta q_2 + \ldots + F_n\,\delta q_n = \sum_{j=1}^{n} F_j\,\delta q_j. \tag{10.7}$$

The F's are known as *generalized forces.* It must be noted that since the q's are not necessarily distances, the F's will not necessarily have the dimensions, MLT^{-2}, of a force.

From equations (10.6) we also see that

$$\frac{dx_1}{dt} = \frac{\partial x_1}{\partial q_1}\frac{dq_1}{dt} + \frac{\partial x_1}{\partial q_2}\frac{dq_2}{dt} + \ldots$$

$$\frac{dx_2}{dt} = \frac{\partial x_2}{\partial q_1}\frac{dq_1}{dt} + \ldots, \quad \text{etc.}$$

or, in condensed notation,

$$\dot{x}_1 = \frac{\partial x_1}{\partial q_1}\dot{q}_1 + \frac{\partial x_2}{\partial q_2}\dot{q}_2 \ldots,$$

$$\dot{x}_2 = \frac{\partial x_2}{\partial q_1}\dot{q}_1 + \ldots, \quad \text{etc.}$$

Therefore, by substituting these into the total kinetic energy of the system,

$$\mathscr{T} = \tfrac{1}{2}m_1\dot{x}_1^2 + \tfrac{1}{2}m_2\dot{x}_2^2 + \ldots,$$

this energy can be expressed in terms of the $\dot{q}$'s, the q's and possibly t:

$$\mathscr{T} \equiv \mathscr{T}(\dot{q}_1, \dot{q}_2, \ldots; q_1, q_2, \ldots; t) \equiv \mathscr{T}(\dot{q}_j; q_j; t).$$

Hence

$$\delta\mathscr{T} = \sum_j \left(\frac{\partial \mathscr{T}}{\partial \dot{q}_j}\delta\dot{q}_j + \frac{\partial \mathscr{T}}{\partial q_j}\delta q_j\right) + \frac{\partial \mathscr{T}}{\partial t}\delta t,$$

where once again the last term is zero in the circumstances we are considering. The quantity $\delta\dot{q}_j$ needs to be identified physically. At any time t the actual 'velocity' of the generalized coordinate is

$$\dot{q}_j = \frac{\mathrm{d}q_j}{\mathrm{d}t}.$$

By definition, this velocity increases by the increment $\delta\dot{q}_j$ when we transfer to the 'velocity' of the same coordinate for the perturbed path. Thus

$$\dot{q}_j + \delta\dot{q}_j = \frac{\mathrm{d}(q_j + \delta q_j)}{\mathrm{d}t} = \frac{\mathrm{d}q_j}{\mathrm{d}t} + \frac{\mathrm{d}(\delta q_j)}{\mathrm{d}t}$$

$$\delta\dot{q}_j = \frac{\mathrm{d}(\delta q_j)}{\mathrm{d}t}.$$

Hence

$$\delta\mathscr{T} = \sum_{j=1}^{n} \left(\frac{\partial \mathscr{T}}{\partial \dot{q}_j}\frac{\mathrm{d}}{\mathrm{d}t}(\delta q_j) + \frac{\partial \mathscr{T}}{\partial q_j}\delta q_j\right). \qquad (10.8)$$

10.5. Lagrange's equations

When the expressions (10.7) and (10.8) are substituted into the condition (10.4), we have

$$\int_{t(1)}^{t(2)} (\delta\mathscr{T} + \delta\mathscr{W})\,\mathrm{d}t = \int_{t(1)}^{t(2)} \sum_j \left[\frac{\partial \mathscr{T}}{\partial \dot{q}_j}\frac{\mathrm{d}}{\mathrm{d}t}(\delta q_j) + \left(\frac{\partial \mathscr{T}}{\partial q_j} + F_j\right)\delta q_j\right]\mathrm{d}t = 0.$$

The first term may be integrated by parts:

$$\int_{t_{(1)}}^{t_{(2)}} \frac{\partial \mathscr{T}}{\partial \dot{q}_j} \frac{\mathrm{d}}{\mathrm{d}t}(\delta q_1)\,\mathrm{d}t = \left[\frac{\partial \mathscr{T}}{\partial \dot{q}_j}\delta q_j\right]_{(1)}^{(2)} - \int_{t_{(1)}}^{t_{(2)}} \frac{\mathrm{d}}{\mathrm{d}t}\left(\frac{\partial \mathscr{T}}{\partial \dot{q}_j}\right)\delta q_j\,\mathrm{d}t.$$

The first term of this is zero since we restrict the perturbations to those leaving the end points unchanged (all the δq_j zero at $t_{(1)}, t_{(2)}$). Hence

$$\int_{t_{(1)}}^{t_{(2)}} \sum_j \left[-\frac{\mathrm{d}}{\mathrm{d}t}\left(\frac{\partial \mathscr{T}}{\partial \dot{q}_j}\right) + \frac{\partial \mathscr{T}}{\partial q_j} + F_j\right]\delta q_j\,\mathrm{d}t = 0. \qquad (10.9)$$

Further progress depends critically upon whether the δq_j are independent. Physically this means that any one of the q's can be varied while the others are kept constant. For example a single particle moving freely in space could have for its generalized coordinates the spherical polar coordinates r, θ, ϕ, and each of these could be varied without affecting the other. If, however, it were constrained to move upon the surface of a sphere of radius a, r could no longer be varied and only variations $\delta\theta$, $\delta\phi$ would be possible. The greatest number of independently variable coordinates is called the number of *degrees of freedom.*

When the generalized coordinates are just sufficient to give the degrees of freedom that a system has, the elaborate expression (10.9) can be reduced considerably. Firstly, since the δq_j are independent, we can choose δq_1, $\delta q_2, \ldots$ in turn to be non-zero while the others are zero. Then

$$\int_{t_{(1)}}^{t_{(2)}} \left[-\frac{\mathrm{d}}{\mathrm{d}t}\left(\frac{\partial \mathscr{T}}{\partial \dot{q}_j}\right) + \frac{\partial \mathscr{T}}{\partial q_j} + F_j\right]\delta q_j\,\mathrm{d}t = 0 \qquad (10.10)$$

for each value of j. Secondly, for any time t between $t_{(1)}$ and $t_{(2)}$ we can choose δq_j to be non-zero in a small interval δt around t and zero elsewhere. The integral (10.10) then becomes approximately

$$\left[-\frac{\mathrm{d}}{\mathrm{d}t}\left(\frac{\partial \mathscr{T}}{\partial \dot{q}_j}\right) + \frac{\partial \mathscr{T}}{\partial q_j} + F_j\right]\delta q_j\,\delta t = 0,$$

and since

$$\delta q_j \neq 0, \qquad \delta t \neq 0,$$

we must have, for each j and all t in the range considered,

$$\frac{\mathrm{d}}{\mathrm{d}t}\left(\frac{\partial \mathscr{T}}{\partial \dot{q}_j}\right) - \frac{\partial \mathscr{T}}{\partial q_j} = F_j. \qquad (10.11)$$

These are *Lagrange's equations.*

10.6. Change of coordinate system

One of the simplest uses of Lagrange's equations is to obtain equations of motion for coordinate systems other than Cartesian. For example, if for a single particle we change from x_1, x_2, x_3 to spherical polar coordinates r, θ, ϕ, we may take these for the generalized coordinates

$$q_1 = r, \qquad q_2 = \theta, \qquad q_3 = \phi.$$

From Figure 10.4 we can see that increments δr, $\delta\theta$, $\delta\phi$ would cause the particle to move distances δr along the radius vector, $r\,\delta\theta$ in the θ direction

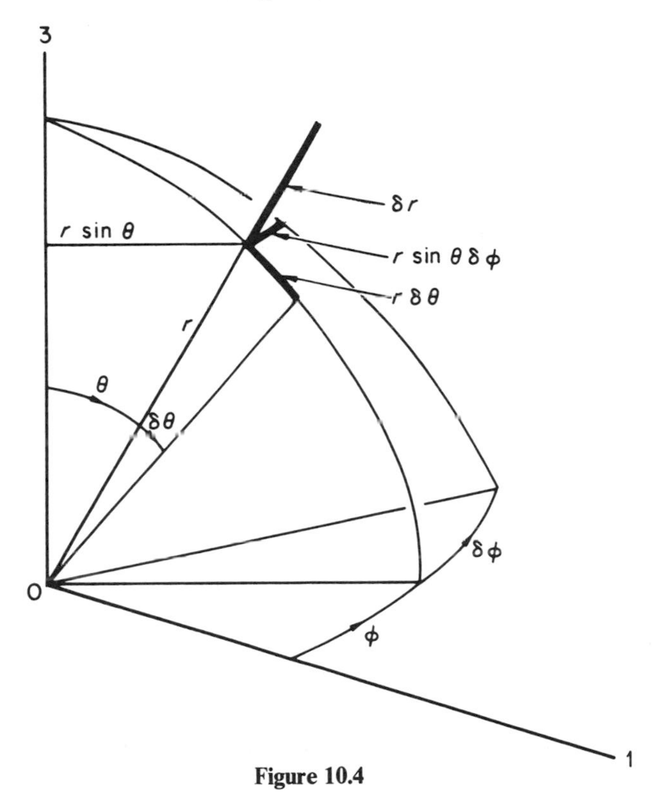

Figure 10.4

and $r \sin\theta\,\delta\phi$ in the ϕ direction. These are all perpendicular to each other and therefore if F_r, F_θ, F_ϕ are the corresponding orthogonal components of the force on the particle, the work done in a general displacement would be

$$\delta\mathscr{W} = F_r\,\delta r + F_\theta\,r\,\delta\theta + F_\phi\,r\sin\theta\,\delta\phi.$$

Since the generalized forces are defined by

$$\delta\mathscr{W} = F_1\,\delta q_1 + F_2\,\delta q_2 + F_3\,\delta q_3,$$

we see that

$$F_1 = F_r, \qquad F_2 = rF_\theta, \qquad F_3 = r\sin\theta\, F_\phi. \tag{10.12}$$

The first is the radial component of the force, while the second and third are the components of the torque about the origin perpendicular to, and along, the O3 axis.

The corresponding orthogonal components of velocity are $\dot{r}$, $r\dot{\theta}$, $r\sin\theta\dot{\phi}$ and the kinetic energy is therefore

$$\mathscr{T} = \frac{m}{2}(\dot{r}^2 + r^2\dot{\theta}^2 + r^2\sin^2\theta\dot{\phi}^2). \tag{10.13}$$

Then

$$\frac{\partial\mathscr{T}}{\partial\dot{q}_1} \equiv \frac{\partial\mathscr{T}}{\partial\dot{r}_1} = m\dot{r}, \qquad \frac{\partial\mathscr{T}}{\partial\dot{q}_2} \equiv \frac{\partial\mathscr{T}}{\partial\dot{\theta}} = mr^2\dot{\theta}, \qquad \frac{\partial\mathscr{T}}{\partial\dot{q}_3} \equiv \frac{\partial\mathscr{T}}{\partial\dot{\phi}} = mr^2\sin^2\theta\dot{\phi}, \tag{10.14}$$

$$\frac{\partial\mathscr{T}}{\partial q_1} \equiv \frac{\partial\mathscr{T}}{\partial r} = mr(\dot{\theta}^2 + \sin^2\theta\dot{\phi}^2), \qquad \frac{\partial\mathscr{T}}{\partial q_2} \equiv \frac{\partial\mathscr{T}}{\partial\theta} = mr^2\sin\theta\cos\theta\dot{\phi},$$

$$\frac{\partial\mathscr{T}}{\partial q_3} \equiv \frac{\partial\mathscr{T}}{\partial\phi} = 0. \tag{10.15}$$

Hence, from equations (10.12), (10.13), (10.14) and (10.15), the first Lagrange's equation,

$$\frac{\mathrm{d}}{\mathrm{d}t}\left(\frac{\partial\mathscr{T}}{\partial\dot{q}_1}\right) - \frac{\partial\mathscr{T}}{\partial q_1} = F_1,$$

becomes

$$\frac{\mathrm{d}}{\mathrm{d}t}(m\dot{r}) - mr(\dot{\theta}^2 + \sin^2\theta\dot{\phi}^2) = F_r, \qquad \text{or} \qquad m\ddot{r} = F_r + mr(\dot{\theta}^2 + \sin^2\theta\dot{\phi}^2) \tag{10.16}$$

—the radial equation of motion, with the 'centrifugal force' term added to the radial component of force. The second Lagrange's equation is

$$\frac{\mathrm{d}}{\mathrm{d}t}(mr^2\dot{\theta}) - mr^2\sin\theta\cos\theta\dot{\phi}^2 = rF_\theta$$

or

$$m(r\ddot{\theta} + 2\dot{r}\dot{\theta} - r\sin\theta\cos\theta\dot{\phi}^2) = F_\theta, \tag{10.17}$$

and the third is

$$\frac{\mathrm{d}}{\mathrm{d}t}(mr^2\sin^2\theta\dot{\phi}) = r\sin\theta F_\phi,$$

which is the equation for rate of change of angular momentum about axis O3. In full this is

$$m(r \sin\theta\ddot{\phi} + 2r\cos\theta\dot{\theta}\dot{\phi} + 2\dot{r}\sin\theta\dot{\phi}) = F_{\phi}. \tag{10.18}$$

The three equations (10.16), (10.17) and (10.18) in principle determine $r(t)$, $\theta(t)$ and $\phi(t)$ when F_r, F_θ and F_ϕ are known.

In the present case we have been able to write down the kinetic energy and the generalized forces in terms of the generalized coordinates directly from considering Figure 10.4. When this cannot easily be done, which is more likely when the generalized coordinates are not orthogonal, the same result can always be achieved systematically, although more lengthily, by substitution into the Cartesian forms. Thus, since

$$x_1 = r\sin\theta\cos\phi, \qquad x_2 = r\sin\theta\sin\phi, \qquad x_3 = r\cos\theta, \tag{10.19}$$

then

$$\left.\begin{aligned}\delta x_1 &= \sin\theta\cos\phi\delta r + r\cos\theta\cos\phi\delta\theta - r\sin\theta\sin\phi\delta\phi,\\ \delta x_2 &= \sin\theta\sin\phi\delta r + r\cos\theta\sin\phi\delta\theta + r\sin\theta\cos\phi\delta\phi,\\ \delta x_3 &= \cos\theta\,\delta r - r\sin\theta\delta\theta.\end{aligned}\right\} \tag{10.20}$$

If we substitute these in the expression

$$\delta\mathscr{W} = F_{x_1}\delta x_1 + F_{x_2}\delta x_2 + F_{x_3}\delta x_3$$

and pick out the coefficient of δr, which is by definition F_1, then

$$F_1 = F_{x_1}\sin\theta\cos\phi + F_{x_2}\sin\theta\sin\phi + F_{x_3}\cos\theta.$$

The right-hand side is the component of $\mathbf{F}$ along the radius vector (i.e. F_r). Hence

$$F_1 = F_r.$$

In a similar way we may show that

$$F_2 = rF_\theta, \qquad F_3 = r\sin\theta F_\phi.$$

On dividing equations (10.20) by δt we obtain expressions for the velocities,

$$\dot{x}_1 = \sin\theta\cos\phi\dot{r} + r\cos\theta\cos\phi\dot{\theta} - r\sin\theta\sin\phi\dot{\phi}, \quad \text{etc.}$$

Substituting these into the kinetic energy,

$$\mathscr{T} = \frac{m}{2}(\dot{x}_1^2 + \dot{x}_2^2 + \dot{x}_3^2),$$

we obtain

$$\mathscr{T} = \frac{m}{2}(\dot{r}^2 + r^2\dot{\theta}^2 + r^2\sin^2\theta\dot{\phi}^2).$$

The expression for the generalized forces and the kinetic energy agree with those obtained above (equations 10.12 and 10.13).

10.7. Constrained systems

When all the N particles of a system can move independently in space there will altogether be $3N$ degrees of freedom and the Lagrange's equations will be the $3N$ equations of motion in any chosen set of generalized coordinates. However, some of the particles may be constrained to slide upon wires or surfaces, and the number of degrees of freedom is thereby reduced. In such cases Lagrange's equations are automatically reduced to the minimum number required to determine the motion.

Suppose, for example, that a bead of mass m slides along a smooth circular wire of radius l. If the plane of the wire and its centre are fixed, the position of the bead can be determined completely by the angle θ of Figure 10.5, and its only degree of freedom is represented by variations in this coordinate.

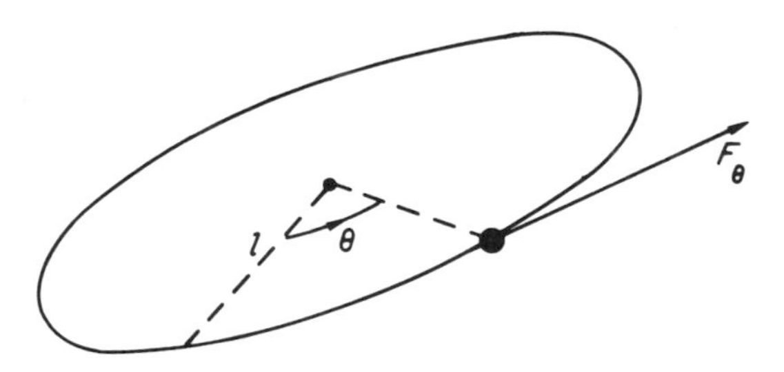

Figure 10.5

Then

$$\mathscr{T} = \tfrac{1}{2}ml^2\dot{\theta}^2,$$

$$\delta\mathscr{W} = F_\theta l\,\delta\theta.$$

In the latter only the tangential component, F_θ, of any external force appears. The normal component and the reaction of the wire on the bead (which, since it is smooth, is necessarily normal) are both perpendicular to the motion allowed by the degree of freedom and will therefore do no work.

The only Lagrange's equation

$$\frac{\mathrm{d}}{\mathrm{d}t}\left(\frac{\partial\mathscr{T}}{\partial\dot{\theta}}\right) - \frac{\partial\mathscr{T}}{\partial\theta} = lF_\theta$$

then gives

$$ml\ddot{\theta} = F_\theta.$$

This is sufficient, given F_θ and two boundary conditions, to determine the motion completely.

In general, since no motion perpendicular to a fixed constraining wire or surface, or in a radial direction at a fixed pivot, is possible, there will be no term in $\delta\mathscr{W}$ representing the work done by the components in such directions of the external force and of the reaction arising from the constraint. When the constraint is smooth the reaction is entirely perpendicular to the motion and disappears completely from $\delta\mathscr{W}$. There will also be no corresponding velocity component contributing to the kinetic energy. Thus the corresponding Lagrange's equation is absent and we are left only with those representing actual motion.

When thc particles constitute a rigid body constraints arise because they must remain at fixed distances from each other. Here, too, Lagrange's equations are the minimum required to determine the motion. For example two equal masses, m, connected by a light rigid rod of length $2l$ and constrained to move on a plane could have their positions determined by the Cartesian components, X_1, X_2, of the mid-point of the rod and by the angle, θ, that gives the orientation of the rod (see Figure 10.6).

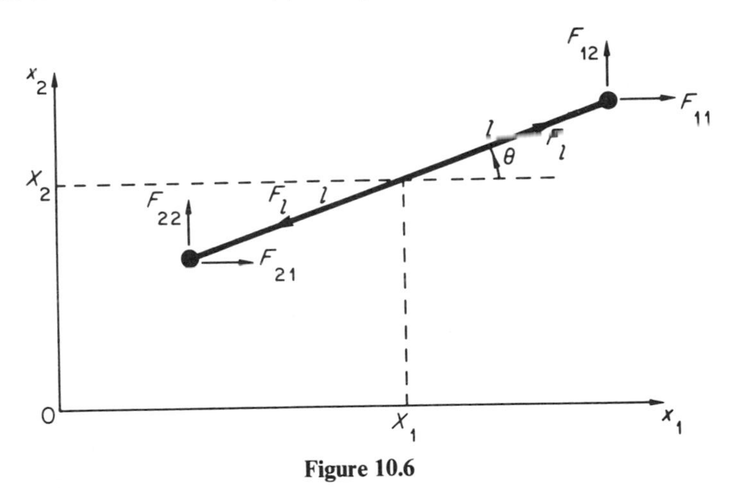

Figure 10.6

Suppose F_{11}, F_{12} and F_{21}, F_{22} are the components of the external forces acting on the particles in the plane of their motion and F_l is the force in the rod. When particle 1 moves there may be a component of motion δl along the direction of the rod and an amount of work $F_l\,\delta l$ will then be done by this force. Since the particles must remain at a fixed distance from each other, the simultaneous motion of particle 2 will necessarily have an equal component of motion, δl, in the same direction. The force exerted by the rod on the second particle will, however, be equal but *opposite* to that on the first and the work done will therefore be equal but of opposite sign.

Thus the contribution to $\delta\mathscr{W}$ from the rod force is zero and we need consider only the external forces:

$$\begin{aligned}\delta\mathscr{W} &= F_{11}\,\delta(X_1 + l\cos\theta) + F_{12}\,\delta(X_2 + l\sin\theta)\\ &\qquad + F_{21}\,\delta(X_1 - l\cos\theta) + F_{22}\,\delta(X_2 - l\sin\theta)\\ &= (F_{11} + F_{21})\,\delta X_1 + (F_{12} + F_{22})\,\delta X_2\\ &\qquad + [(-F_{11} + F_{21})\sin\theta + (F_{12} - F_{22})\cos\theta]l\,\delta\theta\\ &\equiv F_1\,\delta q_1 + F_2\,\delta q_2 + F_3\,\delta q_3.\end{aligned}$$

q_1, q_2 are the coordinates X_1, X_2 and F_1, F_2 are the components of the total external force in the direction of those coordinates. q_3 is the angle θ and F_3 is the total moment, G, of the external forces about the centre of the rod.

The kinetic energy may be written as

$$\mathscr{T} = \mathscr{T}_{\mathrm{cm}} + \mathscr{T}_{\mathrm{rel}}$$

and since the centre of the rod is the centre-of-mass in this example,

$$\mathscr{T} = m(\dot{X}_1^2 + \dot{X}_2^2) + ml^2\dot{\theta}^2.$$

Then the first Lagrange's equation,

$$\frac{\mathrm{d}}{\mathrm{d}t}\left(\frac{\partial\mathscr{T}}{\partial\dot{X}_1}\right) - \frac{\partial\mathscr{T}}{\partial X_1} = F_1,$$

becomes

$$2m\ddot{X}_1 = F_{11} + F_{21}. \tag{10.21}$$

Similarly, the second Lagrange's equation gives

$$2m\ddot{X}_2 = F_{12} + F_{22}. \tag{10.22}$$

The third Lagrange's equation,

$$\frac{\mathrm{d}}{\mathrm{d}t}\left(\frac{\partial\mathscr{T}}{\partial\dot{\theta}}\right) - \frac{\partial\mathscr{T}}{\partial\theta} = F_3, \tag{10.23}$$

becomes

$$2ml^2\ddot{\theta} = G = l[(-F_{11} + F_{21})\sin\theta + (F_{12} - F_{22})\cos\theta]$$

or

$$2ml\ddot{\theta} = (-F_{11} + F_{21})\sin\theta + (F_{12} - F_{22})\cos\theta. \tag{10.24}$$

Equations (10.21) and (10.22) give the motion of the centre-of-mass and equation (10.24) gives the rotation about this.

When the system is composed of many particles kept at *fixed* distances from each other it will similarly be true that since the forces between pairs of particles are equal and opposite (whether these are transmitted by light rods or springs, or, for example, by an electric field between them), the work done by them during any displacements of the system will total zero, and they will therefore not appear in Lagrange's equations.

10.8. Degrees of freedom of a rigid body

When only one particle is involved its maximum number of degrees of freedom is three. Adding a second particle could introduce another three degrees of freedom, but if it is rigidly spaced from the first there will be one equation of constraint reducing these to two. Thus a 'two-particle' rigid body can have at most five degrees of freedom.

A third particle could introduce three more, but if it is fixed relative to the first two the two equations of constraint reduce these to only one, giving a total of six degrees of freedom for a 'three-particle' rigid body. A fourth particle fixed relative to the first three will have three equations of constraint expressing this fact and will therefore introduce no new degrees of freedom; the same will be true for any additional particles. Thus any rigid body has at most six degrees of freedom, although there may be further constraints, such as a fixed point or axis in the body, or motion restricted to sliding on a plane, which will reduce these further.

Another way of looking at this is to say that three coordinates are required to locate a particular point fixed in the body; another two (angles) are required to give the direction of a fixed line in the body passing through this point; and a final sixth coordinate (another angle) is required to determine the orientation of the body about this line.

10.9. Conservative forces; the Lagrangian

Suppose the forces acting on a system are conservative, derivable from a potential energy, $\mathscr{V}$. Then

$$\delta\mathscr{W} = -\delta\mathscr{V} = F_1\,\delta q_1 + F_2\,\delta q_2 + \ldots + F_n\,\delta q_n.$$

Hence if we express $\mathscr{V}$ as a function of the q's, the generalized forces will be

$$F_1 = -\frac{\partial\mathscr{V}}{\partial q_1}, \ldots, \qquad F_n = -\frac{\partial\mathscr{V}}{\partial q_n}.$$

Moreover, $\mathscr{V}$ will be a function only of the positions of the particles, not of their velocities. Hence

$$\frac{\partial\mathscr{V}}{\partial\dot{q}_1} = \ldots = \frac{\partial\mathscr{V}}{\partial\dot{q}_n} = 0.$$

We can therefore rewrite Lagrange's equation as

$$\frac{\mathrm{d}}{\mathrm{d}t}\left(\frac{\partial \mathscr{T}}{\partial \dot{q}_j} - \frac{\partial \mathscr{V}}{\partial \dot{q}_j}\right) - \left(\frac{\partial \mathscr{T}}{\partial q_j} - \frac{\partial \mathscr{V}}{\partial q_j}\right) = 0,$$

or, if we define the *Lagrangian* as

$$\mathscr{L} = \mathscr{T} - \mathscr{V}, \tag{10.25}$$

they take the simple form

$$\frac{\mathrm{d}}{\mathrm{d}t}\left(\frac{\partial \mathscr{L}}{\partial \dot{q}_j}\right) - \frac{\partial \mathscr{L}}{\partial q_j} = 0; \qquad j = 1, \ldots, n. \tag{10.26}$$

According to equation (10.4) Lagrange's equations are equivalent to minimizing the time integral of the Lagrangian:

$$\int_{t(1)}^{t(2)} \delta \mathscr{L}\, \mathrm{d}t = \int_{t(1)}^{t(2)} \delta(\mathscr{T} - \mathscr{V})\, \mathrm{d}t = 0, \tag{10.27}$$

which is known as *Hamilton's principle.*

Although we may wish to incorporate some of the forces in this way into the Lagrangian, there may be some for which this is inconvenient or impossible. If $F'_1, \ldots, F'_n$ are such forces and $\delta\mathscr{W}'$ is the work *they* do in a perturbation of the system, while $F_1, \ldots, F_n$ are those derived from the potential $\mathscr{V}$, then

$$\begin{aligned}
\delta\mathscr{W}' &= F'_1\, \delta q_1 + \ldots + F'_n\, \delta q_n, \\
-\delta\mathscr{V} &= F_1\, \delta q_1 + \ldots + F_n\, \delta q_n, \\
&= -\frac{\partial \mathscr{V}}{\partial q_1}\delta q_1 - \ldots - \frac{\delta \mathscr{V}}{\partial q_n}\delta q_n,
\end{aligned}$$

and the total work done is

$$\delta\mathscr{W} = -\delta\mathscr{V} + \delta\mathscr{W}' = \left(F'_1 - \frac{\partial \mathscr{V}}{\partial q_1}\right)\delta q_1 + \ldots + \left(F'_n - \frac{\partial \mathscr{V}}{\partial q_n}\right)\delta q_n.$$

Hence Lagrange's equations (10.11) may be written

$$\frac{\mathrm{d}}{\mathrm{d}t}\left(\frac{\partial \mathscr{T}}{\partial \dot{q}_j}\right) - \frac{\partial \mathscr{T}}{\partial q_j} = F'_j - \frac{\partial \mathscr{V}}{\partial q_j} \tag{10.28}$$

or

$$\frac{\mathrm{d}}{\mathrm{d}t}\left(\frac{\partial \mathscr{L}}{\partial \dot{q}_j}\right) - \frac{\partial \mathscr{L}}{\partial q_j} = F'_j: \qquad j = 1, \ldots, n, \tag{10.29}$$

where the F'_j are those generalized forces not included in the Lagrangian, $\mathscr{L}$.

10.10. Internal forces; deformable systems

So far we have discussed simple problems for which an elementary approach is as direct a method of solution as Lagrange's equations. Since the latter are derived from the application of Newton's laws to all the particles of a system, it is always true that, in principle, we could start with these laws and arrive at the same results. We have already seen, however, that in general the Lagrangian approach requires fewer equations to be solved.

It is also often the case that, when a convenient description of a system requires angle or distance coordinates of one part of a system relative to another, it may be difficult to be sure about the correct expression of Newton's laws. To illustrate this we shall extend the analysis of a simple pendulum to include the case of a yielding support. The model for this is shown in Figure 10.7, the pivot A moving horizontally on a smooth slide

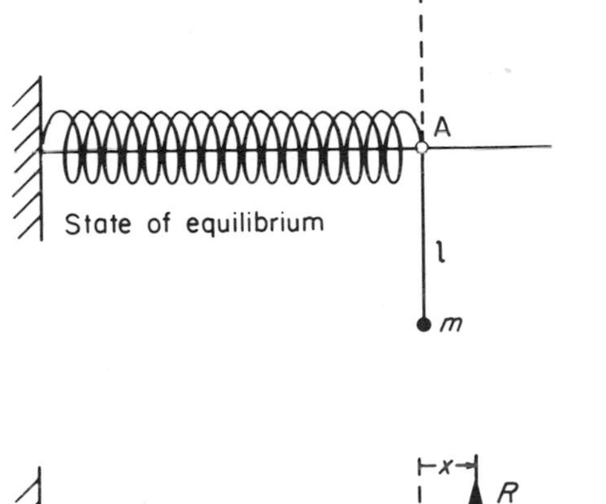

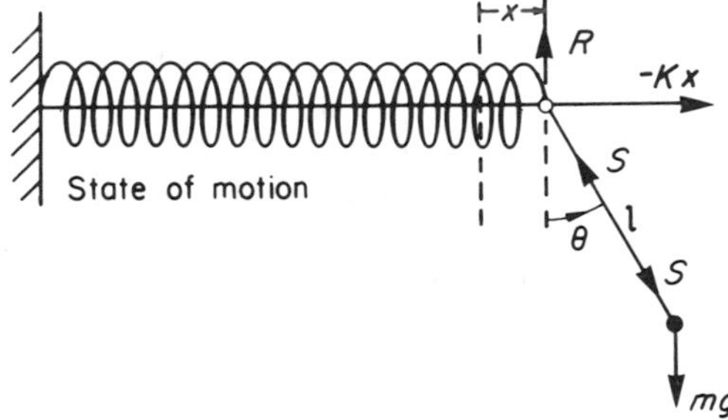

Figure 10.7

under the action of a spring. The whole system of slide and pendulum thus has within it a deformable part, represented by the spring.

The natural and clearest way of describing the system is by the displacement, x, of the pivot from the fixed equilibrium point on the horizontal slide and the inclination, θ, of the pendulum relative to a vertical line through the pivot. The forces acting on the system are the reaction, R, of the slide on the pivot; the gravitational force, mg, on the mass m; the spring force, $-Kx$; and the force, S, in the pendulum rod.

The elementary approach (force = mass × acceleration) might run as follows:

(1) The pivot has no mass. Hence, to avoid infinite acceleration, the forces must balance there.

(2) The mass m will have a horizontal acceleration, $\ddot{x}$, arising from the motion of the pivot, on which is superimposed an acceleration arising from the angular rotation of the pendulum, with components $-l\dot{\theta}^2$ and $l\ddot{\theta}$ along and perpendicular to the rod. Their resultant acceleration is caused by the resultant of the gravitational force and that exerted by the rod.

We have now to decide how best to express these ideas. If we resolve the first vector equation horizontally,

$$S \sin\theta + Kx = 0. \tag{10.30}$$

If we resolve the second one along and perpendicular to the rod,

$$S + mg\cos\theta = m(\ddot{x}\sin\theta - l\dot{\theta}^2), \tag{10.31}$$

$$-mg\sin\theta = m(\ddot{x}\cos\theta + l\ddot{\theta}). \tag{10.32}$$

We cannot avoid bringing in the force S, although it will not figure in the $x(t)$, $\theta(t)$ description of the motion. The first step is therefore to eliminate it. Here it is simple to do so; equations (10.30) and (10.31) give

$$m\sin^2\theta\ddot{x} - ml\sin\theta\dot{\theta}^2 = mg\sin\theta\cos\theta - Kx. \tag{10.33}$$

This and equation (10.32) have to be solved to find $x(t)$ and $\theta(t)$.

Although this process looks quite simple, it has been necessary to exercise some judgment to start with only three equations, and those in a form which readily reduces to the two eventual ones. With a less happy choice we might well have found ourselves with four equations involving components of both S and R, and faced with complicated and error-prone algebra in order to eliminate them.

Such difficulties are dealt with automatically in the Lagrangian method. Here the considerations are:

(1) Kinetic energy. There is only one mass, m, with a velocity arising from the horizontal $\dot{x}$ of the pivot on which is superimposed the $l\dot{\theta}$ of the rod perpendicular to its length.

(2) Work. The gravitational force can be derived from a potential energy, which depends on the height of m relative to some fixed horizontal plane (that through the slide is a convenient choice and gives a potential energy $-mgl\cos\theta$). When the pivot moves a distance δx along the slide

the internal force controlling it will contribute the work $-Kx\,\delta x$. Neither the reaction of the slide nor the force in the rigid rod can do any work and can therefore be ignored.

Subsequent formulation demands no choice or judgment.

$$\mathscr{T} = \frac{m}{2}[(\dot{x} + l\dot{\theta}\cos\theta)^2 + l^2\dot{\theta}^2\sin^2\theta],$$

$$\mathscr{V} = -mgl\cos\theta,$$

$$\delta\mathscr{W}' = -Kx\,\delta x.$$

Hence, with x, θ as the two generalized coordinates,

$$\mathscr{L} = \mathscr{T} - \mathscr{V} = \frac{m}{2}(\dot{x}^2 + 2l\cos\theta\dot{x}\dot{\theta} + l^2\dot{\theta}^2) + mgl\cos\theta, \qquad (10.34)$$

$$\frac{\partial\mathscr{L}}{\partial\dot{x}} = m(\dot{x} + l\cos\theta\dot{\theta}), \qquad \frac{\partial\mathscr{L}}{\partial\dot{\theta}} = m(l\cos\theta\dot{x} + l^2\dot{\theta}),$$

$$\frac{\partial\mathscr{L}}{\partial x} = 0, \qquad \frac{\partial\mathscr{L}}{\partial\theta} = -ml\sin\theta\dot{x}\dot{\theta} - mgl\sin\theta,$$

$$F_1' = -Kx, \qquad F_2' = 0.$$

The two Lagrange's equations are then

$$m(\ddot{x} + l\cos\theta\ddot{\theta} - l\sin\theta\dot{\theta}^2) = -Kx \qquad (10.35)$$

and

$$ml(\cos\theta\ddot{x} - \sin\theta\dot{\theta}\dot{x} + l\ddot{\theta}) + ml(\sin\theta\dot{x}\dot{\theta} + g\sin\theta) = 0$$

or

$$\cos\theta\ddot{x} + l\ddot{\theta} = -g\sin\theta, \qquad (10.36)$$

which are equivalent to equations (10.32) and (10.33).

When x and θ are small enough to neglect second-order quantities, equations (10.35) and (10.36) simplify to

$$\ddot{x} + l\ddot{\theta} = -\frac{Kx}{m},$$

$$\ddot{x} + l\ddot{\theta} = -g\theta.$$

Hence

$$x = \frac{mg\theta}{K}$$

and the two Lagrange's equations become

$$\ddot{\theta}\left(l + \frac{mg}{k}\right) = -g\theta. \tag{10.37}$$

This interesting feature of Lagrange's equations, in which two of them become identical, always occurs when two or more of the generalized coordinates concern a mass-less pivot in the system. Although they may be required to describe the configuration of the system effectively, all of them will not be necessary to describe the dynamics.

Equation (10.37) is the familiar equation for simple harmonic motion. Since the angular frequency is now given by

$$\omega^2 = \frac{g}{l + mg/K},$$

we can see that the yielding pivot increases the effective length of the single pendulum to

$$l' = l + mg/K.$$

This example shows that when any part of a system is deformable, any internal force enters into the Lagrangian description in just the same way as any external force. The distinction between forces, as they are treated by the Lagrangian, is not whether they are external or internal, but whether they are *motion forces* that can do work during a perturbation or *constraint forces* which cannot.

All motion forces have the same standing in Lagrange's equations. In particular, if an internal motion force can be derived from an internal potential energy, then this can be included, if we wish, in the Lagrangian. In the example we are considering the potential energy of the spring is $\frac{1}{2}Kx^2$ and we could have used, instead of equation (10.34), the Lagrangian

$$\mathscr{L} = \frac{m}{2}(\dot{x}^2 + 2l\cos\theta\dot{x}\dot{\theta} + l^2\dot{\theta}^2) + mgl\cos\theta - \tfrac{1}{2}Kx^2.$$

By including the spring force in this way F'_1 would disappear, but we would now have

$$\frac{\partial \mathscr{L}}{\partial x} = -Kx,$$

and the first Lagrange's equation would turn out the same as before (equation 10.35). (See Section 10.17.2, page 320.)

10.11. Internal forces; artificial deformation

In the previous problem we had an example of how forces arising from rigid constraints (the reaction of the slide) and within rigid parts of a system (the force in the rod) are automatically excluded from Lagrange's equations. Can we ever find such forces using Lagrangian methods?

The same problem gives us a clue about this. There was an internal force in the system (the spring force) which *was* involved. This was because the system could be deformed in such a way as to allow this force to do some work. It was a motion force. Any constraint force, internal or external, can be brought back into Lagrange's equations as a motion force by allowing the system to deform in a way that allows the force to do some work. Physically we can imagine doing this by inserting a very strong, but very short, spring at any point at which the force acts.

Thus if we require the reaction, R, of the slide on the pivot we imagine a second short spring lifting the pivot off the slide by a small distance y, as shown in Figure 10.8. It is not necessary to write down the potential energy

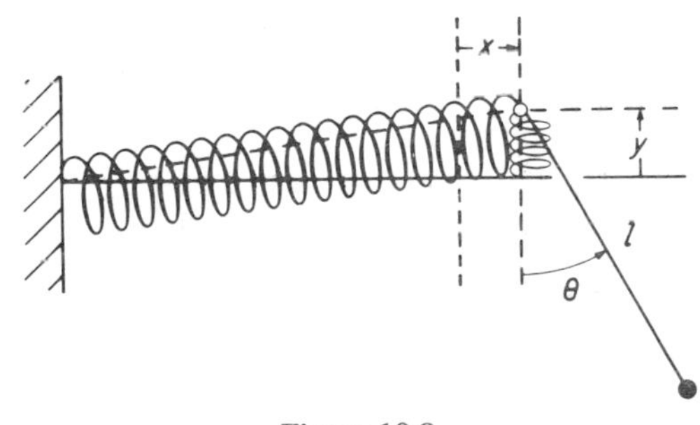

Figure 10.8

in this spring. Indeed, in order to find R it is simpler not to do so but instead to observe that, in addition to the forces accounted for by the potential energy of the first spring and the gravitational potential energy, R will add to the work done an amount

$$\delta \mathscr{W}' = R\,\delta y. \tag{10.38}$$

Variations in y allow another degree of freedom to the system and we must be careful to allow for this in the Lagrangian. The pivot and therefore the bob will have an additional vertical component of velocity, $\dot{y}$, and the kinetic energy will now be

$$\mathscr{T} = \frac{m}{2}[(\dot{x} + l\dot{\theta}\cos\theta)^2 + (\dot{y} + l\dot{\theta}\sin\theta)^2].$$

Raising the pivot a distance y will also raise the bob and increase its gravitational potential energy, although, since the allowed movement is assumed to be small and at right angles to the first spring, the potential energy of that spring will not be changed to first order. Hence

$$\mathscr{V} = mg(y - l\cos\theta) + \frac{Kx^2}{2}.$$

Thus the Lagrangian becomes

$$\mathscr{L} = \mathscr{T} - \mathscr{V} = \frac{m}{2}(\dot{x}^2 + \dot{y}^2 + l^2\dot{\theta}^2 + 2l\cos\theta\dot{x}\dot{\theta} + 2l\sin\theta\dot{y}\dot{\theta})$$

$$+ mg(l\cos\theta - y) - \frac{Kx^2}{2}.$$

This and equation (10.33) give the following Lagrange's equations corresponding to x, θ and y:

$$\frac{\mathrm{d}}{\mathrm{d}t}(m\dot{x} + ml\cos\theta\dot{\theta}) = -Kx, \tag{10.39}$$

$$\frac{\mathrm{d}}{\mathrm{d}t}(ml\cos\theta\dot{x} + ml^2\dot{\theta} + ml\sin\theta\dot{y}) = -mgl\sin\theta - ml\sin\theta\dot{x}\dot{\theta} + ml\cos\theta\dot{y}\dot{\theta}, \tag{10.40}$$

$$\frac{d}{\mathrm{d}t}(m\dot{y} + ml\sin\theta\dot{\theta}) = -mg + R. \tag{10.41}$$

Apparently we now have three complicated equations to solve for x, θ and y. However, we are only concerned with the particular case when the freedom of the pivot to move in the y-direction is reduced to zero, and it is really constrained to move along the horizontal slide. This is the case

$$y = \dot{y} = \ddot{y} = 0.$$

Thus equations (10.39) and (10.40) are just the same as before,

$$m(\ddot{x} + l\cos\theta\ddot{\theta} - l\sin\theta\dot{\theta}^2) = -Kx,$$

$$\cos\theta\ddot{x} + l\ddot{\theta} = -g\sin\theta,$$

while equation (10.41) gives the reaction

$$R = mg + ml(\cos\theta\dot{\theta}^2 + \sin\theta\ddot{\theta}).$$

The second and third terms are just the vertical component of the (mass × acceleration) of the bob, as we might expect.

10.12. Time-dependent motion

Lagrange's equations are sufficiently general to allow $\mathcal{T}$ or $\mathcal{L}$ to depend explicitly upon time, but we have not, so far, exploited this fact. As an example we shall consider a pendulum with a horizontally moving pivot, as in Section 10.10, this time controlled not by a spring but by a mechanism that moves it in some predetermined way, $x(t)$, depending upon time. The kinetic energy will, of course, still involve $\dot{x}$, and will be the same as in Section 10.10:

$$\mathcal{T} = \frac{m}{2}\left[\{\dot{x}(t) + l\dot{\theta}\cos\theta\}^2 + l^2\dot{\theta}^2\sin^2\theta\right],$$

where $\dot{x}(t)$ is written thus to emphasize the fact that $\dot{x}$ is a known function of time. The potential energy is simply

$$\mathcal{V} = -mgl\cos\theta.$$

Then

$$\mathcal{L} = \frac{m}{2}\left[\dot{x}^2(t) + 2l\cos\theta\dot{x}(t)\dot{\theta} + l^2\dot{\theta}^2\right] + mgl\cos\theta,$$

and there is only one Lagrange's equation, which is the same as (10.36)—the second Lagrange's equation of Section 10.10:

$$\cos\theta\,\ddot{x}(t) + l\ddot{\theta} = -g\sin\theta.$$

For small values of θ this is approximately

$$l\ddot{\theta} + g\theta = -\ddot{x}(t).$$

Consequently, when the pivot is moving with a constant velocity, this has no effect upon the 'θ' motion of the pendulum. This is an example of the fact, noted in Chapter 8, that mechanical phenomena are the same in all frames of reference in uniform relative motion. Only an acceleration of the pivot will be noticed. If, for instance, this acceleration is constant,

$$\ddot{x}(t) = a,$$

the oscillations will be about the position of equilibrium given by

$$\tan\theta = -\frac{a}{g},$$

so that the equilibrium angle of inclination in an *accelerated* frame *would* serve to distinguish it from those with constant velocities or different accelerations.

10.13. Charged particle in an electromagnetic field

We saw in Section 10.9 that when the forces acting on a system are conservative it is possible to incorporate the corresponding potential energy into the Lagrangian (equation 10.25),

$$\mathscr{L} = \mathscr{T} - \mathscr{V}, \tag{10.42}$$

and then to rewrite Lagrange's equations,

$$\frac{\mathrm{d}}{\mathrm{d}t}\left(\frac{\partial \mathscr{T}}{\partial \dot{q}_j}\right) - \frac{\partial \mathscr{T}}{\partial q_j} = F_j, \tag{10.43}$$

in the concise form

$$\frac{\mathrm{d}}{\mathrm{d}t}\left(\frac{\partial \mathscr{L}}{\partial \dot{q}_j}\right) - \frac{\partial \mathscr{L}}{\partial q_j} = 0. \tag{10.44}$$

The first term of one such equation is still really only $\mathrm{d}/\mathrm{d}t(\partial \mathscr{T}/\partial \dot{q}_j)$, since $\mathscr{V}$ is a function only of the generalized coordinates q_j and not of their 'velocities' $\dot{q}_j$. For gravitational and electrostatic problems, where a potential energy of this type is relevant, equations (10.42) and (10.44) are applicable.

However, when a charge q is subject to a general electromagnetic field described by the electric field $\mathbf{E}$ and the magnetic induction $\mathbf{B}$ (both of which may be time-varying), the force on it is

$$\mathbf{F} = q\mathbf{E} + q\mathbf{v} \wedge \mathbf{B}, \tag{10.45}$$

where $\mathbf{v}$ is the velocity of the charge. Since this force has a velocity-dependent part we cannot expect it to be derivable from a potential energy of the type we have encountered so far. Even the part $q\mathbf{E}$ cannot be derived from a potential energy when $\mathbf{E}$ varies with time.

Nevertheless, equation (10.43) shows that if we can find a function $\mathscr{V}$ such that

$$F_j = -\frac{\partial \mathscr{V}}{\partial q_j} + \frac{\mathrm{d}}{\mathrm{d}t}\left(\frac{\partial \mathscr{V}}{\partial \dot{q}_j}\right), \tag{10.46}$$

we can define a Lagrangian by equation (10.42) for which equations (10.44) will still be true. It is possible to do this if we use the vector potential $\mathbf{A}$ as well as the scalar potential ϕ to describe the electromagnetic field.

In terms of these potentials,

$$\mathbf{E} = -\mathbf{grad}\,\phi - \frac{\partial \mathbf{A}}{\partial t} \equiv -\nabla\phi - \frac{\partial \mathbf{A}}{\partial t}, \tag{10.47}$$

$$\mathbf{B} = \mathbf{curl}\,\mathbf{A} \equiv \nabla \wedge \mathbf{A}. \tag{10.48}$$

If we consider a single particle of charge q in a Cartesian frame, its coordinates will be x_1, x_2, x_3, and the first component of equation (10.45) will be, using equations (10.47) and (10.48),

$$F_1 = -q\left[\frac{\partial \phi}{\partial x_1} + \frac{\partial A_1}{\partial t}\right] + q\left[v_2\left(\frac{\partial A_2}{\partial x_1} - \frac{\partial A_1}{\partial x_2}\right) - v_3\left(\frac{\partial A_1}{\partial x_3} - \frac{\partial A_3}{\partial x_1}\right)\right]. \tag{10.49}$$

Since x_1, x_2, x_3 occur explicitly only in ϕ and $\mathbf{A}$, we may group the terms on the right-hand side of equation (10.49) as follows:

$$F_1 = -q\frac{\partial}{\partial x_1}(\phi - v_2A_2 - v_3A_3) - q\left(v_2\frac{\partial A_1}{\partial x_2} + v_3\frac{\partial A_1}{\partial x_3} + \frac{\partial A_1}{\partial t}\right).$$

By adding and subtracting $qv_1(\partial A_1/\partial x_1)$ this may be written as

$$\begin{aligned} F_1 &= -q\frac{\partial}{\partial x_1}(\phi - v_1A_1 - v_2A_2 - v_3A_3) \\ &\quad -q\left(v_1\frac{\partial A_1}{\partial x_1} + v_2\frac{\partial A_1}{\partial x_2} + v_3\frac{\partial A_1}{\partial x_3} + \frac{\partial A_1}{\partial t}\right) \\ &= -q\frac{\partial}{\partial x_1}(\phi - \mathbf{v}\,.\,\mathbf{A}) - q\frac{\mathrm{d}A_1}{\mathrm{d}t}, \end{aligned} \tag{10.50}$$

noting that $\mathrm{d}/\mathrm{d}t$ means differentiation following the motion of the particle. (See Sections 10.17.3 and 10.17.4, pages 326 and 327.)

Now in the expression $\phi - \mathbf{v}\,.\,\mathbf{A}$, $\dot{x}_1$ (or v_1) occurs explicitly only in the first term, $\dot{x}_1A_1$, of the scalar product. Hence

$$\frac{\partial}{\partial \dot{x}_1}(\phi - \mathbf{v}\,.\,\mathbf{A}) = -A_1.$$

Then equation (10.50) may be written as

$$F_1 = -q\frac{\partial}{\partial x_1}(\phi - \mathbf{v}\,.\,\mathbf{A}) + q\frac{\mathrm{d}}{\mathrm{d}t}\left[\frac{\partial}{\partial \dot{x}_1}(\phi - \mathbf{v}\,.\,\mathbf{A})\right], \tag{10.51}$$

to which we may add similar equations for the other two components. Comparing equations (10.51) and (10.46) we can see that a 'potential energy' suitable for the Lagrangian is

$$\mathscr{V} = q(\phi - \mathbf{v}\,.\,\mathbf{A}). \tag{10.52}$$

So far we have shown only that the $\mathscr{V}$ so defined will give Lagrange's equations (10.44) *in a Cartesian frame.* However, this is sufficient to show that (equation 10.27)

$$\int_{t(1)}^{t(2)} \delta\mathscr{L}\,\mathrm{d}t = 0. \qquad (10.53)$$

This minimal property of the Lagrangian is no longer tied to any particular coordinate system. We may therefore re-express this property, as in Section 10.9, in the more general form (10.44) of Lagrange's equations. The Lagrangian is, then,

$$\mathscr{L} = \tfrac{1}{2}m\mathbf{v}^2 - q(\phi - \mathbf{v}\,.\,\mathbf{A}). \qquad (10.54)$$

(See Section 10.17.5, page 329.)

10.14. Generalized momenta; constants of motion

For a single particle of mass m moving in a three-dimensional conservative field of force the Lagrangian is, using Cartesian coordinates,

$$\mathscr{L} = \mathscr{T} - \mathscr{V} = \frac{m}{2}(\dot{x}_1^2 + \dot{x}_2^2 + \dot{x}_3^2) - \mathscr{V}(x_1, x_2, x_3).$$

Hence the first Lagrange equation,

$$\frac{\mathrm{d}}{\mathrm{d}t}\left(\frac{\partial\mathscr{L}}{\partial\dot{x}_1}\right) - \frac{\partial\mathscr{L}}{\partial x_1} = 0,$$

is simply the first component of Newton's second law:

$$\frac{\mathrm{d}}{\mathrm{d}t}(m\dot{x}_1) + \frac{\partial\mathscr{V}}{\partial x_1} = 0$$

or

$$\frac{\mathrm{d}}{\mathrm{d}t}(p_1) = F_1,$$

where p_1 and F_1 are the first components of the momentum and force. The two remaining Lagrange equations give the other two components of Newton's law.

When $\mathscr{L}$ does not contain x_1, which in this case implies that $\mathscr{V}$ is independent of x_1 and that there is no component of force in direction 1, then

$$\frac{\mathrm{d}}{\mathrm{d}t}(m\dot{x}_1) = -\frac{\partial\mathscr{V}}{\partial x_1} = 0.$$

Thus the component of momentum, p_1, is constant.

By analogy with these results the quantities $\partial\mathscr{L}/\partial\dot{q}_j$, obtained from the Lagrangian when it is expressed in generalized coordinates, are called the *generalized momenta.* From the Lagrange equation,

$$\frac{\mathrm{d}}{\mathrm{d}t}\left(\frac{\partial\mathscr{L}}{\partial\dot{q}_j}\right) - \frac{\partial\mathscr{L}}{\partial q_j} = 0,$$

we can see that when q_j is absent from the Lagrangian the corresponding momentum,

$$p_j = \frac{\partial\mathscr{L}}{\partial\dot{q}_j},$$

is a *constant of the motion.* (The p_j are sometimes called *canonical* or *conjugate* momenta. A coordinate q_j that is absent from the Lagrangian is called a *cyclic* coordinate.)

As one example let us consider the Lagrangian of two particles of equal mass m, connected by a spring of natural length $2l$, moving in a plane under no external forces. As Lagrangian coordinates we shall use X_1, X_2, which define the spring centre; r, the distance of either particle from this centre; and θ, the orientation of the spring axis (see Figure 10.9).

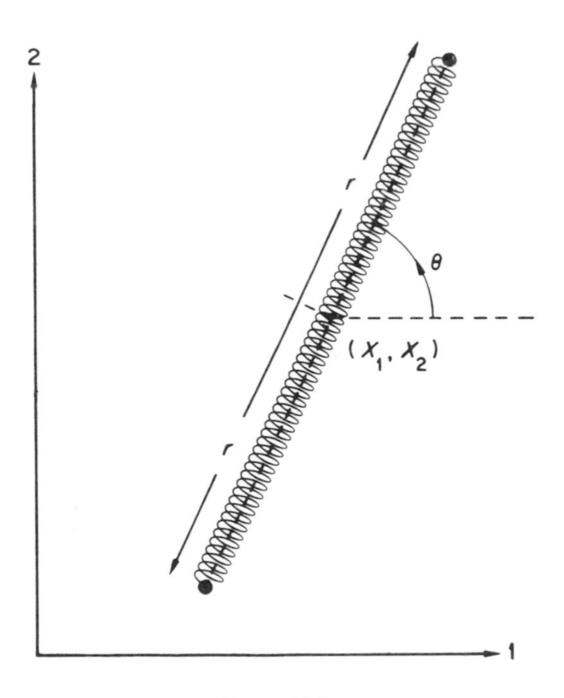

Figure 10.9

Since the spring centre is the centre-of-mass of the system, the total energy is

$$\mathcal{T} = \mathcal{T}_{\text{cm}} + \mathcal{T}_{\text{rel}}$$
$$= m(\dot{X}_1^2 + \dot{X}_2^2) + m(\dot{r}^2 + r^2\dot{\theta}^2).$$

The potential energy is that stored in the spring,

$$\mathcal{V} = \tfrac{1}{2}K(2r - 2l)^2 = 2K(r - l)^2,$$

where K is the strength constant of the spring. Then the Lagrangian is

$$\mathcal{L} = \mathcal{T} - \mathcal{V} = m(\dot{X}_1^2 + \dot{X}_2^2 + \dot{r}^2 + r^2\dot{\theta}^2) - 2K(r - l)^2.$$

In this X_1, X_2 and θ are cyclic coordinates and we therefore immediately have the three constants of motion:

$$\frac{\partial \mathcal{L}}{\partial \dot{X}_1} = 2m\dot{X}_1 = \text{constant},$$

$$\frac{\partial \mathcal{L}}{\partial \dot{X}_2} = 2m\dot{X}_2 = \text{constant},$$

$$\frac{\partial \mathcal{L}}{\partial \dot{\theta}} = 2mr^2\dot{\theta} = \text{constant}.$$

The first two express conservation of total momentum and the third conservation of angular momentum relative to the centre-of-mass for a system free of external forces.

For another example consider a charged particle in an electromagnetic field. In three-dimensional Cartesian coordinates the Lagrangian (10.54) becomes

$$\mathcal{L} = \tfrac{1}{2}m(\dot{x}_1^2 + \dot{x}_2^2 + \dot{x}_3^2) - q\phi + q(\dot{x}_1A_1 + \dot{x}_2A_2 + \dot{x}_3A_3).$$

ϕ, A_1, A_2, A_3 may all be functions of x_1, x_2, x_3 and t, but not of $\dot{x}_1$, $\dot{x}_2$, $\dot{x}_3$. Hence the three generalized momenta are

$$p_1 = \frac{\partial \mathcal{L}}{\partial \dot{x}_1} = m\dot{x}_1 + qA_1, \qquad p_2 = m\dot{x}_2 + qA_2, \qquad p_3 = m\dot{x}_3 + qA_3.$$

Thus if x_1 is a cyclic coordinate the corresponding constant of motion is not the (mechanical) momentum $m\dot{x}_1$, but the generalized momentum $m\dot{x}_1 + qA_1$.

10.15. Hamilton's equations

The laws and concepts we have used so far have all been based upon a description of a system in terms of the positions of all its parts as functions of time. Thus $x_1, x_2, \ldots, x_{3N}$ would describe the coordinates of an N-particle system and F_i, the force component corresponding to the ith coordinate, could depend upon all these coordinates, their speeds and possibly explicitly upon time. There will then be $3N$ equations of motion of the form:

$$m_i\ddot{x}_i = F_i(x_1, x_2, \ldots, x_{3N}; \dot{x}_1, \dot{x}_2, \ldots, \dot{x}_{3N}; t). \tag{10.55}$$

'Solving' a problem about the system usually means finding solutions of the equations (10.55) in the form:

$$x_i \equiv x_i(t), \qquad i = 1, 2, \ldots, 3N.$$

Each of these solutions contains, in general, two constants which can be chosen to fit two boundary conditions such as the initial values and speeds of each coordinate.

In using generalized coordinates and recasting Newton's laws as Lagrange's equations, which are also second-order differential equations, we have maintained the same basic description. From this viewpoint, velocity and momentum appear as derived, or secondary, dynamical quantities. There are, however, considerable advantages in considering position and momentum, or, more generally, generalized coordinates and generalized momenta, on an equal footing, particularly when the transitions from classical mechanics to statistical and quantum mechanics are to be made.

Let us consider first a simple harmonic oscillator in one dimension. Starting with the equation of motion,

$$m\ddot{x} = -Kx, \tag{10.56}$$

we may write the solution as

$$x = A\cos(\omega_0 t + \alpha), \qquad \omega_0^2 = \frac{K}{m}.$$

From this we may derive the velocity,

$$\dot{x} = -\omega_0 A \sin(\omega_0 t + \alpha), \tag{10.57}$$

and the momentum is then

$$p = m\dot{x} = -\omega_0 m A \sin(\omega_0 t + \alpha). \tag{10.58}$$

Now the total energy is

$$\mathscr{E} = \mathscr{T} + \mathscr{V} = \tfrac{1}{2}m\dot{x}^2 + \tfrac{1}{2}Kx^2.$$

If, using equation (10.58), we rewrite this as

$$\mathscr{E} = \frac{p^2}{2m} + \frac{Kx^2}{2}, \tag{10.59}$$

we can think of x as a 'position coordinate' and p as a 'momentum coordinate', each contributing in a similar way (as its square) to the total energy, with $K/2$ and $1/(2m)$ as the corresponding constants. From equation (10.59) we can see that these 'coordinates' define an ellipse with semi-axes $\sqrt{2\mathscr{E}/K}$, $\sqrt{2m\mathscr{E}}$, as shown in Figure 10.10. The two-dimensional x, p plane of Figure 10.10 is

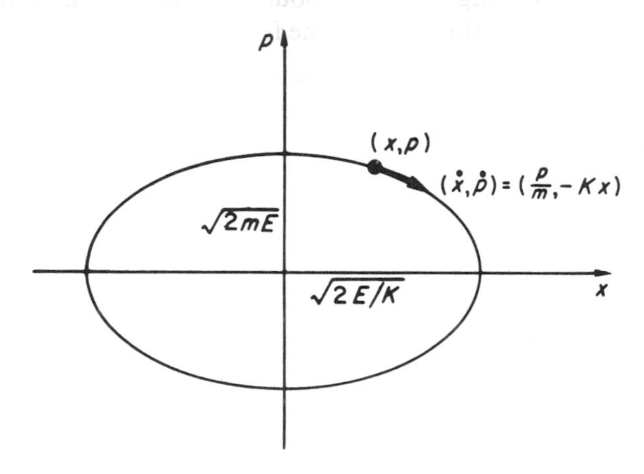

Figure 10.10

known as *phase-space*. Each point in phase-space on the ellipse defined by a particular value of $\mathscr{E}$ gives a possible pair of simultaneous values of x and p. Moreover, the way in which the point moves around the ellipse is determined by equation (10.59), since, using equations (10.58) and (10.56), its velocity components in phase-space may be expressed as

$$\dot{x} = \frac{p}{m} = \frac{\partial \mathscr{E}}{\partial p}, \tag{10.60}$$

$$\dot{p} = -Kx = -\frac{\partial \mathscr{E}}{\partial x}. \tag{10.61}$$

We can therefore regard the expression (10.59) for the total energy as the fundamental description of the system (note that it contains the same two constants, m and K, as equation (10.56)) and derive from it, using the two first-order equations (10.60) and (10.61), the development of the motion from any given pair of values, x and p. The three equations (10.59), (10.60) and (10.61) are thus equivalent in every way to the equations (10.56), (10.57) and (10.58). We may also note that the law of energy conservation can be deduced from the 'new' equations as follows:

$$\begin{aligned}\frac{\mathrm{d}\mathscr{E}}{\mathrm{d}t} &= \frac{\partial \mathscr{E}}{\partial x}\frac{\mathrm{d}x}{\mathrm{d}t} + \frac{\partial \mathscr{E}}{\partial p}\frac{\mathrm{d}p}{\mathrm{d}t} \\ &= \frac{\partial \mathscr{E}}{\partial x}\frac{\partial \mathscr{E}}{\partial p} - \frac{\partial \mathscr{E}}{\partial p}\frac{\partial \mathscr{E}}{\partial x} = 0.\end{aligned}$$

It is not necessary for the total energy to be constant in adopting this new viewpoint. Consider, for example, the system discussed in Section 10.17.4, page 327. The total energy is

$$\mathscr{E} = \mathscr{T} + \mathscr{V}$$

$$= \mathscr{T}_0 + \tfrac{1}{2}m(\dot{x}_1^2 + \dot{x}_2^2 + \dot{x}_3^2) + \frac{q^2}{4\pi\varepsilon_0}\frac{1}{[(x_1 - ut)^2 + x_2^2 + x_3^2]^{1/2}},$$

where $\mathscr{T}_0$ is the constant kinetic energy of the charged particle which is kept moving at a constant velocity by the external force. Rewritten in terms of the momenta and position coordinates,

$$\mathscr{E} = \mathscr{T}_0 + \frac{1}{2m}(p_1^2 + p_2^2 + p_3^2) + \frac{q^2}{4\pi\varepsilon_0}\frac{1}{[(x_1 - ut)^2 + x_2^2 + x_3^2]^{1/2}}. \tag{10.62}$$

In this case the condition of the system is described by a 'point' $(x_1, x_2, x_3, p_1, p_2, p_3)$ in six-dimensional phase-space and by t also.

The definition of the momenta and their rates of change,

$$p_1 = m\dot{x}_1, \qquad \dot{p}_1 = -\frac{\partial \mathscr{V}}{\partial x_1}, \quad \text{etc.,}$$

together with equation (10.62), show that

$$\dot{x}_1 = \frac{\partial \mathscr{E}}{\partial p_1}, \qquad \dot{x}_2 = \frac{\partial \mathscr{E}}{\partial p_2}, \qquad \dot{x}_3 = \frac{\partial \mathscr{E}}{\partial p_3}, \tag{10.63}$$

$$\dot{p}_1 = -\frac{\partial \mathscr{E}}{\partial x_1}, \qquad \dot{p}_2 = -\frac{\partial \mathscr{E}}{\partial x_2}, \qquad \dot{p}_3 = -\frac{\partial \mathscr{E}}{\partial x_3}, \tag{10.64}$$

since the third term in $\mathscr{E}$ is $\mathscr{V}(x_1, x_2, x_3, t)$ and is the only one containing the space coordinates.

Thus we have six first-order equations, (10.63) and (10.64), of the same form as (10.60) and (10.61), which determine the motion from the expression (10.62) for the total energy. Since $\mathscr{E}$ now contains t explicitly, its variation with time is given by

$$\begin{aligned}\frac{d\mathscr{E}}{dt} &= \sum_{i=1}^{3}\frac{\partial \mathscr{E}}{\partial x_i}\frac{dx_i}{dt} + \sum_{i=1}^{3}\frac{\partial \mathscr{E}}{\partial p_i}\frac{dp_i}{dt} + \frac{\partial \mathscr{E}}{\partial t}\\ &= \sum_{i=1}^{3}\left[\frac{\partial \mathscr{E}}{\partial x_i}\frac{\partial \mathscr{E}}{\partial p_i} - \frac{\partial \mathscr{E}}{\partial p_i}\frac{\partial \mathscr{E}}{\partial x_i}\right] + \frac{\partial \mathscr{E}}{\partial t}\\ &= \frac{\partial \mathscr{E}}{\partial t}.\end{aligned} \tag{10.65}$$

Rather surprisingly it is only the explicit time dependence of $\mathscr{E}$ that matters; gains in kinetic and potential energy arising from the motion of the particle that is free to move exactly cancel. From equations (10.65) and (10.62) we find

$$\frac{\mathrm{d}\mathscr{E}}{\mathrm{d}t} = \frac{q^2}{4\pi\varepsilon_0}\frac{(x_1 - ut)u}{[(x_1 - ut)^2 + x_2^2 + x_3^2]^{3/2}},$$

in agreement with the result obtained in Section 10.17.4.

We shall now see how to extend the ideas arising from these simple cases to the more general system in which the forces may be velocity-, as well as time-, dependent, and for which generalized coordinates are used. In Section 10.13 we showed that it may still be possible to construct a suitable Lagrangian, $\mathscr{L}$, in terms of which the motion of the system is described by Lagrange's equations,

$$\frac{\mathrm{d}}{\mathrm{d}t}\left(\frac{\partial \mathscr{L}}{\partial \dot{q}_j}\right) - \frac{\partial \mathscr{L}}{\partial q_j} = 0.$$

In terms of the generalized momenta,

$$p_j = \frac{\partial \mathscr{L}}{\partial \dot{q}_j}, \tag{10.66}$$

Lagrange's equations may be written as

$$\dot{p}_j = \frac{\partial \mathscr{L}}{\partial q_j}. \tag{10.67}$$

It is no longer necessarily true that the total energy is the most suitable basic description of the system. Instead we use the *Hamiltonian*, $\mathscr{H}$, defined by

$$\mathscr{H} = \sum_j p_j\dot{q}_j - \mathscr{L}. \tag{10.68}$$

In this expression, as it stands, the summation is written in terms solely of the p_j and the $\dot{q}_j$, while $\mathscr{L}$ has been considered so far as a function of the q_j, the $\dot{q}_j$ and possibly of t. Hence the most general variation of $\mathscr{H}$ must be written as

$$\delta\mathscr{H} = \sum_j p_j\,\delta\dot{q}_j + \sum_j \dot{q}_j\,\delta p_j - \sum_j \frac{\partial \mathscr{L}}{\partial \dot{q}_j}\delta\dot{q}_j - \sum_j \frac{\partial \mathscr{L}}{\partial q_j}\delta q_j - \frac{\partial \mathscr{L}}{\partial t}\delta t.$$

In view of equation (10.66) this reduces to

$$\delta\mathscr{H} = \sum_j \dot{q}_j\,\delta p_j - \sum_j \frac{\partial \mathscr{L}}{\partial q_j}\delta q_j - \frac{\partial \mathscr{L}}{\partial t}\delta t. \tag{10.69}$$

In the preceding section we saw that $\mathscr{E}$ might be considered as a function of position and momentum coordinates, and possibly of t. We can do the same for the Hamiltonian, since equations (10.66) enable us, in principle, to solve for the $\dot{q}_j$ in terms of the q_j, p_j and t. When this is done we have a form of the Hamiltonian in which only the q_j, p_j and possibly t occur explicitly:

$$\mathscr{H} \equiv \mathscr{H}(q_1, \ldots, q_n; p_1, \ldots, p_n; t). \tag{10.70}$$

For this the most general variation is

$$\delta\mathscr{H} = \sum_j \frac{\partial\mathscr{H}}{\partial q_j}\delta q_j + \sum_j \frac{\partial\mathscr{H}}{\partial p_j}\delta p_j + \frac{\partial\mathscr{H}}{\partial t}\delta t. \tag{10.71}$$

If the generalized coordinates have been chosen so that independent variations in all the q_j and p_j, and in t, are physically possible, we may equate the coefficients of each δq_j, δp_j and δt in equations (10.69) and (10.71), thus obtaining the sets of equations:

$$\dot{q}_j = \frac{\partial\mathscr{H}}{\partial p_j}, \qquad -\frac{\partial\mathscr{L}}{\partial q_j} = \frac{\partial\mathscr{H}}{\partial q_j}, \qquad j = 1, 2, \ldots, n.$$

The first of these sets is analogous to equations (10.63), while the second becomes analogous to equations (10.64) if we use Lagrange's equations in the form (10.67). We then have the $2n$ first-order equations:

$$\dot{q}_j = \frac{\partial\mathscr{H}}{\partial p_j}, \qquad \dot{p}_j = -\frac{\partial\mathscr{H}}{\partial q_j}, \qquad j = 1, 2, \ldots, n, \tag{10.72}$$

to describe in $2n$-dimensional generalized phase-space the motion of the point that represents the way in which the system develops. The fundamental description of the system is the Hamiltonian, defined by equation (10.68), but written, as in the expression (10.70), in terms of the *generalized coordinates* q_j, the *generalized momenta* and, possibly, time, t.

Equations (10.72) are known as Hamilton's equations. We may note that if they are used to simplify the time derivative of the Hamiltonian it becomes

$$\begin{aligned}\frac{d\mathscr{H}}{dt} &= \sum_j \frac{\partial\mathscr{H}}{\partial q_j}\frac{dq_j}{dt} + \sum_j \frac{\partial\mathscr{H}}{\partial p_j}\frac{dp_j}{dt} + \frac{\partial\mathscr{H}}{\partial t} \\ &= \sum_j \left[\frac{\partial\mathscr{H}}{\partial q_j}\frac{\partial\mathscr{H}}{\partial p_j} - \frac{\partial\mathscr{H}}{\partial p_j}\frac{\partial\mathscr{H}}{\partial q_j}\right] + \frac{\partial\mathscr{H}}{\partial t} \\ &= \frac{\partial\mathscr{H}}{\partial t},\end{aligned} \tag{10.73}$$

which is the generalized form of the result (10.65).

10.16. Motion in phase-space; Liouville's theorem

The Hamiltonian formulation is, in practice, far more useful as a link with other aspects of physical theory than as a means of investigating problems of the type discussed in this book. One such link is with statistical mechanics and thermodynamics, where the 'system' consists of large numbers of particles or parts (for example the molecules of a gas).

For such systems we cannot hope to solve in detail the equation of motion of all its parts. In any case, measurements of the individual parts are impracticable, and our understanding of the systems comes from such concepts as density, pressure and temperature. These are statistical or cooperative concepts, based upon the averaged or integrated properties of individual parts, which, in turn, are derived from their positions and momenta. Now the latter may all be represented by a point or points in phase-space. Consequently we may expect concepts and laws concerning them which are useful for describing multi-particle systems to correspond to general patterns of behaviour in phase-space of the points representing them.

Let us first note that, once the Hamiltonian for a system is known, equations (10.72) uniquely determine the 'velocity' of the representative point in phase-space, in magnitude and direction. Thus, given a starting point at any time, the change in position during a succeeding small interval, dt, is uniquely determined. Proceeding step by step in this way, the whole path of the representative point may be traced out. If it returns to its initial position its direction at the second time of passing may be different when $\mathscr{H}$ has an explicit dependence upon t, since the $\partial\mathscr{H}/\partial p_j$ and the $\partial\mathscr{H}/\partial q_j$ may then have a similar dependence. However, when $\mathscr{H}$ is a function only of the p_j and the q_j, equations (10.72) show that the $\dot{q}_j$ and the $\dot{p}_j$ are always the same for any one position in phase-space, no matter when the representative point reaches it. Thus if the point passes through any point twice it will continue to trace out the *same* closed trajectory indefinitely in phase-space.

When we have a number of systems that are similar (i.e. described by the *same* Hamiltonian) but differing slightly in condition (i.e. with slightly different positions and momenta), they may be described by a set of neighbouring representative points in the *same* phase-space. Figure 10.11 illustrates

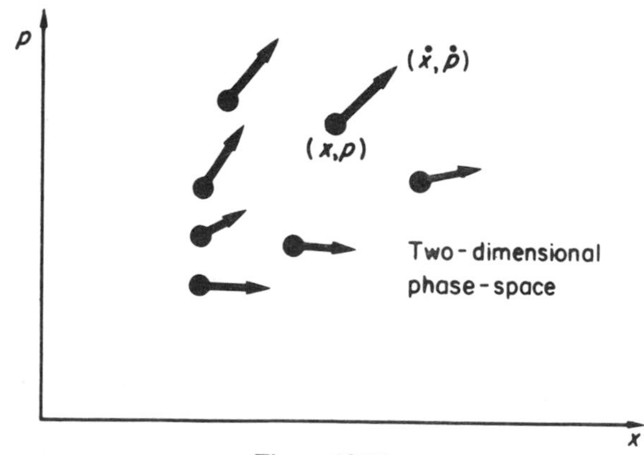

Figure 10.11

this for two dimensions, although it must be remembered that we are normally dealing with a multi-dimensional phase-space.

One of the 'patterns of behaviour' mentioned above concerns the way in which these points move as a group through the phase-space. Suppose ρ is the density of the representative points. By this we mean that if a 'cell of volume $d\sigma$' in phase-space is defined by the intervals $dq_1, \ldots, dq_n, dp_1, \ldots, dp_n$ in the coordinates, then the number of points in the cell is

$$N = \rho \, d\sigma = \rho dq_1 \ldots dq_n \, dp_1 \ldots dp_n.$$

Now the flux of these points is $\rho\mathbf{v}$, where $\mathbf{v}$ is the 'velocity' with which these points are moving in phase-space. This velocity has the $2n$ components $\dot{q}_1, \ldots, \dot{q}_n, \dot{p}_1, \ldots, \dot{p}_n$. The rate at which points accumulate in any cell must be equal to the net flux of particles into it. The *equation of continuity* expressing this is

$$\frac{\partial \rho}{\partial t} + \sum_j \frac{\partial(\rho \dot{q}_j)}{\partial q_j} + \sum_j \frac{\partial(\rho \dot{p}_j)}{\partial p_j} = 0 \tag{10.74}$$

or

$$\frac{\partial \rho}{\partial t} + \sum_j \left(\dot{q}_j \frac{\partial \rho}{\partial q_j} + \dot{p}_j \frac{\partial \rho}{\partial p_j} \right) + \sum_j \rho \left(\frac{\partial \dot{q}_j}{\partial q_j} + \frac{\partial \dot{p}_j}{\partial p_j} \right) = 0.$$

(See Section 10.17.6, page 332.) From Hamilton's equations (10.72),

$$\frac{\partial \dot{q}_j}{\partial q_j} + \frac{\partial \dot{p}_j}{\partial p_j} = \frac{\partial^2 \mathscr{H}}{\partial q_j \, \partial p_j} - \frac{\partial^2 \mathscr{H}}{\partial p_j \, \partial q_j} = 0.$$

Hence

$$\frac{\partial \rho}{\partial t} + \sum_j \left(\dot{q}_j \frac{\partial \rho}{\partial q_j} + \dot{p}_j \frac{\partial \rho}{\partial p_j} \right) = 0. \tag{10.75}$$

This tells us that ρ remains constant, *following the motion* of the representative points. Thus, although they may move relative to each other, the density of any group of representative points remains constant in phase-space. *They move as if they were point molecules in a moving incompressible fluid.* This is *Liouville's theorem.* Note that by itself the theorem does not imply that the density is constant throughout phase-space. (See Section 10.17.7, page 334.)

10.17. Comments and worked examples

10.17.1. See Section 10.1

In one dimension the potential energy near any point x_0 may be written as

$$\mathscr{V}(x_0 + \delta x) = \mathscr{V}(x_0) + \delta x \left(\frac{d\mathscr{V}}{dx} \right)_{x_0} + \tfrac{1}{2} \delta x^2 \left(\frac{d^2 \mathscr{V}}{dx^2} \right)_{x_0} + \ldots.$$

When the first-order term, $\delta x(\mathrm{d}\mathscr{V}/\mathrm{d}x)_{x_0}$, is zero then the force acting at x_0 is

$$F_{x_0} = -\left(\frac{\mathrm{d}\mathscr{V}}{\mathrm{d}x}\right)_{x_0} = 0.$$

Thus if the particle is once stationary at x_0 it will remain so. When the second-order term, $\delta x^2(\mathrm{d}^2\mathscr{V}/\mathrm{d}x^2)_{x_0}$, is positive then

$$\left(\frac{\mathrm{d}F}{\mathrm{d}x}\right)_{x_0} = -\left(\frac{\mathrm{d}^2\mathscr{V}}{\mathrm{d}x^2}\right)_{x_0} < 0.$$

Thus F will be negative to the right of x_0 and positive to the left, which means that if the particle is displaced from x_0 the force will act to bring it back towards x_0.

In more than one dimension,

$$\begin{aligned}\mathscr{V}(x_{10} + \delta x_1, x_{20} + \delta x_2, \ldots) &= \mathscr{V}(x_{10}, x_{20}, \ldots)\\ &+ \left(\delta x_1 \frac{\partial \mathscr{V}}{\partial x_1} + \delta x_2 \frac{\partial \mathscr{V}}{\partial x_2} + \ldots\right)_{x_{10}, x_{20}, \ldots}\\ &+ \left(\tfrac{1}{2}\delta x_1^2 \frac{\partial^2 \mathscr{V}}{\partial x_1^2} + \tfrac{1}{2}\delta x_2^2 \frac{\partial^2 \mathscr{V}}{\partial x_2^2} + \delta x_1\,\delta x_2 \frac{\partial^2 \mathscr{V}}{\partial x_1\,\partial x_2} + \ldots\right)_{x_{10}, x_{20}, \ldots} + \ldots.\end{aligned}$$

For $x_{10}, x_{20}, \ldots$ to be an equilibrium position the condition is the simple extension of that above—the first-order term must be zero for all values of $\delta x_1, \delta x_2, \ldots$:

$$\frac{\partial \mathscr{V}}{\partial x_1} = \frac{\partial \mathscr{V}}{\partial x_2} = \ldots = 0 \quad \text{at} \quad x_{10}, x_{20}, \ldots.$$

For stability, however, we must have the second-order term positive for all possible values of $\delta x_1, \delta x_2, \ldots$:

$$\tfrac{1}{2}\delta x_1^2 \frac{\partial^2 \mathscr{V}}{\partial x_1^2} + \tfrac{1}{2}\delta x_2^2 \frac{\partial^2 \mathscr{V}}{\partial x_2^2} + \delta x_1\,\delta x_2 \frac{\partial^2 \mathscr{V}}{\partial x_1\,\partial x_2} + \ldots > 0 \quad \text{at} \quad x_{10}, x_{20}, \ldots.$$

For this to hold it is *necessary* that

$$\frac{\partial^2 \mathscr{V}}{\partial x_1^2} > 0, \quad \frac{\partial^2 \mathscr{V}}{\partial x_2^2} > 0, \ldots \quad \text{at} \quad x_{10}, x_{20}, \ldots,$$

but these inequalities are not *sufficient* to ensure stability.

10.17.2. See Section 10.10

Example 33

A cylindrical satellite of mass M and radius a is spinning about its axis with angular speed ω. It carries within it two hollow antennae, each of mass

m and length $2a$ lying one within the other along a diameter. These are driven out symmetrically in opposite directions at constant speed, stretching a spring of negligible mass and natural length $2a$ which joins their ends. Figure 10.12 shows the system partly extended.

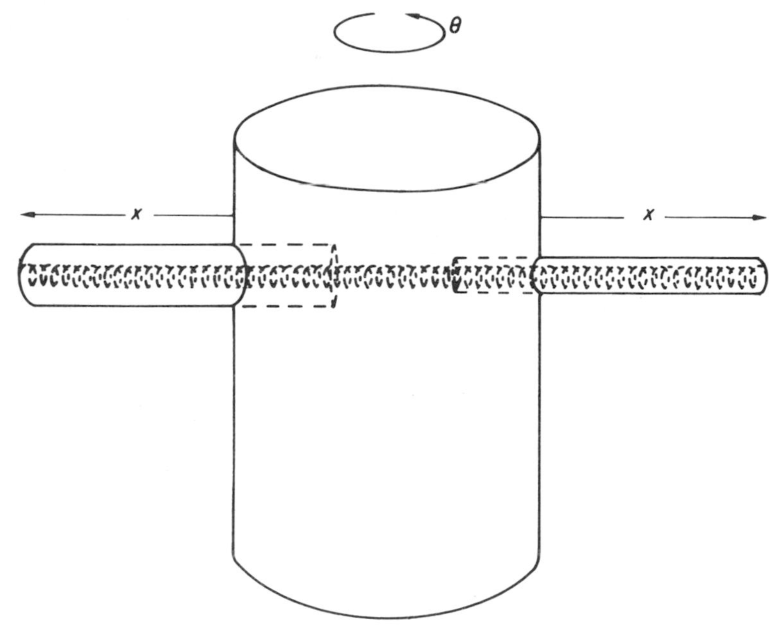

Figure 10.12

(a) Show that it is possible to choose the strength constant, K, of the spring so that no net work is done by the motor driving the antennae in fully extending them a distance $2a$ out of the satellite. Determine this value of K.

(b) With this value of K show that if u is the speed relative to the satellite at which each antennae moves, the maximum power delivered by the motor driving them is

$$P_{\max} = \frac{192m^2\omega^2 au(4m + 3M)}{(52m + 3M)^2}.$$

(a) The effect of the spring is to pull the antennae into the satellite. The effect of the centrifugal force arising from the rotation is to make the antennae fly out. Clearly K can be chosen sufficiently large for the former effect to predominate at all stages of the extension and the motor will be required to do work on the system throughout. With a sufficiently small K the latter effect predominates and work will be done against the motor throughout.

Assuming that the work done depends continuously upon K, there will be some intermediate value for which the net work done is zero. Note that this may mean that work is done by the motor during part of the extension and *against* it for the rest.

To investigate the process in more detail let us set up the Lagrangian for the system. This is not necessary for part (a) of the problem, but we shall use Lagrange's equation for part (b). Since there are no external forces the centre-of-mass of the whole system moves with constant velocity, and we shall therefore use a frame of reference based on this. Because of the symmetrical motion of the antennae their joint centre-of-mass remains throughout at a fixed point on the axis of the cylinder. The same is true of the centre-of-mass of the cylinder itself, and since the only initial rotation is about this axis, both centres-of-mass will therefore remain fixed in the frame we have chosen.

The configuration of the system may be fixed by θ, the angle of rotation, and x, the extension of either antenna, and we shall use these as the generalized coordinates.

The centre-of-mass of each antenna has velocity components $\dot{x}$ along, and $x\dot{\theta}$ perpendicular to, its length. The antenna has a moment of inertia $\frac{1}{3}ma^2$ about its centre-of-mass and is rotating about it with angular speed $\dot{\theta}$. Hence the kinetic energy of each antenna is

$$\begin{aligned}\mathcal{T}_\mathrm{a} &= \mathcal{T}_\mathrm{cm} + \mathcal{T}_\mathrm{rel}\\ &= \tfrac{1}{2}m(\dot{x}^2 + x^2\dot{\theta}^2) + \tfrac{1}{6}ma^2\dot{\theta}^2.\end{aligned}$$

The satellite has a moment of inertia $\frac{1}{2}Ma^2$ about its centre-of-mass and rotates about this with angular speed $\dot{\theta}$. Since this centre-of-mass is stationary, the kinetic energy of the satellite is

$$\mathcal{T}_\mathrm{s} = \tfrac{1}{4}Ma^2\dot{\theta}^2.$$

The potential energy of the spring is

$$\mathcal{V} = \tfrac{1}{2}K(2x)^2 = 2Kx^2.$$

The Lagrangian of the system is therefore

$$\begin{aligned}\mathcal{L} &= \mathcal{T}_\mathrm{s} + 2\mathcal{T}_\mathrm{a} - \mathcal{V}\\ &= (mx^2 + \tfrac{1}{3}ma^2 + \tfrac{1}{4}Ma^2)\dot{\theta}^2 + m\dot{x}^2 - 2Kx^2. \qquad (10.76)\end{aligned}$$

The remaining forces acting are the reactions between the antennae and their guides in the satellite, which are constraint forces and may therefore be omitted from Lagrange's equation, and the forces extending the antennae, which are motion forces and must appear. Suppose F is the force on each antennae. During a variation, δx, *each* antennae will extend this amount, and the work done will therefore be

$$\delta \mathscr{W} = 2F\,\delta x.$$

Lagrange's equations are then

$$\frac{\mathrm{d}}{\mathrm{d}t}\left(\frac{\partial \mathscr{L}}{\partial \dot{\theta}}\right) - \frac{\partial \mathscr{L}}{\partial \theta} = 0, \tag{10.77}$$

$$\frac{\mathrm{d}}{\mathrm{d}t}\left(\frac{\partial \mathscr{L}}{\partial \dot{x}}\right) - \frac{\partial \mathscr{L}}{\partial x} = 2F. \tag{10.78}$$

Since θ is absent from the Lagrangian, equation (10.77) gives

$$\frac{\partial \mathscr{L}}{\partial \dot{\theta}} = \text{constant},$$

which simply expresses conservation of angular momentum in the absence of external forces:

$$2(mx^2 + \tfrac{1}{3}ma^2 + \tfrac{1}{4}Ma^2)\dot{\theta} = J = \text{constant}. \tag{10.79}$$

Equation (10.78) gives

$$2m\ddot{x} - 2mx\dot{\theta}^2 + 4Kx = 2F. \tag{10.80}$$

The first term is zero when the antennae move at constant speed, and $\dot{\theta}$ may be substituted from equation (10.79). Thence

$$F = 2Kx - \frac{mJ^2x}{4(A + mx^2)^2}, \tag{10.81}$$

where

$$A = (\tfrac{1}{3}m + \tfrac{1}{4}M)a^2. \tag{10.82}$$

Hence the work done as both antennae are fully extended is

$$\begin{aligned}
\mathscr{W} = 2\int_0^{2a} F\,\mathrm{d}x &= 4K\int_0^{2a} x\,\mathrm{d}x - \frac{mJ^2}{2}\int_0^{2a}\frac{x\,\mathrm{d}x}{(A + mx^2)^2} \\
&= 2K[x^2]_0^{2a} + \frac{J^2}{4}\left[\frac{1}{A + mx^2}\right]_0^{2a} \\
&= 8Ka^2 - \frac{J^2}{4}\left(\frac{1}{A} - \frac{1}{A + 4ma^2}\right) \\
&= 8Ka^2 - \frac{mJ^2a^2}{A(A + 4ma^2)}.
\end{aligned}$$

The strength constant for which this is zero is

$$K = \frac{mJ^2}{8A(A + 4ma^2)}. \tag{10.83}$$

We may calculate the constant J by substituting into equation (10.79) the initial conditions

$$x = 0, \qquad \dot{\theta} = \omega,$$

thus obtaining

$$J = (\tfrac{2}{3}m + \tfrac{1}{2}M)a^2\omega. \tag{10.84}$$

Then, from equations (10.82), (10.83) and (10.84),

$$K = \frac{m(\frac{2}{3}M + \frac{1}{2}M)^2\omega^2}{8(\frac{1}{3}m + \frac{1}{4}M)(\frac{13}{3}m + \frac{1}{4}M)} = \frac{m\omega^2}{2}\frac{4m + 3M}{52m + 3M}.$$

(b) The power delivered by the motor in driving both antennae at a radial speed u is

$$P = 2Fu,$$

which, from equation (10.81), is

$$P = \left[4Kx - \frac{mJ^2x}{2(A + mx^2)^2}\right]u. \tag{10.85}$$

$P(x)$ is sketched in Figure 10.13, from which it is clear that the physical maximum occurs at the maximum extension,

$$x = 2a.$$

Hence

$$P_{\max} = \left[8Ka - \frac{mJ^2a}{(A + 4ma^2)^2}\right]u.$$

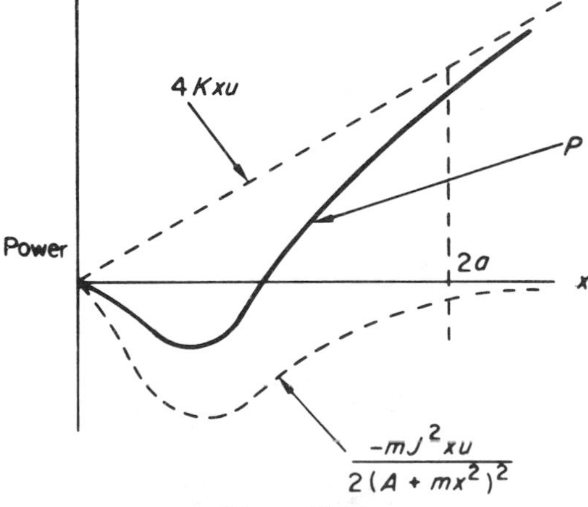

Figure 10.13

Substituting from equations (10.82), (10.83) and (10.84), we obtain

$$P_{\text{max}} = \frac{192m^2\omega^2 au(4m + 3M)}{(52m + 3M)^2} \tag{10.86}$$

at the position of furthest extent,

$$x = 2a.$$

An alternative approach to (a) is to note that if no net work is to be done by the motors the total mechanical energy of the system,

$$\begin{aligned} \mathscr{E} &= \mathscr{T}_{\text{s}} + 2\mathscr{T}_{\text{a}} + \mathscr{V} \\ &= (mx^2 + \tfrac{1}{3}ma^2 + \tfrac{1}{4}Ma^2)\dot{\theta}^2 + m\dot{x}^2 + 2Kx^2, \end{aligned} \tag{10.87}$$

will be the same in the initial and final conditions. Since there are no external forces the angular momentum will remain constant throughout:

$$J = 2(mx^2 + \tfrac{1}{3}ma^2 + \tfrac{1}{4}Ma^2)\dot{\theta} = \text{constant.} \tag{10.88}$$

($mx^2 + \frac{1}{3}ma^2$ is the moment of inertia of each antenna and $\frac{1}{2}Ma^2$ the moment of inertia of the satellite, about the axis of revolution.)

If the antennae start and finish at rest relative to the satellite, the initial condition is

$$x = 0, \qquad \dot{x} = 0, \qquad \dot{\theta} = \omega \tag{10.89}$$

and the final condition is

$$x = 2a, \qquad \dot{x} = 0 \quad 0 = \omega_f, \text{ say.} \tag{10.90}$$

Substituting these values into equations (10.87) and (10.88),

$$\begin{aligned} (\tfrac{1}{3}m + \tfrac{1}{4}M)a^2\omega^2 &= (\tfrac{13}{3}m + \tfrac{1}{4}M)a^2\omega_f^2 + 8Ka^2, \\ (\tfrac{1}{3}m + \tfrac{1}{4}M)a^2\omega &= (\tfrac{13}{3}m + \tfrac{1}{4}M)a^2\omega_f. \end{aligned}$$

On eliminating ω_f from these equations we obtain

$$\begin{aligned} K &= \frac{\omega^2}{96}\left[4m + 3M - \frac{(4m + 3M)^2}{52m + 3M}\right] \\ &= \frac{m\omega^2}{2}\,\frac{4m + 3M}{52m + 3M}. \end{aligned}$$

Not only does this method give the same result as before, but it is a more satisfactory way of analysing the physical process since the initial and final conditions (10.89) and (10.90) describe antennae *at rest* relative to the satellite at the beginning and end. However, in taking $\ddot{x}$ as zero in equation (10.80) we imply that each antenna has the same *non-zero* radial speed in both conditions. This would give a constant value to the term $m\dot{x}^2$ in equation

(10.87) and would make no difference to the total energy balance. Nevertheless, the fact that in practice the antennae *must* be brought to rest means that there must be a short period of deceleration just before they are fully extended when the motor would be shut off. It is therefore strictly incorrect to use the expression (10.85) for the power delivered right to the end of the motion, and the result (10.86) is only true for small values of u. It is left to the reader to estimate the distance over which the deceleration occurs (use equations (10.79), (10.80) and (10.81) with $F = 0$, $\ddot{x} \neq 0$ and $x \simeq 2a$).

10.17.3. See Section 10.13

We must remember that to describe a moving particle fully we must specify its position (x_1, x_2, x_3) *and* the time (t) when it is at that position. Hence any quantity, Φ, which depends upon the motion of the particle may be expected, in general, to be a function of all four of the variables:

$$\Phi \equiv \Phi(x_1, x_2, x_3, t).$$

When the particle changes its position by $\delta x_1, \delta x_2, \delta x_3$ in a time δt, the corresponding change in Φ will then be

$$\delta\Phi = \Phi(x_1 + \delta x_1, x_2 + \delta x_2, x_3 + \delta x_3, t + \delta t) - \Phi(x_1, x_2, x_3, t).$$

Expanding $\Phi(x_1 + \delta x_1, \ldots)$ as a Taylor series, we then have

$$\delta\Phi = \frac{\partial\Phi}{\partial x_1}\delta x_1 + \frac{\partial\Phi}{\partial x_2}\delta x_2 + \frac{\partial\Phi}{\partial x_3}\delta x_3 + \frac{\partial\Phi}{\partial t} + \ldots.$$

Hence the time rate of change of Φ is

$$\begin{aligned}\frac{d\Phi}{dt} = \lim_{\delta t\to 0}\frac{\delta\Phi}{\delta t} &= \frac{\partial\Phi}{\partial x_1}\frac{dx_1}{dt} + \frac{\partial\Phi}{\partial x_2}\frac{dx_2}{dt} + \frac{\partial\Phi}{\partial x_3}\frac{dx_3}{dt} + \frac{\partial\Phi}{\partial t} \\ &= v_1\frac{\partial\Phi}{\partial x_1} + v_2\frac{\partial\Phi}{\partial x_2} + v_3\frac{\partial\Phi}{\partial x_3} + \frac{\partial\Phi}{\partial t} \\ &= \mathbf{v}\,.\,\nabla\Phi + \frac{\partial\Phi}{\partial t},\end{aligned} \tag{10.91}$$

where $\mathbf{v}$ is the particle velocity.

Thus the rate of change *following the motion of the particle* consists of two parts. The first, $\mathbf{v}\,.\,\nabla\Phi$, arises from the movement of the particle itself and can contribute to $d\Phi/dt$ even when, at any fixed point in space, Φ is constant. The second, $\partial\Phi/\partial t$, arises directly from the time variation of Φ at any fixed point and will contribute to $d\Phi/dt$ even when the particle is stationary.

10.17.4. See Section 10.13

Example 34

An external force keeps a particle of charge q moving at a constant velocity $\mathbf{u}$. A second particle of charge q is free to move under the influence of the first. Determine the rate at which the potential energy of the system is changing, and show that this is zero when the relative velocity of the two particles is perpendicular to the line joining them. Account for the total energy balance of the system.

Choose Cartesian axes such that the first particle moves along axis 1, passing the origin at time zero. Then its position at any time t in the frame of reference is $(ut, 0, 0)$ where u is a constant speed, while the other particle has position (x_1, x_2, x_3) and velocity (v_1, v_2, v_3), as shown in Figure 10.14. Then the potential energy is

$$\mathscr{V}(x_1, x_2, x_3, t) = \frac{q^2}{4\pi\varepsilon_0} \frac{1}{[(x_1 - ut)^2 + x_2^2 + x_3^2]^{1/2}}. \tag{10.92}$$

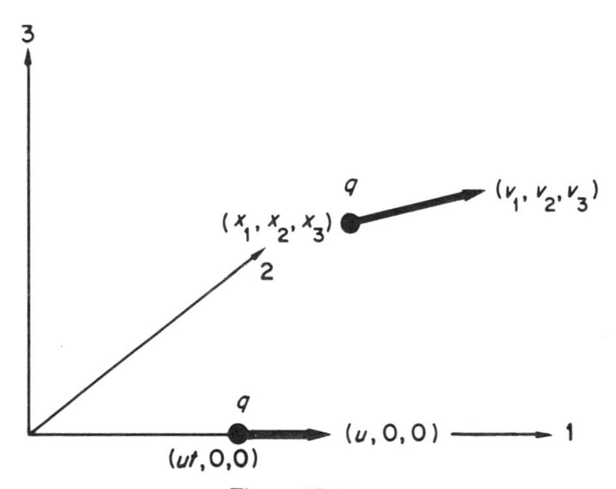

Figure 10.14

From equation (10.91) of the preceding section,

$$\frac{d\mathscr{V}}{dt} = \mathbf{v} \,.\, \nabla\mathscr{V} + \frac{\partial\mathscr{V}}{\partial t} \tag{10.93}$$

$$= v_1 \frac{\partial\mathscr{V}}{\partial x_1} + v_2 \frac{\partial\mathscr{V}}{\partial x_2} + v_3 \frac{\partial\mathscr{V}}{\partial x_3} + \frac{\partial\mathscr{V}}{\partial t} \tag{10.94}$$

$$= -q^2 \frac{v_1(x_1 - ut) + v_2 x_2 + v_3 x_3 - u(x_1 - ut)}{4\pi\varepsilon_0[(x_1 - ut)^2 + x_2^2 + x_3^2]^{3/2}}. \tag{10.95}$$

Now the components of the relative velocity vector are $v_1 - u, v_2, v_3$ and the components of the vector joining the particles are $x_1 - ut, x_2, x_3$. The condition for these two vectors to be perpendicular is

$$(v_1 - u)(x_1 - ut) + v_2 x_2 + v_3 x_3 = 0.$$

When this is true, equation (10.95) shows that

$$\frac{\mathrm{d}\mathscr{V}}{\mathrm{d}t} = 0.$$

(This result could be obtained more readily by using a frame moving with the driven particle, but this method would not demonstrate the analysis of the preceding section.)

The kinetic energy of the first particle is constant since it moves with constant velocity. That of the second particle is

$$\mathscr{T} = \tfrac{1}{2} m \mathbf{v}^2,$$

where m is its mass. Then

$$\frac{\mathrm{d}\mathscr{T}}{\mathrm{d}t} = m\mathbf{v} \cdot \frac{\mathrm{d}\mathbf{v}}{\mathrm{d}t}. \tag{10.96}$$

Now if $\mathbf{F}$ is the force exerted on the second particle by the first,

$$\mathbf{F} = m \frac{\mathrm{d}\mathbf{v}}{\mathrm{d}t},$$

and at any time it is derivable from the electrostatic potential energy, $\mathscr{V}$:

$$\mathbf{F} = -\nabla \mathscr{V}.$$

Hence equation (10.96) may be written as

$$\frac{\mathrm{d}\mathscr{T}}{\mathrm{d}t} = \mathbf{v} \cdot \mathbf{F},$$

and equation (10.93) as

$$\frac{\mathrm{d}\mathscr{V}}{\mathrm{d}t} = -\mathbf{v} \cdot \mathbf{F} + \frac{\partial \mathscr{V}}{\partial t}.$$

Thus the rate of change of the total energy of the system is

$$\frac{\mathrm{d}(\mathscr{T} + \mathscr{V})}{\mathrm{d}t} = \frac{\partial \mathscr{V}}{\partial t} = \frac{q^2}{4\pi\varepsilon_0} \frac{(x_1 - ut)u}{[(x_1 - ut)^2 + x_2^2 + x_3^2]^{3/2}}. \tag{10.97}$$

Now the first particle is subject to the force $-\mathbf{F}$ exerted on it by the second and to some external force $\mathbf{F}_{\text{ext}}$. Since it is not accelerated,

$$\mathbf{F}_{\text{ext}} - \mathbf{F} = 0.$$

The only source of energy that could account for the change in $\mathscr{T} + \mathscr{V}$ is the work done on the first particle by the external force. Since this particle has a velocity component (u) only along the axis 1, only the corresponding component of the external force,

$$F_{\text{ext}\,1} = F_1 = \frac{q^2}{4\pi\varepsilon_0} \frac{x_1 - ut}{[(x_1 - ut)^2 + x_2^2 + x_3^2]^{3/2}},$$

will contribute to the work done, and will do so at a rate

$$\frac{\mathrm{d}\mathscr{W}}{\mathrm{d}t} = F_{\text{ext}\,1}u = \frac{q^2}{4\pi\varepsilon_0} \frac{(x_1 - ut)u}{[(x_1 - ut)^2 + x_2^2 + x_3^2]^{3/2}}. \qquad (10.98)$$

From the expressions (10.97) and (10.98) we can see that the changing energy of the system is exactly accounted for by the work delivered by the external force.

It should be noted that we have neglected any magnetic interaction between the moving charges. This is justified when their speeds are small compared with that of light. In general, however, a time-dependent potential energy such as (10.92) is inadequate to describe fully an electromagnetic field, and the magnetic vector potential is also required (see Section 10.13).

10.17.5. See Section 10.13

Example 35

In a photo-electronic tube electrons are emitted from a cathode at zero potential with what may be assumed to be zero velocity. Under the influence of stationary electric and magnetic fields they travel along the tube and are collected at a target at potential ψ_t. The fields have rotational symmetry about an axis passing through the centres of the cathode and of the target.

An electron emitted from the cathode at a distance ρ_0 from this axis reaches the target at a distance ρ. Show that the least value of ψ which would enable the electron to do this is given by

$$A - \frac{\rho_0}{\rho}A_0 = \pm\sqrt{\frac{2m\psi_{\text{min}}}{e}},$$

where A_0, A are respectively the magnitudes of the vector potential at the points of emission and collection of the electron.

Electrons are emitted from a cathode of radius 1 mm and, after travelling through a constant axial magnetic field of 10^{-2} Tesla, are used to scan a target at $+1$ Volt relative to the cathode. Show that the smallest spot into which the electron beam may be focused at the target has a radius of 0·72 mm.

$$\left[\text{For an electron } \sqrt{\frac{2m}{e}} = 3{\cdot}37 \times 10^{-6}\,\text{m}^{-1}\,\text{s}\,\text{V}^{1/2}.\right]$$

In cylindrical polar coordinates ρ, ϕ, z, the three velocity components are $\dot{\rho}$, $\rho\dot{\phi}$, $\dot{z}$. From equations (10.54), the Lagrangian for the electron is therefore

$$\begin{aligned}\mathscr{L} &= \tfrac{1}{2}m\mathbf{v}^2 + e\psi - e\mathbf{v}\,.\,\mathbf{A}\\ &= \tfrac{1}{2}m(\dot{\rho}^2 + \rho^2\dot{\phi}^2 + \dot{z}^2) + e\psi - e(\dot{\rho}A_\rho + \rho\dot{\phi}A_\phi + \dot{z}A_z),\end{aligned}$$

where ψ, A are the electric (scalar) and magnetic (vector) potentials. Since there is cylindrical symmetry,

$$\psi \equiv \psi(\rho, z) \tag{10.99}$$

and

$$A_\rho = A_z = 0, \qquad A_\phi \equiv A(\rho, z). \tag{10.100}$$

The form (10.99) ensures that

$$E_\phi = -\frac{1}{\rho}\frac{\partial\psi}{\partial\phi} = 0,$$

while the relationship

$$\mathbf{B} = \mathbf{curl\ A},$$

applied to the components (10.100), shows that

$$B_\rho = \frac{1}{\rho}\left[\frac{\partial A_z}{\partial\phi} - \frac{\partial(\rho A_\phi)}{\partial z}\right] = -\frac{\partial A(\rho, z)}{\partial z}, \tag{10.101}$$

$$B_\phi = \frac{\partial A_\rho}{\partial z} - \frac{\partial A_z}{\partial\rho} = 0, \tag{10.102}$$

$$B_z = \frac{1}{\rho}\left[\frac{\partial(\rho A_\phi)}{\partial\rho} - \frac{\partial A_\rho}{\partial\phi}\right] = \frac{1}{\rho}\frac{\partial}{\partial\rho}[\rho A(\rho, z)]. \tag{10.103}$$

The fields **E** and **B**, therefore, have the required symmetry. A and ψ are time-independent. Hence

$$\mathscr{L} = \tfrac{1}{2}m(\dot{\rho}^2 + \rho^2\dot{\phi}^2 + \dot{z}^2) + e\psi(\rho, z) - e\rho\dot{\phi}A(\rho, z). \tag{10.104}$$

Since ϕ does not appear in $\mathscr{L}$ we have immediately one constant of motion:

$$\frac{\mathrm{d}}{\mathrm{d}t}\left(\frac{\partial\mathscr{L}}{\partial\dot{\phi}}\right) = \frac{\mathrm{d}}{\mathrm{d}t}(m\rho^2\dot{\phi} - e\rho A) = 0, \tag{10.105}$$

$$\rho^2\dot{\phi} - \frac{e}{m}\rho A = \beta = \text{constant}. \tag{10.106}$$

We could attempt to combine all the Lagrange equations to obtain other conservation laws. However, it is simpler to observe that since **B** is a stationary field and the force it exerts on the electron is perpendicular to the electron's velocity, no work will be done by it. Thus there is only interchange of energy between the kinetic energy of the electron and the potential energy it has in the electric field. Conservation of energy may then be written as

$$\mathscr{T} + \mathscr{V} = \tfrac{1}{2}m(\dot{\rho}^2 + \rho^2\dot{\phi}^2 + \dot{z}^2) - e\psi = \gamma = \text{constant.} \qquad (10.107)$$

If the electron emerges from the cathode at zero velocity at a distance ρ_0 from the axis, where A_ϕ has the value A_0 and ψ is zero, equations (10.106) and (10.107) relate the fields to the position and velocity at any point along the electron trajectory according to the equations

$$\rho^2\dot{\phi} - \frac{e}{m}\rho A_\phi = -\frac{e}{m}\rho_0 A_0, \qquad (10.108)$$

$$\dot{\rho} + \rho^2\dot{\phi}^2 + \dot{z}^2 \frac{2e}{m} - \psi = 0. \qquad (10.109)$$

We can see from equation (10.108) that the rotational velocity component is given by

$$\rho\dot{\phi} = \frac{e}{m}\left(A_\phi - \frac{\rho_0}{\rho}A_0\right). \qquad (10.110)$$

Thus if the electron reaches the target at a radial distance ρ, its rotational velocity component is determined entirely by the magnetic field. In addition to this, there may be radial and axial components ($\dot{\rho}$ and $\dot{z}$), which, as equation (10.109) shows, can only serve to increase the potential required to collect the electron. This potential will be at a minimum ψ_{min} when

$$\dot{\rho} = \dot{z} = 0.$$

In this case equations (10.108) and (10.109) show that

$$\frac{2e}{m}\psi_{\text{min}} = \rho^2\dot{\phi}^2 = \frac{e^2}{m^2}\left(A - \frac{\rho_0}{\rho}A_0\right)^2.$$

Hence

$$A - \frac{\rho_0}{\rho}A_0 = \pm\sqrt{\frac{2m}{e}\psi_{\text{min}}}. \qquad (10.111)$$

A constant axial field has components

$$B_\rho = B_\phi = 0, \qquad B_z = B = \text{constant},$$

and inspection of equations (10.101), (10.102) and (10.103) shows that a suitable vector potential is then

$$A_\phi = \tfrac{1}{2}\rho B.$$

Substituting into equation (10.111) we have

$$\tfrac{1}{2}B\left(\rho - \frac{\rho_0^2}{\rho}\right) = \pm\sqrt{\frac{2m}{e}\psi_{\min}}.$$

If we are interested in bringing the electrons into a small spot,

$$\rho < \rho_0,$$

and the negative square root is relevant:

$$\frac{\rho_0^2}{\rho} - \rho = \frac{2}{B}\sqrt{\frac{2m}{e}\psi_{\min}}.$$

Solving this for the positive value of ρ,

$$\rho = \frac{l}{2}\left(\sqrt{1 + \frac{4\rho_0^2}{l^2}} - 1\right),$$

where

$$l = \frac{2}{B}\sqrt{\frac{2m}{e}\psi_{\min}}.$$

For the values given,

$$\rho_0 = 1\ \text{mm}, \qquad l = 2 \times 10^2 \times 3{\cdot}37 \times 10^{-6} = 6{\cdot}7 \times 10^{-4}\ \text{m} = 0{\cdot}67\ \text{mm}.$$

Then

$$\rho = \frac{0{\cdot}67}{2}\left(\sqrt{1 + \frac{4}{0{\cdot}67^2}} - 1\right) = \frac{0{\cdot}67 \times 2{\cdot}13}{2} = 0{\cdot}72\ \text{mm}.$$

10.17.6. See Section 10.16

The corresponding results for an actual fluid may be shown as follows. Suppose ρ is the density of its molecules (i.e., the number per unit volume) and $\dot{\mathbf{x}}$ its velocity, with components $\dot{x}_1, \dot{x}_2, \dot{x}_3$. All these quantities may depend upon position and time. Let us consider the rate at which molecules enter a small cell of sides $\mathrm{d}x_1, \mathrm{d}x_2, \mathrm{d}x_3$ centred about the point x_1, x_2, x_3 (see Figure 10.15).

The rate at which they cross face A into the cell is

$$R_{\mathrm{A}} = (\mathrm{d}x_2\,\mathrm{d}x_3\rho\dot{x}_1)_{x_1 - \frac{1}{2}\mathrm{d}x_1, x_2, x_3}.$$

It is the value of $\rho\dot{x}_1$ at $x_1 - \frac{1}{2}\mathrm{d}x_1, x_2, x_3$ that is required, since this is the mid-point of face A. The rate across face B (still *into* the cell) is

$$R_{\mathrm{B}} = -(\mathrm{d}x_2\,\mathrm{d}x_3\rho\dot{x}_1)_{x_1 + \frac{1}{2}\mathrm{d}x_1, x_2, x_3}.$$

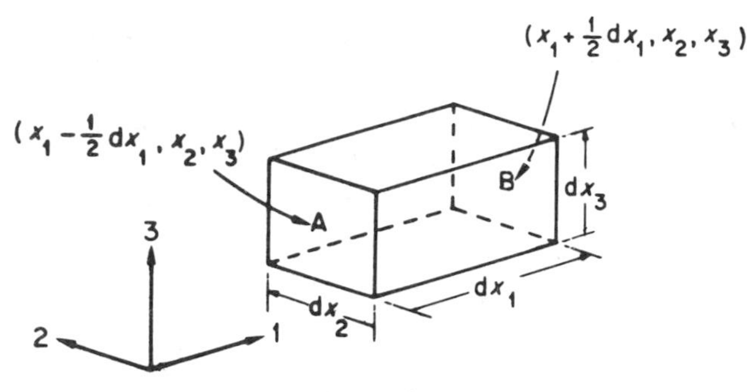

Figure 10.15

The net rate into the cell from these contributions is then

$$R_A + R_B = -dx_2\,dx_3\,dx_1\frac{\partial(\rho\dot{x}_1)}{\partial x_1} = -\sigma\frac{\partial(\rho\dot{x}_1)}{\partial x_1},$$

where

$$\sigma = dx_1\,dx_2\,dx_3$$

is the volume of the cell. Adding the contributions from the three pairs of opposite faces, the total rate is

$$R = -\sigma\left[\frac{\partial(\rho\dot{x}_1)}{\partial x_1} + \frac{\partial(\rho\dot{x}_2)}{\partial x_2} + \frac{\partial(\rho\dot{x}_3)}{\partial x_3}\right]. \tag{10.112}$$

Since the cell is fixed in space, this rate may also be written as

$$R = \sigma\frac{\partial\rho}{\partial t}. \tag{10.113}$$

Equating (10.112) and (10.113) we obtain the equation of continuity,

$$\frac{\partial\rho}{\partial t} + \frac{\partial(\rho\dot{x}_1)}{\partial x_1} + \frac{\partial(\rho\dot{x}_2)}{\partial x_2} + \frac{\partial(\rho\dot{x}_3)}{\partial x_3} = 0 \tag{10.114}$$

or, in vector form,

$$\frac{\partial\rho}{\partial t} + \operatorname{div}(\rho\dot{\mathbf{x}}) \equiv \frac{\partial\rho}{\partial t} + \nabla\,.\,(\rho\dot{\mathbf{x}}) = 0. \tag{10.115}$$

Equations (10.114) or (10.115) express the very general result that molecules are neither created nor destroyed as the fluid moves, and apply equally well, for example, to a non-uniform compressible gas as to a uniform incompressible liquid. On differentiating the products we obtain

$$\left\{\frac{\partial \rho}{\partial t} + \dot{x}_1 \frac{\partial \rho}{\partial x_1} + \dot{x}_2 \frac{\partial \rho}{\partial x_2} + \dot{x}_3 \frac{\partial \rho}{\partial x_3}\right\} + \rho\left(\frac{\partial \dot{x}_1}{\partial x_1} + \frac{\partial \dot{x}_2}{\partial x_2} + \frac{\partial \dot{x}_3}{\partial x_3}\right) = 0 \qquad (10.116)$$

or

$$\left\{\frac{\partial \rho}{\partial t} + \dot{\mathbf{x}} \,.\, \mathbf{grad}\, \rho\right\} + \rho \operatorname{div} \dot{\mathbf{x}} \equiv \left\{\frac{\partial \rho}{\partial t} + \dot{\mathbf{x}} \,.\, \nabla \rho\right\} + \rho \nabla \,.\, \dot{\mathbf{x}} = 0. \qquad (10.117)$$

The terms in the brackets { } describe the change in density following the motion of the molecules (see Section 10.17.3). By definition this is zero for an incompressible fluid, so for such a fluid

$$\frac{\partial \dot{x}_1}{\partial x_1} + \frac{\partial \dot{x}_2}{\partial x_2} + \frac{\partial \dot{x}_3}{\partial x_3} = \operatorname{div} \dot{\mathbf{x}} \equiv \nabla \,.\, \dot{\mathbf{x}} = 0. \qquad (10.118)$$

In phase-space the zero divergence of the velocity of the representative points is ensured by Hamilton's equations.

10.17.7. See Section 10.16

We could imagine a volume of incompressible fluid with constant density and composed of molecules of one type in which were embedded a number of pockets of another incompressible fluid with a different constant density and composed of molecules of another type (oil drops in water, for example). Provided the fluids did not diffuse into each other, the equations (10.74) and (10.75) would still apply to each fluid and therefore throughout the whole volume (except at the boundaries between the different fluids), although the density would clearly not be constant throughout.

10.18. Problems

10.1. A governor consists of two equal masses m connected to four light rods of equal length b, smoothly pivoted at their ends and rotating as a whole with angular velocity $\dot{\phi}$ about a vertical shaft, as shown in Figure 10.16. P is a point fixed in the shaft, Q a sliding collar, and R and S are the two masses. Between P and Q is a spring of natural length b and strength constant K. θ is the inclination of the rods to the vertical.

Describe carefully how energy is stored in the various parts of the governor when it is in its most general state of motion, and show that the Lagrangian may be written as

$$\mathscr{L} = mb^2(\dot{\theta}^2 + \dot{\phi}^2 \sin^2 \theta) - 2mgb \cos \theta - \tfrac{1}{2}Kb^2(2 \cos \theta - 1)^2.$$

If θ_0 is the inclination of the rods when the governor is in stable static equilibrium, and θ_1 when it is in stable equilibrium at a constant angular velocity ω, prove that

$$\omega^2 = \frac{2K}{m}\left(\frac{\cos \theta_0}{\cos \theta_1} - 1\right).$$

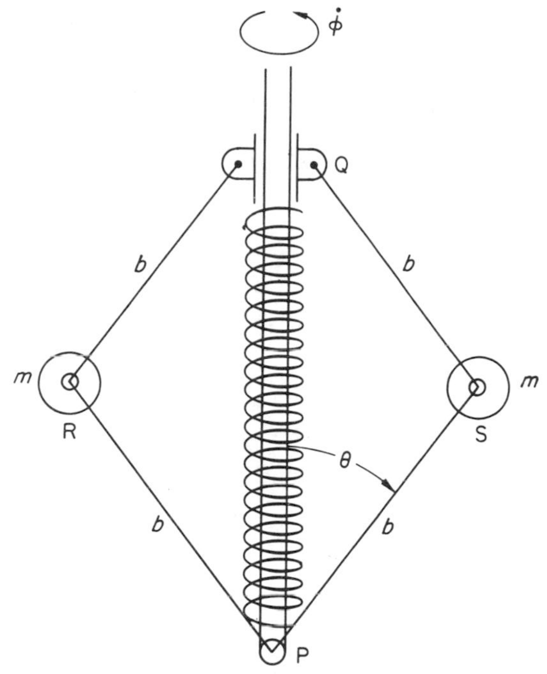

Figure 10.16

10.2. A smooth rod of mass m is free to rotate horizontally about a pivot at one end. On the rod slides a trolley of mass $4m/9$. Initially the rod is rotating with the trolley at its moving end, and a force is applied to the trolley along the axis of the rod, causing it to move with constant speed towards the pivot. Show that the force required reaches a maximum when the trolley is half-way along the rod.

10.3. The horizontal boom of a tower crane is rotated at constant angular velocity while the carriage bearing the load is driven outwards at constant speed along the boom. (The load is assumed not to swing relative to the carriage.) If P_1 is the power delivered by the motor rotating the boom and P_2 that by the motor moving the carriage outwards, show that

$$\frac{P_1}{P_2} = -2,$$

explaining the significance of the minus sign.

Calculate the change in energy of the system (boom, carriage and load) as the carriage moves from the axis of rotation to the extremity of the jib, and show that this is equal to the net work done by the two motors.

If the boom rotates one turn in 30 s and the carriage moves at a speed of $0{\cdot}5$ m s^{-1}, calculate the maximum torque that must be applied to a boom of length 30 m when the mass of the carriage plus load is 3000 kg. Hence show that if the torque is transmitted through a pinion engaging a horizontal gearwheel of diameter 2 m attached to the boom, the force between the teeth of the gears is roughly 2 tons weight.

10.4. Obtain Lagrange's equations for the system shown in Figure 10.9 (page 311) when the two masses are acted upon by external forces $\mathbf{F}_1$, $\mathbf{F}_2$, both these forces and the spring being coplanar. Show that as the strength constant of the spring tends to infinity the equations become those appropriate to a light rigid rod with equal masses at its ends, and determine the stress in the rod when it is in its general state of motion.

10.5. An inextensible string passes over a pulley, supporting a mass m_1 at one end and a second pulley at the other. Another inextensible string passes over the second pulley, supporting masses m_2 and m_3 at its ends, as shown in Figure 10.17. Each pulley has a mass m, moment of inertia I and radius a.

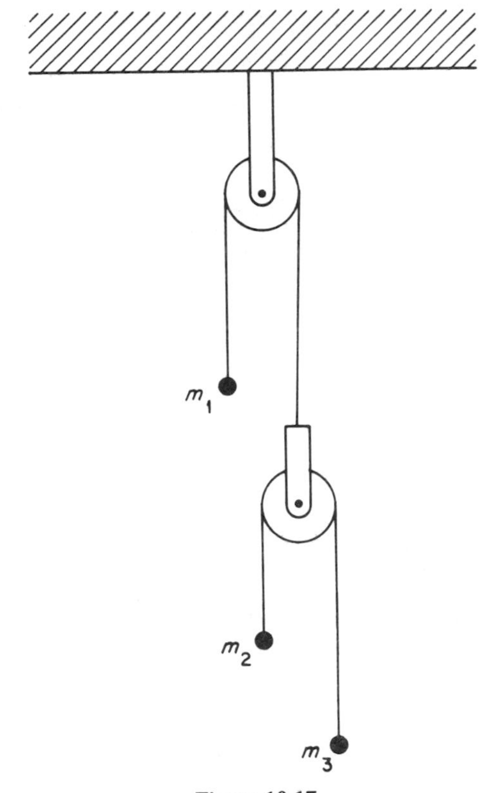

Figure 10.17

Show that m_1 will remain at rest if

$$m_1 = m_2 + m_3 + m - \frac{(m_2 - m_3)^2}{m_2 + m_3 + I/a^2},$$

and that m_2 will remain at rest if

$$m_2 = \frac{(m_1 - m)(m_3 + I/a^2) + 2m_1 m_3}{m_1 + 4m_3 + m + 2I/a^2}.$$

10.6. A uniform vertical cylinder of mass M and radius R has light rigid rods pivoted at opposite ends of a diameter so that they may rotate in a vertical plane. At the end of each rod is a mass m. Gravity acts upon the system, and the inclination of the rods, $\theta(t)$, is made to vary in some prescribed way with time while the cylinder is free to rotate about its vertical axis, as shown in Figure 10.18.

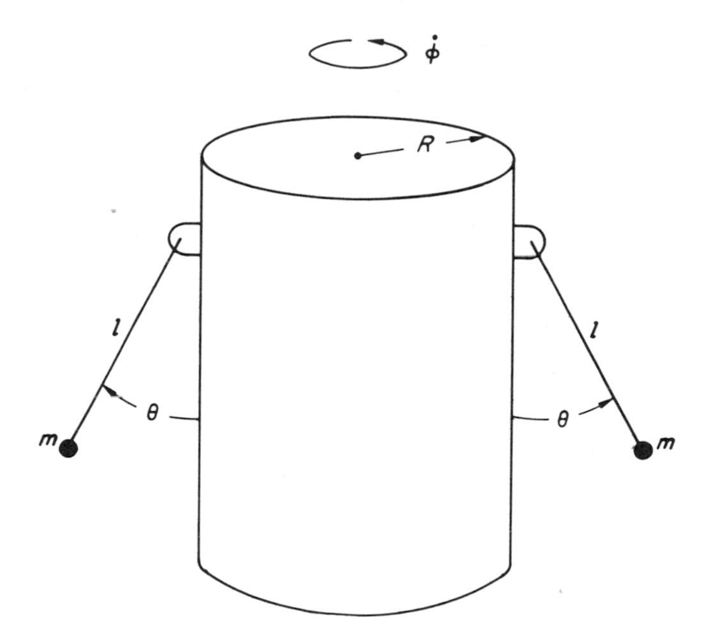

Figure 10.18

Show that if the rods are lowered from the horizontal at a constant rate

$$\dot{\theta} = -\beta,$$

then the power required to maintain the movement of the rods at any inclination is

$$P = 2ml\beta[\cos\theta(R + l\sin\theta)\dot{\phi}^2 - g\sin\theta],$$

and that angular speed $\dot{\phi}$ is increased by the factor

$$\gamma = \frac{1 + \dfrac{4m}{M}\left(\dfrac{R + l}{R}\right)^2}{1 + 4m/M},$$

as the rods move from the horizontal to the vertical position. Using masses and dimensions that would make this structure a model for a typical spinning skater, determine what increase in his spin rate could be achieved by lowering his arms from the horizontal to the vertical. Could he do better than this?

10.7. The *spherical pendulum* is physically similar to the simple pendulum, consisting of a point mass m supported by a light inextensible string or rod of length a from a fixed point. It is, however, free to move in space so that not only does the inclination of the string vary, but also the orientation of the vertical plane through it. Show that angles θ, ϕ describing the inclination and orientation are suitable generalized coordinates and that in terms of them the Lagrangian is

$$\mathscr{L} = \frac{ma^2}{2}(\dot{\theta}^2 + \sin^2\theta\dot{\phi}^2) + mga\cos\theta.$$

Use Lagrange's equations to show that, provided the string remains taut,

$$\sin^2\theta\dot{\phi} = \beta = \text{constant},$$

and that θ varies according to the equation

$$\frac{\dot{\theta}^2}{2} + \frac{\beta^2}{2\sin^2\theta} - \frac{g\cos\theta}{a} = \gamma = \text{constant}.$$

Determine the condition under which the pendulum rotates at a constant inclination, θ_c, and show that if it is perturbed by a sudden blow towards the axis of rotation, the period of oscillation about θ_c is

$$\tau = 2\pi\left(\frac{a}{g}\frac{\cos\theta_c}{1 + 3\cos^2\theta_c}\right)^{1/2}.$$

10.8. According to the previous problem oscillations about the inclination $\theta_c = 0$ would apparently have a period

$$\tau = \pi\left(\frac{a}{g}\right)^{1/2}.$$

$\theta_c = 0$, however, corresponds to the vertically downward equilibrium position for the pendulum, and oscillations about this are known to have a period

$$\tau = 2\pi\left(\frac{a}{g}\right)^{1/2}.$$

What is the explanation of this apparent contradiction?

10.9. Use the technique of artificially deforming the string of a simple pendulum to determine the tension in it. Show that if the amplitude of the swing is $\frac{1}{10}$ radian, then the tension varies from 1 per cent above mg to $\frac{1}{2}$ per cent below, m being the mass of the bob.

10.10. A wedge of mass M and included angle α lies on a smooth horizontal surface. Under the influence of gravity a uniform sphere of mass m rolls down the inclined plane of the wedge without slipping. x is the horizontal displacement of the wedge and y is the displacement of the point of contact of the sphere measured along the inclined plane from some point on it.

If the wedge and sphere are initially at rest and their initial positions are taken as the zeros for x and y, show that the subsequent motion is given by

$$x = -\frac{m}{M'}\frac{g}{4}\sin 2\alpha\, t^2,$$

$$y = \frac{m + M}{M'}\frac{g}{2}\sin\alpha\, t^2,$$

where

$$M' = \frac{7M}{5} + m\left(\sin^2\alpha + \frac{2}{5}\right).$$

10.11. (a) When the included angle of the wedge in the preceding problem is a right-angle, the 'inclined' plane is vertical and $\alpha = \pi/2$. Substituting this value in the results obtained we find that the wedge remains stationary:

$$x = 0,$$

while the sphere falls vertically downwards:

$$y = \tfrac{5}{14}gt^2.$$

However, we would expect any mass falling under gravity to have the motion

$$y = \tfrac{1}{2}gt^2.$$

Explain, in detail, how this apparent contradiction arises.

(b) Suppose the angle of the wedge is within the limits

$$\pi/2 < \alpha < \pi.$$

Describe the physical situation implied by this range of values, and show that, although x and y now *both* increase with t, the centre-of-mass remains stationary.

10.12. What additional coordinate may be introduced to enable the reaction between the wedge and the horizontal plane in problem 10.8 to be calculated? Show that this reaction is

$$R = (M + m)\left(1 - \frac{m}{M'}\sin^2\alpha\right)g.$$

How do you account for the value

$$R = \left(M + \frac{2m}{7}\right)g$$

when the wedge is right-angled?

10.13. A horizontal force drives the wedge of problem 10.8 in such a way that the sphere remains stationary relative to the wedge. Show that

$$F = (M + m)g\tan\alpha.$$

10.14. A spring of strength constant K is stretched between fixed points and has two masses m attached at its points of trisection. These masses oscillate free of external forces along the spring axis, their displacements from their positions of equilibrium being x and y. Show that the Lagrangian may be written as

$$\mathscr{L} = \tfrac{1}{2}m(\dot{x}^2 + \dot{y}^2) - K(x^2 + y^2 - xy).$$

Show that both masses will oscillate with the *same* angular frequency ω provided this has one of the values $(K/m)^{1/2}$, $(3K/m)^{1/2}$. Determine the relative amplitudes and phases of x and y for these two cases.
(*Hint*: substitute complex solutions with time dependence $\exp(j\omega t)$ for both x and y in Lagrange's equations, and show that their amplitudes will both be zero unless ω has one of the suggested values.)

10.15. By changing to the generalized coordinates

$$X = x + y, \qquad Y = x - y,$$

show that the Lagrangian for the preceding problem becomes identical with that for two *independent* oscillators of angular frequencies $(K/m)^{1/2}$ and $(3K/m)^{1/2}$, and that Lagrange's equations are then the simple harmonic equations

$$\ddot{X} + \frac{K}{m}X = 0, \qquad \ddot{Y} + \frac{3K}{m}Y = 0.$$

Hence show that any free oscillation of the system may be described as a sum of these two oscillations, with suitably chosen amplitudes and phases for X and Y. (The two single frequency solutions of problem 10.12 are known as *normal modes* and the generalized coordinates which separate them, as shown in problem 10.13, are called *normal coordinates.*)

10.16. In problem 10.12 the masses are held stationary at the values

$$x = a, \qquad y = 0,$$

and then released. Show that their subsequent motion is

$$x = \tfrac{1}{2}a[\cos(K/m)^{1/2}t + \cos(3K/m)^{1/2}t],$$

$$y = \tfrac{1}{2}a[\cos(K/m)^{1/2}t - \cos(3K/m)^{1/2}t].$$

Can the first mass ever exactly regain its initial displacement?

10.17. In Figure 10.19 RQ and SP are light rigid rods of length a pivoted at T. OR and

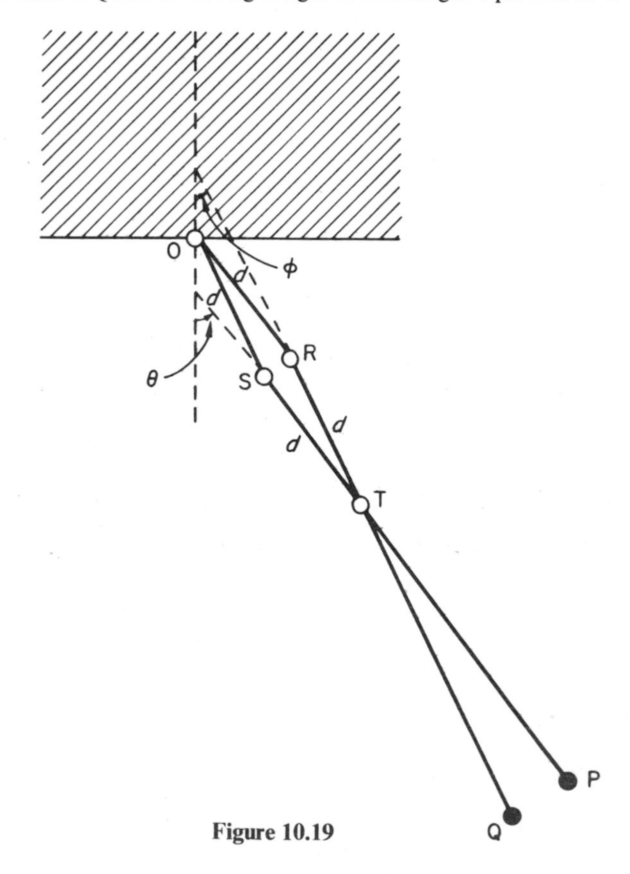

Figure 10.19

OS are two light rigid rods pivoted at R and S and at the fixed point O. d is the length of each of OR, OS, RT and ST. Equal masses, m, are attached to P and Q. The whole system is allowed to move in a vertical plane under gravity.

Show that if the inclinations of SP and RQ to the vertical are θ, ϕ respectively, then the Lagrangian for *small* oscillations is

$$\mathscr{L} = \tfrac{1}{2}m(a^2 + d^2)(\dot{\theta}^2 + \dot{\phi}^2) + 2mad\dot{\theta}\dot{\phi} - \tfrac{1}{2}mg(a + d)(\theta^2 + \phi^2).$$

Show that if θ and ϕ are both to oscillate with the same frequency, this must have one of the values $[g/(a + d)]^{1/2}$, $[g(a + d)/(a - d)]^{1/2}$. What is the nature of the oscillations in these two cases? Show that the electrical analogue is two circuits *inductively* coupled.

Chapter 11

Special Relativity

Once physical phenomena and the results of experiments concerning them have been described mathematically it is possible to extend this description in a precise, systematic and self-consistent way. We have seen in the preceding chapters how the three Newtonian laws may be developed in this manner to give concepts and predictions applicable to a much greater range of experiments than those concerned merely with the motion of single particles.

One important aspect of this process of extension and generalization has already been pointed out. When experimental observations appear the same to a large class of observers, they become 'laws'. Newton's laws have this status because they can be shown to be the same for all inertial observers. In demonstrating this, and in much of the foregoing discussion, we have constantly implied some method of measuring physical quantities (for example the components of vectors) and of relating the measurements that different observers make of the same quantity.

When a vector is measured in various frames two assumptions are involved. Firstly, it is assumed that *instantaneous* perpendiculars from the vector to the axes can be constructed. Secondly, it is assumed that the *same* scales exist on all the axes concerned so that the points of intersection of the perpendiculars with the axes can be used to measure the components in any frame. It is only when the scales are the same that any meaning can be attached to comparing the magnitudes in different frames.

How do these assumptions stand up to experimental tests? In many cases extremely well, which is why the assumptions are usually not explicitly mentioned but regarded rather as self-evident truths. However, if we consider the first more carefully we realize that there is no way of physically connecting two points, or sending a message between them, instantaneously. *A light signal is the fastest method known to us*, and this implies that any method of projecting vectors on to axes will take a finite time. In consequence, any *experimentally constructed* components will lag behind the vector they represent when that vector is varying in length or direction in the frame under consideration.

This fact does not necessarily invalidate the concept of instantaneous projection, since we might argue that, once the speed of light is known, we could allow for the experimental time lag and predict what the components *would be* if the vectors could be projected on to the axes instantaneously. Whether or not we accept such an argument, it is clear that any formulation of the laws of physics must

(1) require that the speed of light be determined in any frame of reference, and
(2) accord with the fact that the speed of light is the upper limit for the speed of any material particle or the transmission of any information.

The second assumption (that scales of measurement remain the same for various frames of reference) is the basis for virtually all engineering design. Parts of a machine which fit together when they are stationary still do so when they move relative to each other (allowing, of course, for any strains in the material that might arise from the motion). Since their sizes and shapes are unaltered we may call this experimental fact *geometrical invariance.* However, although this is a very widely tested fact, it is tested only within the range of speeds available to engineering practice, and we must bear this in mind in future discussion. (See Section 11.12.1, page 371.)

11.1. Geometrical invariance

Assuming that instantaneous projection can be imagined if not actually performed, let us see how the 'fact' of geometrical invariance enters into the description of mechanics that we have used so far. Given any two positions in space we can consider the vector, $\Delta\mathbf{x}$, joining them as a *displacement vector* or *space interval*, defined by the elements of the 3×1 column matrix, (Δx), in a frame S. These elements need not be small, although we may later take the interval as an infinitesimal one when velocities are to be considered.

The methods of Chapter 9 show that the corresponding space intervals (Δx), $(\Delta x')$ in two frames with different orientations are related by

$$(\Delta x') = (A)(\Delta x), \qquad (\Delta x) = (\tilde{A})(\Delta x').$$

The properties of the transformation matrix (A) are such that the scalar product of two vectors, $\Delta\mathbf{x}$ and $\Delta\mathbf{y}$, is the same whichever frame it is calculated in:

$$\begin{aligned}\Delta x'_1\,\Delta y'_1 &+ \Delta x'_2\,\Delta y'_2 + \Delta x'_3\,\Delta y'_3\\ &= (\widetilde{\Delta x'})(\Delta y')\\ &= (\widetilde{\Delta x})(\tilde{A})(A)(\Delta y)\\ &= (\widetilde{\Delta x})(\Delta y) = \Delta x_1\,\Delta y_1 + \Delta x_2\,\Delta y_2 + \Delta x_3\,\Delta y_3.\end{aligned} \tag{11.1}$$

By taking

$$\Delta\mathbf{x} = \Delta\mathbf{y},$$

this result shows that

$$\Delta x_1'^2 + \Delta x_2'^2 + \Delta x_3'^2 = \Delta x_1^2 + \Delta x_2^2 + \Delta x_3^2 \tag{11.2}$$

(i.e., the length of the vector $\Delta\mathbf{x}$ is the same in both frames). If we write this as $\Delta x, \Delta x'$ in the two frames, equation (11.2) is equivalent to

$$\Delta x' = \Delta x.$$

Similarly,

$$\Delta y' = \Delta y.$$

Hence if $\theta_{yx}, \theta'_{yx}$ are the angles between the pair of vectors as measured in the two frames, equation (11.2) is the same as

$$\Delta x' \, \Delta y' \cos \theta'_{yx} = \Delta x \, \Delta y \cos \theta_{yx},$$

and therefore

$$\theta'_{yx} = \theta_{yx}.$$

Thus the transformation matrix (A) is such that distances and angles appear the same in the two frames. They are *invariants* of the transformation. We can think of them as predictions derived from the transformation—predictions which are confirmed experimentally and which therefore give support to the validity of the transformation.

Suppose, instead, that we start with these invariance properties as experimental observations and ask which transformations (A) will satisfy them. If lengths and angles are to be the same for S and S', reversal of the argument above shows that the scalar product of two displacement vectors must be invariant:

$$(\Delta x')(\Delta y') = (\widetilde{\Delta x})(\tilde{A})(A)(\Delta y) = (\widetilde{\Delta x})(\Delta y).$$

Thus (A) must satisfy

$$(\tilde{A})(A) = (1). \tag{11.3}$$

If we were to write out equation (11.3) in full we should find relationships among the elements a_{ij} of (A) which would enable us to give them some direct physical meaning.

To demonstrate this simply we shall take the special case in which the third component of the vectors is unchanged. (A) will then be of the form:

$$(A) = \begin{pmatrix} a_{11} & a_{12} & 0 \\ a_{21} & a_{22} & 0 \\ 0 & 0 & 1 \end{pmatrix},$$

so that

$$(\tilde{A})(A) = \begin{pmatrix} a_{11}^2 + a_{21}^2 & a_{11}a_{12} + a_{21}a_{22} & 0 \\ a_{12}a_{11} + a_{22}a_{21} & a_{12}^2 + a_{22}^2 & 0 \\ 0 & 0 & 1 \end{pmatrix}.$$

Thus if equation (11.3) is to be true, the a_{ij} must satisfy

$$\left.\begin{aligned} a_{11}^2 + a_{21}^2 &= 1, \\ a_{12}^2 + a_{22}^2 &= 1, \\ a_{11}a_{12} + a_{21}a_{22} &= 0. \end{aligned}\right\} \tag{11.4}$$

The form of equations (11.4) suggests a cosine and sine type of solution, and it is easy to check that such a one is

$$a_{11} = \cos\phi, \qquad a_{12} = \sin\phi, \qquad a_{21} = -\sin\phi, \qquad a_{22} = \cos\phi.$$

Then

$$(A) = \begin{pmatrix} \cos\phi & \sin\phi & 0 \\ -\sin\phi & \cos\phi & 0 \\ 0 & 0 & 1 \end{pmatrix}.$$

On calculating the inverse we find

$$(A)^{-1} = \begin{pmatrix} \cos\phi & -\sin\phi & 0 \\ \sin\phi & \cos\phi & 0 \\ 0 & 0 & 1 \end{pmatrix}.$$

Thus

$$(A)^{-1} = (\tilde{A}),$$

and the reverse transformation is therefore

$$(\Delta x) = (A)^{-1}(\Delta x') = (\tilde{A})(\Delta x').$$

As yet we know nothing about ϕ except that it is a real angle. To interpret it physically consider the vector

$$\Delta\mathbf{x} \underset{S'}{\leftrightarrow} (\Delta x') = \begin{pmatrix} \Delta x' \\ 0 \\ 0 \end{pmatrix}$$

which is parallel with the axis 1′. For S this will appear as

$$\Delta\mathbf{x} \underset{S'}{\leftrightarrow} (\Delta x) = (\tilde{A})(\Delta x') = \begin{pmatrix} \Delta x' \cos\phi \\ \Delta x' \sin\phi \\ 0 \end{pmatrix}$$

which is at an angle ϕ to the axis 1. Hence S will consider the S' axis 1′ to be rotated through an angle ϕ relative to his own axis 1. By taking $\Delta\mathbf{x}$ parallel in turn with the axes 2′ and 3′, S will find that the axis 2′ is also rotated through the angle ϕ relative to axis 2, while axis 3′ remains parallel with axis 3 (see Figure 11.1).

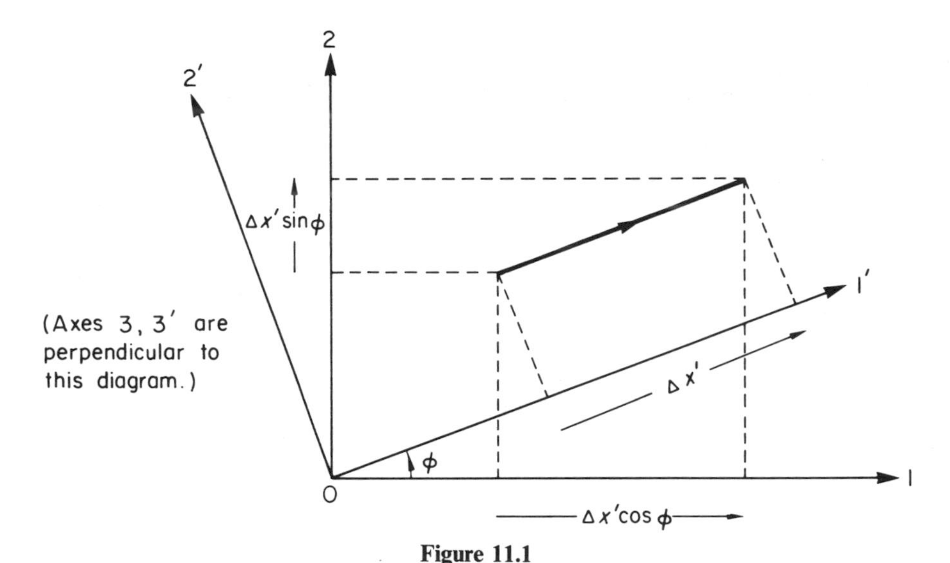

Figure 11.1

S will therefore consider that the S' frame is an orthogonal one like his own, but rotated through the angle ϕ in the 12 plane. By taking $\Delta\mathbf{x}$ parallel in turn with the axes 1, 2 and 3 we can similarly show that S' will consider that the S frame is an orthogonal one, but rotated through the angle $-\phi$ in the 1′2′ plane. Both observers are therefore in agreement over the meaning of the transformation (A).

11.2. The breakdown of classical mechanics

Lengths and angles are not the only physical quantities that have invariance properties. When we constantly assume quantities to be invariant, and find that these assumptions give equations that accord with experiments, we tend to think that the invariance is obvious and hardly worth mentioning. Mass and electric charge can be added to length and angle as examples of quantities we treat as invariants long before we grace them with that title and realize its full implications.

Sometimes, however, it is difficult to extract an invariance principle from a set of observations, especially when they concern phenomena that are not common experience and when the invariance appears to contradict what is otherwise held to be obvious.

In this category are experiments concerning the speed of light. We have already seen that it is essential to know this in any frame if we are to be able to relate, in a systematic *experimental* way, physical phenomena in different frames. Although from common experience it appears to be infinite, as far back as 1676 Römer determined a finite velocity of about 220,000 km s^{-1}. There was no reason to suppose, however, that it was different from any other velocities encountered in nature except that it was so large compared with them that it could often be treated as infinite in calculations.

Nevertheless, later experiments, in which interference effects were used to compare very accurately the velocity of light in various directions and for various observers, showed that it had very different properties from the much smaller velocities of ordinary mechanical experiments. This is not the place to describe in detail the nature of electromagnetic radiation and how it appears to different observers. All that need concern us here can be summarized in one surprisingly simple invariance principle:

The speed of light, in a vacuum, is the same for all inertial observers.

(This speed is denoted by c, and has the value $2{\cdot}998 \times 10^8$ m s^{-1}.)

To see one of the implications of this let us consider a light signal traversing a *physical* distance $\Delta x'$ which appears stationary to S' (for example from one end to the other of a measuring rod fixed in the S' frame). If this traversal takes a time $\Delta t'$, then S' will observe the speed of light as

$$c' = \frac{\Delta x'}{\Delta t'}. \tag{11.5}$$

Suppose S' is moving with a constant speed U relative to S in the direction of the measuring rod. Then for S the light signal has to travel not only the length of the rod during the traversal, but also the distance moved by the rod (see Figure 11.2). Thus if S says the light signal traverses the moving rod, of length Δx, in a time Δt at a speed c, then

$$c\,\Delta t = \Delta x + U\,\Delta t,$$

so that

$$c = \frac{\Delta x}{\Delta t} + U. \tag{11.6}$$

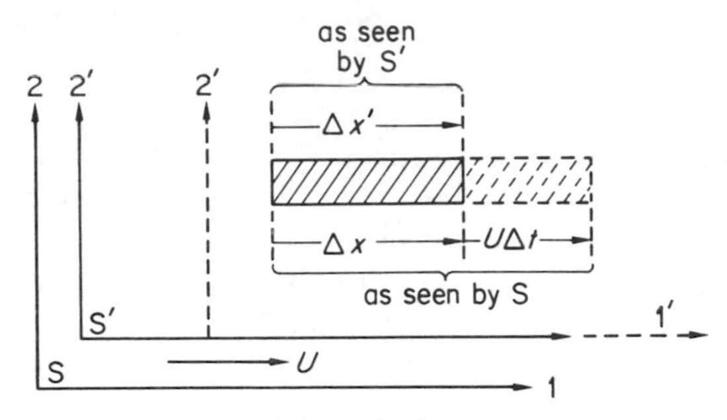

Figure 11.2

What exactly do we mean by Δx and Δt—the length and time intervals observed by S? The first is simple. In the preceding section we pointed out the importance of length invariance, which is equivalent here to the equation

$$\Delta x = \Delta x'. \tag{11.7}$$

The relationship between Δt and $\Delta t'$, however, is something we have not discussed so far. This is because we have always assumed that time is something all observers agree about. Of course, they may choose different time origins, just as they may choose different origins for their space axes, but we have never doubted the invariance of the time intervals:

$$\Delta t = \Delta t'. \tag{11.8}$$

We have thought of this as 'more obvious' even than length invariance. For one thing there is only one 'time axis' so we do not have to bother about time intervals of the same length, but 'pointing in different directions'. Time and length invariance then give, from equations (11.5) to (11.8),

$$c = c' + U. \tag{11.9}$$

This is, of course, a particular case of the law for compounding velocities that we derived in Section 8.2. The Galilean transformation plus time invariance lead inevitably to the result (11.9). But this contradicts the principle that the speed of light is invariant. Where have we gone wrong?

We have forgotten the warning of Chapter 1. Classical mechanics, like any other branch of science, is derived from a restricted range of phenomena. In particular, they involve objects in laboratories, rockets and planets, none travelling with speeds much above 10^5 m s^{-1}. What right have we to assume that the laws of classical mechanics will apply when the speeds are 10^8 m s^{-1} or more? None, unless experiments at such speeds show them still to be true.

11.3. Space-time invariance

It is clear that we shall have to revise our ideas about the invariance of distance and time intervals, at any rate when large speeds are involved. Even before this, however, we would do well to avoid any traps over hidden assumptions about time by mentioning it explicitly in any description of physical processes. Thus, instead of describing, as we have done so far, a particle as having a position (x_1, x_2, x_3) at time t, we shall refer to this as an *event* in the history of the particle defined by the four quantities (x_1, x_2, x_3, t). Time is then put on level terms with space and may be considered a 'co-ordinate', although this does not mean that it necessarily has the same properties as a space coordinate.

The two positions at two different times along the path of the particle then become two events in its history, which determine a *space-time interval* $(\Delta x_1, \Delta x_2, \Delta x_3, \Delta t)$. Instead of treating separately the space and time parts as seen by two observers and finding out how they are related,

$$(\Delta x_1, \Delta x_2, \Delta x_3) \rightarrow (\Delta x'_1, \Delta x'_2, \Delta x'_3),$$

$$\Delta t \rightarrow \Delta t',$$

we consider the relationship between the space-time intervals:

$$(\Delta x_1, \Delta x_2, \Delta x_3, \Delta t) \rightarrow (\Delta x'_1, \Delta x'_2, \Delta x'_3, \Delta t').$$

We are therefore led to think in terms of a *space-time vector* defined by *four* components. Following the arguments of Section 11.1 we may expect the transformation relating the vector in two frames to be determined, in part at least, by the invariance properties of this vector.

To see what these are we first take the 'particle' to be a light signal, so that the intervals refer to the passage of a light wave as seen by S and S'. If they are both to find the same speed, c, for this, then

$$\frac{(\Delta x_1^2 + \Delta x_2^2 + \Delta x_3^2)^{1/2}}{\Delta t} = c = \frac{(\Delta x_1'^2 + \Delta x_2'^2 + \Delta x_3'^2)^{1/2}}{\Delta t'}$$

or

$$\Delta x_1^2 + \Delta x_2^2 + \Delta x_3^2 - c^2\,\Delta t^2 = 0 = \Delta x_1'^2 + \Delta x_2'^2 + \Delta x_3'^2 - c^2\,\Delta t'^2. \quad (11.10)$$

These equations are beginning to look like those of the invariance law,

$$\Delta x_1^2 + \Delta x_2^2 + \Delta x_3^2 = \Delta x_1'^2 + \Delta x_2'^2 + \Delta x_3'^2, \quad (11.11)$$

except that they are restricted to the value zero and that the fourth term is subtracted from, and not added to, the others. We might expect some restriction in value since we are dealing with a special type of space-time interval—one obtained from the motion of a light signal. We could use two events in the history of a material particle to define an interval in each frame. From these we could obtain the (average) speeds v, v^*:

$$\frac{(\Delta x_1^2 + \Delta x_2^2 + \Delta x_3^2)^{1/2}}{\Delta t} = v,$$

$$\frac{(\Delta x_1'^2 + \Delta x_2'^2 + \Delta x_3'^2)^{1/2}}{\Delta t'} = v^*.$$

From these equations we can write expressions of the type (11.11):

$$\Delta x_1^2 + \Delta x_2^2 + \Delta x_3^2 - c^2\Delta t^2 = (v^2 - c^2)\Delta t^2, \tag{11.12}$$

$$\Delta x_1'^2 + \Delta x_2'^2 + \Delta x_3'^2 - c^2\Delta t'^2 = (v'^2 - c^2)\,\Delta t'^2. \tag{11.13}$$

There does not now appear to be any simple relationship between these two quantities since v and v^* are not necessarily the same. However, experiments involving particles (usually nuclear particles) at very high speeds yield another fundamental observation which was commented on earlier:

> *No material particles are observed with speeds greater than, or equal to, that of light.*

This refers to all observers moving with any constant relative velocities (except that their speeds, too, are always found to be less than that of light). Thus when either of the expressions (11.12) or (11.13) is known to refer to the passage of a material particle it will be negative and will imply that the other is negative also.

Suppose we go one step further and assume that they are *equal*. This assumption would then be the invariance law,

$$\Delta x_1^2 + \Delta x_2^2 + \Delta x_3^2 - c^2\,\Delta t^2 = \Delta x_1'^2 + \Delta x_2'^2 + \Delta x_3'^2 - c^2\,\Delta t'^2, \tag{11.14}$$

similar in form to the law (11.11) except for the negative sign of the last term.

There is no good reason to be surprised at this change of sign. The signs of the three space terms must be the same since no direction in free space has any special properties; space is 'isotropic'. Time, however, does have different properties, and it is not to be wondered at that the sign of its corresponding term is different. It *must* be different if the proposed law (11.14) is to include the special case (11.10).

11.4. The space-time four-vector

Assuming the law (11.14) to be true, how can we express this in a way that systematically extends the matrix description of mechanics developed in

Chapter 8? The invariant derived from the three-component column matrix (or *three-vector*), describing an interval between two *positions*, or *space interval*,

$$(\Delta x) = \begin{pmatrix} \Delta x_1 \\ \Delta x_2 \\ \Delta x_3 \end{pmatrix},$$

was its *length*, whose square was written as the (matrix) product of the vector and its transpose:

$$(\widetilde{\Delta x})(\Delta x) = (\Delta x_1\ \Delta x_2\ \Delta x_3)\begin{pmatrix} \Delta x_1 \\ \Delta x_2 \\ \Delta x_3 \end{pmatrix} = \Delta x_1^2 + \Delta x_2^2 + \Delta x_3^2. \tag{11.15}$$

To correspond with (11.15) we shall need the four-component column matrix (or *four-vector*), describing an interval between two *events*, or *space-time interval*,

$$(\Delta x) = \begin{pmatrix} \Delta x_1 \\ \Delta x_2 \\ \Delta x_3 \\ \Delta x_4 \end{pmatrix} \equiv \begin{pmatrix} \Delta x_1 \\ \Delta x_2 \\ \Delta x_3 \\ c\Delta t \end{pmatrix}. \tag{11.16}$$

Note that we have now written the fourth component as $c\,\Delta t$. This ensures that each component has the same dimension (length) and that the squares of each are the terms of the expression $\Delta x_1^2 + \Delta x_2^2 + \Delta x_3^2 - c^2\,\Delta t^2$ which occurs in the proposed law (11.14). From now on we shall refer to this expression as the 'length' of the four-vector, meaning by this either the quantity as just written *or* its square root; the intention will be clear from the context. This length cannot simply be written as $(\widetilde{\Delta x})(\Delta x)$, where this product follows the normal rules of matrix multiplication, since this would give a positive fourth term. There are various ways of overcoming this difficulty. One is to redefine the four-vector as

$$(\Delta x) = \begin{pmatrix} \Delta x_1 \\ \Delta x_2 \\ \Delta x_3 \\ \Delta x_4 \end{pmatrix} \equiv \begin{pmatrix} \Delta x_1 \\ \Delta x_2 \\ \Delta x_3 \\ jc\Delta t \end{pmatrix}, \tag{11.17}$$

where

$$j = \sqrt{-1}.$$

Then $(\widetilde{\Delta x})(\Delta x)$, calculated in the normal way, will give the negative sign, $j^2 = -1$, for the last term. This can only be regarded as a mathematical trick; j has no physical significance here, as it does, for example, in alternating-current circuit theory where it describes a phase difference. It is simply an automatic way of remembering where to put the minus sign in equation (11.14).

We could remember this instead by always putting in a symbol, (G) say, between the factors to indicate the change of sign:

$$(\widetilde{\Delta x})(G)(\Delta x) = \Delta x_1^2 + \Delta x_2^2 + \Delta x_3^2 - \Delta x_4^2. \tag{11.18}$$

(G) may be regarded as changing either

$$(\widetilde{\Delta x}) = (\Delta x_1 \; \Delta x_2 \; \Delta x_3 \; \Delta x_4)$$

to

$$(\widetilde{\Delta x})(G) = (\Delta x_1 \; \Delta x_2 \; \Delta x_3 \; -\Delta x_4),$$

or

$$(\Delta x) = \begin{pmatrix} \Delta x_1 \\ \Delta x_2 \\ \Delta x_3 \\ \Delta x_4 \end{pmatrix} \quad \text{to} \quad (G)(\Delta x) = \begin{pmatrix} \Delta x_1 \\ \Delta x_2 \\ \Delta x_3 \\ -\Delta x_4 \end{pmatrix}.$$

Either product, $[(\widetilde{\Delta x})(G)](\Delta x)$ or $(\widetilde{\Delta x})[(G)(\Delta x)]$ will then give the same result (11.18) according to the usual law of matrix multiplication.

Once we have got this far we can make the whole process of deriving the invariant (11.18) follow normal matrix laws by writing (G) as the simple 4×4 matrix:

$$(G) = \begin{pmatrix} 1 & 0 & 0 & 0 \\ 0 & 1 & 0 & 0 \\ 0 & 0 & 1 & 0 \\ 0 & 0 & 0 & -1 \end{pmatrix}.$$

It does not matter whether $(\widetilde{\Delta x})$ and (G) are multiplied together first, or (G) and (Δx). In either case the effect is to change the sign of the last element of $(\widetilde{\Delta x})$ or of (Δx) and thus give the required invariant,

$$(\widetilde{\Delta x})(G)(\Delta x) = \Delta x_1^2 + \Delta x_2^2 + \Delta x_3^2 - \Delta x_4^2. \tag{11.19}$$

Although the introduction of (G) may appear an unnecessary complication, it has important advantages. Once we have seen that its form has the property required it is rarely, if ever, necessary to write out (G) in full. There is no more to remember and write than when introducing j into the definition of the four-vector, so that, compared with that method, the use of (G) is not a complication at all. From a physical point of view it much more satisfactorily separates the components of the four-vector, which are four real measurements made of an event, from the properties of space and time which are represented by (G). This matrix is called the space-time *metric* (or *metric tensor*).

The corresponding matrix for three-vectors,

$$(\widetilde{\Delta x})(G)(\Delta x) = \Delta x_1^2 + \Delta x_2^2 + \Delta x_3^2,$$

would be the metric

$$(G) = \begin{pmatrix} 1 & 0 & 0 \\ 0 & 1 & 0 \\ 0 & 0 & 1 \end{pmatrix}.$$

Since this is a unit matrix it can be, and nearly always is, omitted from any product. However, the fact that the metric here, if we wish to write it out explicitly, *is* a unit matrix expresses very clearly the isotropic nature of space referred to above.

11.5. The Lorentz transformation

As we saw in Chapter 8, grouping a set of observations together may be convenient, but the group does not have any special physical significance unless it satisfies some simple transformation law relating its values for two observers. In Section 11.1 the length and angle invariance properties of three-vectors were used to show what the transformation should be. We shall use exactly the same approach to discover the transformation for four-vectors, but starting now with the requirement that it is the length of a four-vector, as described in equation (11.19), that is to be invariant.

Here, too, we choose a special case; this time one in which Δx_2 and Δx_3 are unchanged. (We must allow at least two of the components to change, otherwise they would all remain unchanged; the time component must be one of the two that change, otherwise we shall simply be back to the rotation in space described by the (A) of Chapter 8.) The transformation will then be of the form:

$$\begin{pmatrix} \Delta x_1' \\ \Delta x_2' \\ \Delta x_3' \\ \Delta x_4' \end{pmatrix} = \begin{pmatrix} \lambda_{11} & 0 & 0 & \lambda_{14} \\ 0 & 1 & 0 & 0 \\ 0 & 0 & 1 & 0 \\ \lambda_{41} & 0 & 0 & \lambda_{44} \end{pmatrix} \begin{pmatrix} \Delta x_1 \\ \Delta x_2 \\ \Delta x_3 \\ \Delta x_4 \end{pmatrix}$$

or, more concisely,

$$(\Delta x') = (\Lambda)(\Delta x). \tag{11.20}$$

For the four-vector length to be invariant we must have

$$(\widetilde{\Delta x'})(G)(\Delta x') = (\widetilde{\Delta x})(\tilde{\Lambda})(G)(\Lambda)(\Delta x) = (\widetilde{\Delta x})(G)(\Delta x).$$

Hence the transformation (Λ) must satisfy

$$(\tilde{\Lambda})(G)(\Lambda) = (G). \tag{11.21}$$

Now

$$(G)(\Lambda) = \begin{pmatrix} 1 & 0 & 0 & 0 \\ 0 & 1 & 0 & 0 \\ 0 & 0 & 1 & 0 \\ 0 & 0 & 0 & -1 \end{pmatrix}\begin{pmatrix} \lambda_{11} & 0 & 0 & \lambda_{14} \\ 0 & 1 & 0 & 0 \\ 0 & 0 & 1 & 0 \\ \lambda_{41} & 0 & 0 & \lambda_{44} \end{pmatrix} = \begin{pmatrix} \lambda_{11} & 0 & 0 & \lambda_{14} \\ 0 & 1 & 0 & 0 \\ 0 & 0 & 1 & 0 \\ -\lambda_{41} & 0 & 0 & -\lambda_{44} \end{pmatrix}.$$

(Note that (G) multiplying any matrix from the left changes the sign of its last row; when multiplying from the right it changes the sign of the last column—the obvious extension of its effect upon single column or row matrices. This observation enables us still to avoid writing out (G) in full.) Hence

$$(\tilde{\Lambda})(G)(\Lambda) = \begin{pmatrix} \lambda_{11} & 0 & 0 & \lambda_{41} \\ 0 & 1 & 0 & 0 \\ 0 & 0 & 1 & 0 \\ \lambda_{14} & 0 & 0 & \lambda_{44} \end{pmatrix}\begin{pmatrix} \lambda_{11} & 0 & 0 & \lambda_{14} \\ 0 & 1 & 0 & 0 \\ 0 & 0 & 1 & 0 \\ -\lambda_{41} & 0 & 0 & -\lambda_{44} \end{pmatrix}$$

$$= \begin{pmatrix} \lambda_{11}^2 - \lambda_{41}^2 & 0 & 0 & \lambda_{11}\lambda_{14} - \lambda_{41}\lambda_{44} \\ 0 & 1 & 0 & 0 \\ 0 & 0 & 1 & 0 \\ \lambda_{14}\lambda_{11} - \lambda_{44}\lambda_{41} & 0 & 0 & \lambda_{14}^2 - \lambda_{44}^2 \end{pmatrix},$$

so that if equation (11.21) is to be satisfied we must have

$$\left.\begin{aligned} \lambda_{11}^2 - \lambda_{41}^2 &= 1, \\ \lambda_{14}^2 - \lambda_{44}^2 &= -1, \\ \lambda_{11}\lambda_{14} - \lambda_{41}\lambda_{44} &= 0. \end{aligned}\right\} \tag{11.22}$$

The form of these equations leads us to consider cosh and sinh solutions (instead of the cosine and sine solutions of Section 11.1), and it is easy to see that one such solution is

$$\lambda_{11} = \cosh\chi, \qquad \lambda_{41} = -\sinh\chi, \qquad \lambda_{14} = -\sinh\chi, \qquad \lambda_{44} = \cosh\chi.$$

Then

$$(\Lambda) = \begin{pmatrix} \cosh\chi & 0 & 0 & -\sinh\chi \\ 0 & 1 & 0 & 0 \\ 0 & 0 & 1 & 0 \\ -\sinh\chi & 0 & 0 & \cosh\chi \end{pmatrix} \tag{11.23}$$

and its inverse is

$$(\Lambda)^{-1} = \begin{pmatrix} \cosh\chi & 0 & 0 & \sinh\chi \\ 0 & 1 & 0 & 0 \\ 0 & 0 & 1 & 0 \\ \sinh\chi & 0 & 0 & \cosh\chi \end{pmatrix}, \tag{11.24}$$

giving the inverse transformation

$$(\Delta x) = (\Lambda)^{-1}(\Delta x'). \tag{11.25}$$

Note that, unlike the three-vector transformation, (Λ) and $(\Lambda)^{-1}$ are both symmetric and are not the transpose of each other. It is still true, however, that we obtain one from the other by changing the sign of χ, just as (A) and $(A)^{-1}$ are obtained from each other by changing the sign of ϕ.

We were able to interpret the ϕ of Section 11.1 by choosing three-vectors parallel with particular axes. We shall do the same here, although we will have to be very careful about what is meant by a four-vector 'parallel with an axis'. Formally it is easy to write down such a vector. For example if it is parallel with the axis $1'$ it will be of the form:

$$(\Delta x') = \begin{pmatrix} \Delta x'_1 \\ 0 \\ 0 \\ 0 \end{pmatrix}.$$

However, what is the physical interpretation of this? Remembering that $(\Delta x')$ is the space-time interval between two events *as observed by* S', we note the following:

(1) Since $\Delta x'_2 = \Delta x'_3 = 0$, but $\Delta x'_1 \neq 0$, the two events occur in the S' frame at the ends of a line parallel with the axis $1'$.

(2) Since $\Delta x'_4 = c\,\Delta t' = 0$, S' observes the two events *at the same time.*

To give a physical example of such an interval we might imagine that two spark gaps are fixed at the ends of a bar which is parallel with the axis 1′. Two wires, whose lengths are equal according to S', lead from these to a single switch. When the switch is closed the two spark gaps will fire simultaneously in the S' frame. The two events which define the interval are then the positions and times of the two sparks.

According to equation (11.25) the interval as observed by S is

$$\begin{pmatrix} \Delta x_1 \\ \Delta x_2 \\ \Delta x_3 \\ c\Delta t \end{pmatrix} = \begin{pmatrix} \cosh\chi & 0 & 0 & \sinh\chi \\ 0 & 1 & 0 & 0 \\ 0 & 0 & 1 & 0 \\ \sinh\chi & 0 & 0 & \cosh\chi \end{pmatrix} \begin{pmatrix} \Delta x'_1 \\ 0 \\ 0 \\ 0 \end{pmatrix}.$$

From this we see that

$$\Delta x_2 = \Delta x_3 = 0, \qquad \Delta x_1 = \cosh\chi\,\Delta x'_1 \neq 0,$$

which shows that S observes the two events at the ends of a line which is also parallel with the axis 1. Hence the axes 1′ and 1 are parallel. (Note that $\Delta x_1 \neq \Delta x'_1$ and that $\Delta t \neq 0$, so that S does not observe the events as simultaneous ones or find them the same distance apart as does S'. We shall refer to this later; see Sections 11.12.3, page 372, and 11.12.4, page 373.)

When the interval is taken parallel in turn with the axes 2′ and 3′ a similar argument shows that these axes are parallel respectively with the axes 2 and 3. (In these cases $\Delta t = 0$ also, so that the events are simultaneous for both observers.)

Finally, let us take an interval parallel with the axis 4′. This will have the form:

$$(\Delta x') = \begin{pmatrix} 0 \\ 0 \\ 0 \\ \Delta x'_4 \end{pmatrix} = \begin{pmatrix} 0 \\ 0 \\ 0 \\ c\Delta t' \end{pmatrix}.$$

Since $\Delta x'_1 = \Delta x'_2 = \Delta x'_3 = 0$, the two events occur *in the same position* in the S' frame. They refer, therefore, to two different times in the history of a point which is *fixed in the S' frame*. Equations (11.25) and (11.24) show that S will observe the interval as

$$\begin{pmatrix} \Delta x_1 \\ \Delta x_2 \\ \Delta x_3 \\ c\Delta t \end{pmatrix} = \begin{pmatrix} \cosh\chi & 0 & 0 & \sinh\chi \\ 0 & 1 & 0 & 0 \\ 0 & 0 & 1 & 0 \\ \sinh\chi & 0 & 0 & \cosh\chi \end{pmatrix} \begin{pmatrix} 0 \\ 0 \\ 0 \\ c\Delta t' \end{pmatrix} = \begin{pmatrix} \sinh\chi\, c\Delta t' \\ 0 \\ 0 \\ \cosh\chi\, c\Delta t' \end{pmatrix}.$$

He therefore sees a change of position for the point,

$$\Delta x_1 = \sinh\chi \, c\Delta t', \qquad \Delta x_2 = \Delta x_3 = 0,$$

in a time

$$\Delta t = \cosh\chi \, \Delta t'.$$

S would therefore say that a point fixed in the S' frame has velocity components

$$U_1 = \frac{\Delta x_1}{\Delta t} = c\tanh\chi, \qquad U_2 = U_3 = 0$$

in his own frame. In other words, S will say that the S' framc has axes parallel with his (as we saw above), but is moving with the velocity

$$U = c\tanh\chi$$

in the direction of the axis 1. In a similar way, the transformation (11.20) and (11.23) shows that S' will consider the frame S to have parallel axes and to be moving with the velocity $-U$ in the direction of the axis $1'$.

Since

$$\cosh\chi = \frac{1}{(1-\tanh^2\chi)^{1/2}}, \qquad \sinh\chi = \frac{\tanh\chi}{(1-\tanh^2\chi)^{1/2}},$$

we can write the transformation and its inverse in more directly physical terms as

$$(\Lambda) = \begin{pmatrix} \Gamma & 0 & 0 & -B\Gamma \\ 0 & 1 & 0 & 0 \\ 0 & 0 & 1 & 0 \\ -B\Gamma & 0 & 0 & \Gamma \end{pmatrix}, \qquad (\Lambda)^{-1} = \begin{pmatrix} \Gamma & 0 & 0 & B\Gamma \\ 0 & 1 & 0 & 0 \\ 0 & 0 & 1 & 0 \\ B\Gamma & 0 & 0 & \Gamma \end{pmatrix} \tag{11.26}$$

where

$$B = \frac{U}{c}, \qquad \Gamma = \frac{1}{(1-B^2)^{1/2}} = \frac{1}{(1-U^2/c^2)^{1/2}}. \tag{11.27}$$

(See Section 11.12.2, page 372.) These matrices constitute the *Lorentz transformation.* In component form the transformation is

$$\Delta x_1' = \frac{\Delta x_1 - U\,\Delta t}{(1-U^2/c^2)^{1/2}}, \qquad \Delta x_1 = \frac{\Delta x_1' + U\,\Delta t'}{(1-U^2/c^2)^{1/2}}, \tag{11.28}$$

$$\Delta x_2' = \Delta x_2, \tag{11.29}$$

$$\Delta x_3' = \Delta x_3, \tag{11.30}$$

$$\Delta t' = \frac{\Delta t - U\,\Delta x_1/c^2}{(1-U^2/c^2)^{1/2}}, \qquad \Delta t = \frac{\Delta t' + U\,\Delta x_1'/c^2}{(1-U^2/c^2)^{1/2}}. \tag{11.31}$$

(See Sections 11.12.3, 11.12.4 and 11.12.5, pages 372, 373 and 374.)

11.5.1. Lengths and scalar products

The derivation of the Lorentz transformation has ensured that the 'length' of a four-vector is invariant. The equivalent mathematical statement is that if (a), (a') are four-vectors such that

$$(a') = (\Lambda)(a) \quad \text{or} \quad (a) = (\Lambda)^{-1}(a'), \tag{11.32}$$

then

$$(\tilde{a}')(G)(a') = (\tilde{a})(G)(a). \tag{11.33}$$

This result follows from the fact that (Λ) is chosen to satisfy

$$(\tilde{\Lambda})(G)(\Lambda) = (G). \tag{11.34}$$

We have seen that (Λ) is symmetrical,

$$(\tilde{\Lambda}) = (\Lambda), \tag{11.35}$$

so the relationship (11.34) may be rewritten as

$$(\Lambda)(G)(\Lambda) = (G). \tag{11.36}$$

It is similarly true that

$$(\Lambda)^{-1}(G)(\Lambda)^{-1} = (G). \tag{11.37}$$

Corresponding to the squared 'length' of a single four-vector we may consider the 'scalar product' of two four-vectors, (a) and (b), defined as

$$(\tilde{a})(G)(b) = a_1b_1 + a_2b_2 + a_3b_3 - a_4b_4. \tag{11.38}$$

Using equations (11.32) and (11.36), we see that the scalar product of the transformed vectors, (a') and (b'), is

$$(\tilde{a}')(G)(b') = (\tilde{a})(\tilde{\Lambda})(G)(\Lambda)(b) = (\tilde{a})(G)(b). \tag{11.39}$$

The scalar product is therefore an invariant.

There is also a result involving 'mixed' scalar products which is often useful in calculations:

$$\begin{aligned} a_1b'_1 &+ a_2b'_2 + a_3b'_3 + a_4b'_4 \\ &= (\tilde{a})(b') = (\tilde{a}')(\tilde{\Lambda})^{-1}(\Lambda)(b) = (\tilde{a}')(\Lambda)^{-1}(\Lambda)(b) \\ &= (\tilde{a}')(b) \\ &= a'_1b_1 + a'_2b_2 + a'_3b_3 + a'_4b_4. \end{aligned} \tag{11.40}$$

(Note the positive sign of the last term in these expressions.)

11.6. The momentum-energy four-vector

Classical mechanics does not stop short at position or displacement three-vectors. We have seen, for example, that velocity, acceleration and momentum are three-vectors that obey the same transformation laws between frames of different orientation, preserving their magnitudes and relative directions.

The four-vector is unlikely to be a useful concept unless it, too, can be used for other dynamical quantities and their corresponding invariants can be shown to have physical significance. Let us first look at the velocity. If the displacement four-vector is shrunk to the infinitesimal displacement between neighbouring events along the path of a particle,

$$(\mathrm{d}x) = \begin{pmatrix} \mathrm{d}x_1 \\ \mathrm{d}x_2 \\ \mathrm{d}x_3 \\ \mathrm{d}x_4 \end{pmatrix} = \begin{pmatrix} \mathrm{d}x_1 \\ \mathrm{d}x_2 \\ \mathrm{d}x_3 \\ c\,\mathrm{d}t \end{pmatrix}, \tag{11.41}$$

the velocity components observed by S would be the first three components of

$$\frac{1}{\mathrm{d}t}(\mathrm{d}x) = \begin{pmatrix} \mathrm{d}x_1/\mathrm{d}t \\ \mathrm{d}x_2/\mathrm{d}t \\ \mathrm{d}x_3/\mathrm{d}t \\ \mathrm{d}x_4/\mathrm{d}t \end{pmatrix} = \begin{pmatrix} \mathrm{d}x_1/\mathrm{d}t \\ \mathrm{d}x_2/\mathrm{d}t \\ \mathrm{d}x_3/\mathrm{d}t \\ c \end{pmatrix} = \begin{pmatrix} v_1 \\ v_2 \\ v_3 \\ c \end{pmatrix} = (v), \quad \text{say.} \tag{11.42}$$

However, this is not a four-vector, since we have divided the four-vector $(\mathrm{d}x)$ by a quantity $\mathrm{d}t$ which is *not* the same for two observers S and S'. For the latter observer the velocity components would be given by

$$\frac{1}{\mathrm{d}t'}(\mathrm{d}x') = \begin{pmatrix} \mathrm{d}x'_1/\mathrm{d}t \\ \mathrm{d}x'_2/\mathrm{d}t \\ \mathrm{d}x'_3/\mathrm{d}t \\ c \end{pmatrix} = \begin{pmatrix} v_1^* \\ v_2^* \\ v_3^* \\ c \end{pmatrix} = (v^*), \quad \text{say.} \tag{11.43}$$

Now if we apply the Lorentz transformation to equation (11.42), we obtain

$$(\Lambda)(v) = (\Lambda)\frac{1}{\mathrm{d}t}(\mathrm{d}x) = \frac{1}{\mathrm{d}t}(\Lambda)(\mathrm{d}x) = \frac{1}{\mathrm{d}t}(\mathrm{d}x').$$

Hence, using equation (11.43),

$$(\Lambda)(v) = \frac{dt'}{dt}(v^*).$$

Since dt and dt' are, in general, *not* equal, then

$$(\Lambda)(v) \neq (v^*).$$

(v) and (v^*) therefore cannot be four-vectors. This is the reason why (v^*) rather than (v') is used for $(1/dt')(dx')$. Primes are used here to show that the quantity is the Lorentz transformation of the corresponding unprimed quantity, just as in Chapter 8 they were used for quantities related by the orthogonal transformation.

If we are to obtain one four-vector from another by multiplying or dividing by a scalar quantity, *this must be the same for both observers.* Thus to derive a four-vector with the dimensions of velocity from the four-vector (11.41) we must divide by an *invariant time*. We can obtain this from the invariant

$$dx_1^2 + dx_2^2 + dx_3^2 - c^2\,dt^2 = dx_1'^2 + dx_2'^2 + dx_3'^2 - c^2\,dt'^2$$

by first rewriting it in the form

$$\left(1 - \frac{v^2}{c^2}\right) dt^2 = \left(1 - \frac{v^{*2}}{c^2}\right) dt'^2, \tag{11.44}$$

where v and v^* are the speeds of the particle according to the two observers:

$$v^2 = v_1^2 + v_2^2 + v_3^2 \quad \text{and} \quad v^{*2} = v_1^{*2} + v_2^{*2} + v_3^{*2}.$$

We may now define the dimensionless quantities

$$\gamma = \left(1 - \frac{v^2}{c^2}\right)^{-1/2} \quad \text{and} \quad \gamma^* = \left(1 - \frac{v^{*2}}{c^2}\right)^{-1/2}, \tag{11.45}$$

and hence rewrite equation (11.44) as

$$\frac{dt}{\gamma} = \frac{dt'}{\gamma^*}.$$

The two sides of this equation are quantities found separately by the two observers, with the dimension of time, and therefore define an invariant quantity of the type we are seeking. It is known as the (infinitesimal) *proper time interval* $d\tau$:

$$d\tau = \begin{cases} \dfrac{dt}{\gamma} = dt\left(1 - \dfrac{v^2}{c^2}\right)^{-1/2} \\[2ex] \dfrac{dt'}{\gamma^*} = dt'\left(1 - \dfrac{v^{*2}}{c^2}\right)^{-1/2}. \end{cases} \tag{11.46}$$

From these equations we can see that the proper time interval approaches the ordinary time interval in the classical limit of small speeds:

$$\frac{d\tau}{dt}, \frac{d\tau}{dt'} \rightarrow 1 \quad \text{as} \quad \frac{v}{c}, \frac{v^*}{c} \rightarrow 0.$$

Thus if we divide (dx), (dx') by $d\tau$ rather than by dt and dt', we shall obtain another pair of corresponding four-vectors and these will give the classical components of velocity (and c) in the low-speed limit. We may call this the *proper velocity* (w), defined in the two frames as

$$(w) = \frac{1}{d\tau}(dx) \equiv \left(\frac{dx}{d\tau}\right) = \begin{pmatrix} \gamma\, dx_1/dt \\ \gamma\, dx_2/dt \\ \gamma\, dx_3/dt \\ \gamma c \end{pmatrix} = \begin{pmatrix} \gamma v_1 \\ \gamma v_2 \\ \gamma v_3 \\ \gamma c \end{pmatrix},$$

$$(w') = \frac{1}{d\tau}(dx') \equiv \left(\frac{dx'}{d\tau}\right) = \begin{pmatrix} \gamma^*\, dx'_1/dt' \\ \gamma^*\, dx'_2/dt' \\ \gamma^*\, dx'_3/dt' \\ \gamma^* c \end{pmatrix} = \begin{pmatrix} \gamma^* v_1^* \\ \gamma^* v_2^* \\ \gamma^* v_3^* \\ \gamma^* c \end{pmatrix},$$

When the proper velocity four-vector refers to the motion of a point particle, we can obtain another four-vector, (p), with the dimensions of momentum, simply by multiplying by a mass with the same value, m_0, for S and S':

$$(p) = m_0(w) = \begin{pmatrix} m_0\gamma v_1 \\ m_0\gamma v_2 \\ m_0\gamma v_3 \\ m_0\gamma c \end{pmatrix}, \qquad (p') = m_0(w') = \begin{pmatrix} m_0\gamma^* v_1^* \\ m_0\gamma^* v_2^* \\ m_0\gamma^* v_3^* \\ m_0\gamma^* c^* \end{pmatrix}. \tag{11.47}$$

This is known as the *momentum-energy four-vector*. In the next three sections we shall see the reasons for this name, and give a definite meaning to m_0.

11.6.1. Relativistic momentum

When the speed, v, of a particle in the S frame is small we may write for the first component, p_1, of the momentum-energy four-vector

$$\begin{aligned} p_1 = m_0\gamma v_1 &= m_0\left(1 - \frac{v^2}{c^2}\right)^{-1/2} v_1 \\ &= m_0 v_1\left(1 + \frac{1}{2}\frac{v^2}{c^2} + \ldots\right). \end{aligned} \tag{11.48}$$

Thus when v is small enough for $(v/c)^2$ to be neglected in this expansion, p_1 is the first component of the classical momentum, provided we identify m_0 with the inertial mass which we have earlier taken to be an intrinsic property of the particle. The second and third components will have similar properties, and we may therefore group the first three components of the four-vector as a three-vector,

$$\begin{pmatrix} p_1 \\ p_2 \\ p_3 \end{pmatrix} \underset{S}{\leftrightarrow} \mathbf{p},$$

which, as

$$\frac{v}{c} \rightarrow 0,$$

approaches the classical momentum. A similar classical limit holds in the S' frame as

$$\frac{v^*}{c} \rightarrow 0.$$

From the Lorentz transformation

$$(p') = (\Lambda)(p),$$

we have

$$\begin{aligned} p'_1 &= \Gamma(p_1 - Bp_4) \\ &= \Gamma(m_0\gamma v_1 - Bm_0\gamma c) \\ &= m_0\left(1 - \frac{U^2}{c^2}\right)^{-1/2}\left(1 - \frac{v^2}{c^2}\right)^{-1/2}(v_1 - U) \\ &= m_0(v_1 - U)\left(1 + \frac{1}{2}\frac{U^2 + v^2}{c^2} + \ldots\right). \end{aligned}$$

Thus when U^2/c^2 and v^2/c^2 can be neglected,

$$p'_1 = m_0(v_1 - U).$$

In the same limit,

$$p'_2 = m_0 v_2, \qquad p'_3 = m_0 v_3,$$

which is just the classical result from the Galilean transformation relating the S and S' frames. It is on this basis that we may consider the first three components of the momentum-energy four-vector to constitute a three-vector, $\mathbf{p}$, known as the *relativistic momentum.*

11.6.2. Relativistic energy

The fourth component of the momentum-energy vector is

$$p_4 = m_0\gamma c = m_0 c\left(1 - \frac{v^2}{c^2}\right)^{-1/2}$$

$$= m_0 c\left(1 + \frac{1}{2}\frac{v^2}{c^2} + \ldots\right)$$

$$= m_0 c + \frac{T_{cl}}{c} + \ldots, \tag{11.49}$$

where T_{cl} is the classical kinetic energy. We therefore define the *relativistic energy* as

$$E = p_4 c = m_0\gamma c^2 = m_0 c^2 + T_{cl} + \ldots. \tag{11.50}$$

For small speeds this is the classical kinetic energy plus the constant m_0c^2.
E is often called the *total relativistic energy* to distinguish it from the *relativistic kinetic energy*, defined by

$$T_{rl} = E - m_0c^2 = m_0c^2(\gamma - 1). \tag{11.51}$$

'Total' is used here to denote the inclusion of the term m_0c^2, not the kinetic plus potential energy of earlier chapters. Since m_0c^2 remains even when the particle is at rest in the frame of reference, this quantity is known as the *rest energy*, and m_0 as the *rest mass.*

If we collect together the results of this and the preceding section we see that the four-vector may be written in any of the following ways:

$$(p) = m_0(w) = \begin{pmatrix} m_0\gamma v_1 \\ m_0\gamma v_2 \\ m_0\gamma v_3 \\ m_0\gamma c \end{pmatrix} = \begin{pmatrix} p_1 \\ p_2 \\ p_3 \\ E/c \end{pmatrix} = \begin{pmatrix} \mathbf{p} \\ E/c \end{pmatrix}. \tag{11.52}$$

Equations (11.52) are the justification for calling (p) the (relativistic) momentum-energy four-vector.

11.6.3. Invariant mass

Since we have constructed momentum-energy in such a way as to make it a relativistic four-vector, it will necessarily have an invariant 'length'. This is given by

$$p_1^2 + p_2^2 + p_3^2 - \frac{E^2}{c^2} = m_0^2\gamma^2(v_1^2 + v_2^2 + v_3^2 - c^2)$$

$$= m_0^2\gamma^2(v^2 - c^2)$$

$$= -m_0^2c^2. \tag{11.53}$$

Hence

$$\frac{E^2}{c^4} - \frac{p^2}{c^2} = m_0^2,$$

and the invariance of the length is simply another way of stating that the same mass, m_0, is associated with the particle by all observers whose frames are connected by a Lorentz transformation. As we have seen in the preceding sections, this mass is identical with the inertial mass with which we are familiar in non-relativistic mechanics. (See Section 11.12.6, page 376.)

11.6.4. The classical limit

It may seem inconsistent that to show the correspondence between classical and relativistic momentum we considered only the first term in the expansion (11.48), while for energy we took the first two terms in the expansion (11.50). This choice is, in fact, consistent as the following argument shows.

Substituting

$$\gamma = \left(1 - \frac{v^2}{c^2}\right)^{-1/2} = 1 + \frac{1}{2}\frac{v^2}{c^2} + \ldots$$

into equation (11.52) enables us to write

$$(p) = m_0 c \begin{pmatrix} \gamma v_1/c \\ \gamma v_2/c \\ \gamma v_3/c \\ \gamma \end{pmatrix} = m_0 c \begin{pmatrix} v_1/c + \frac{1}{2}v_1 v^2/c^3 + \ldots \\ v_2/c + \frac{1}{2}v_2 v^2/c^3 + \ldots \\ v_3/c + \frac{1}{2}v_3 v^2/c^3 + \ldots \\ 1 + \frac{1}{2}v^2/c^2 + \ldots \end{pmatrix}. \qquad (11.54)$$

Each of the velocity components v_1, v_2, v_3 has a smaller magnitude than the speed v. Thus when this speed is small enough for v^3/c^3 and smaller terms to be neglected, the expression (11.54) reduces to

$$(p) = m_0 c \begin{pmatrix} v_1/c \\ v_2/c \\ v_3/c \\ 1 + \frac{1}{2}v^2/c^2 \end{pmatrix} = \begin{pmatrix} m_0 v_1 \\ m_0 v_2 \\ m_0 v_3 \\ m_0 c + \frac{1}{2}m_0 v^2/c \end{pmatrix} = \begin{pmatrix} \mathbf{p}_{\mathrm{cl}} \\ m_0 c + T_{\mathrm{cl}}/c \end{pmatrix}. \qquad (11.55)$$

It is therefore the *same* condition,

$$\left(\frac{v}{c}\right)^3 \ll 1, \qquad (11.56)$$

that gives all the classical values in (11.52).

11.7. Conservation of momentum and energy

When particles collide in the absence of external forces, or during a time so short that the effect of external forces is negligible, the classical momentum three-vector is conserved. Thus if particles $a, b, \ldots$ with masses $m_a, m_b, \ldots$ have momenta $\mathbf{p}_{ai}, \mathbf{p}_{bi}, \ldots$ initially and, after a collision or series of collisions, the momenta become finally $\mathbf{p}_{af}, \mathbf{p}_{bf}, \ldots$, then

$$\mathbf{p}_{ai} + \mathbf{p}_{bi} + \ldots = \mathbf{p}_{af} + \mathbf{p}_{bf} + \ldots. \tag{11.57}$$

If the collisions are elastic then, provided the particles are sufficiently far apart in the initial and final condition, the kinetic energy will be conserved. Thus if $T_{ai}, T_{bi}, \ldots$ and $T_{af}, T_{bf}, \ldots$ are the individual kinetic energies before and after the collisions,

$$T_{ai} + T_{bi} + \ldots = T_{af} + T_{bf} + \ldots. \tag{11.58}$$

None of the masses is changed, since in classical elastic collisions no particles break up or join together. Hence, adding the rest energies to both sides of equation (11.58),

$$m_a c^2 + T_{ai} + m_b c^2 + T_{bi} + \ldots = m_a c^2 + T_{af} + m_b c^2 + T_{bf} + \ldots. \tag{11.59}$$

We saw from equation (11.55) that the three components of classical momentum, $\mathbf{p}_{cl}$, together with the term $m_0 c + T_{cl}/c$ are identical with the relativistic momentum-energy four-vector in the low-speed limit. Hence equations (11.57) and (11.59) are equivalent to the four-vector conservation equation

$$[(p_a)_i + (p_b)_i + \ldots]_{lsp} = [(p_a)_f + (p_b)_f + \ldots]_{lsp}, \tag{11.60}$$

where the suffix '*lsp*' means that all the four-vectors are to be evaluated *in the low-speed limit.*

What if the collisions are not low-speed ones? Some of the most important practical aspects of relativity come from observations that in this case conservation of classical momentum and energy is no longer true, but *conservation of the corresponding relativistic quantities is.* In other words, we can remove the low-speed limit restriction on equation (11.60) and write more generally

$$(p)_i = (p)_f, \tag{11.61}$$

where $(p)_i$ is the total or resultant momentum-energy four-vector before the collisions,

$$(p)_i = (p_a)_i + (p_b)_i + \ldots, \tag{11.62}$$

and $(p)_f$ is the total four-vector afterwards,

$$(p)_f = (p_a)_f + (p_b)_f + \ldots. \tag{11.63}$$

Equation (11.61) expresses a conservation law for S. Suppose S' is another observer moving with a constant velocity $\mathbf{U}$ relative to S. We can choose axes 1, 1′ for S and S' to lie along the direction of $\mathbf{U}$. Then the Lorentz transformation will give these four-vectors as seen by S':

$$(p')_i = (\Lambda)(p)_i, \qquad (p')_f = (\Lambda)(p)_f. \tag{11.64}$$

Equations (11.61) and (11.64) show that S' will observe the same conservation law:

$$(p')_i = (p')_f. \tag{11.65}$$

Any law like this, equating four-vectors, will be true in any frames connected by Lorentz tranformations (i.e. the law will be the same for all observers who have constant velocities relative to each other). We therefore lose no generality by choosing a frame of reference in which the algebra or arithmetic is as simple as possible.

11.8. Change of rest mass

Suppose two particles of rest mass m_a and m_b collide elastically. Conservation of relativistic momentum-energy,

$$(p_a)_i + (p_b)_i = (p_a)_f + (p_b)_f, \tag{11.66}$$

is, as we have just seen, equivalent for small speeds to

$$m_a\mathbf{v}_{ai} + m_b\mathbf{v}_{bi} = m_a\mathbf{v}_{af} + m_b\mathbf{v}_{bf} \tag{11.67}$$

and

$$m_ac + \tfrac{1}{2}m_av_{ai}^2/c + m_bc + \tfrac{1}{2}m_bv_{bi}^2/c = m_ac + \tfrac{1}{2}m_av_{af}^2/c + m_bc + \tfrac{1}{2}m_bv_{bf}^2/c. \tag{11.68}$$

Equation (11.67) expresses conservation of classical momentum, while equation (11.68) yields

$$\tfrac{1}{2}m_av_{ai}^2 + \tfrac{1}{2}m_bv_{bi}^2 = \tfrac{1}{2}m_av_{af}^2 + m_bv_{bf}^2, \tag{11.69}$$

which is conservation of classical kinetic energy.

Conservation of kinetic energy alone is a rather limited law, and arises only because we are considering initial and final conditions in which the particles are so far apart that any internal potential energy, representing the forces between them, can be ignored. It could not, for example, apply to the case in which a single particle breaks into two parts which then move apart. We saw, in Chapter 5, that this always involves a gain in kinetic energy.

However, since the concept of momentum-energy conservation has proved so fruitful up to now, let us see what the implications are if we assume it to be true even for a process of the type just mentioned—one which in the classical context we should have called inelastic. Suppose the initial rest mass of a particle is m_0 and that it breaks into fragments with rest masses m_a and m_b, with speeds v_{af} and v_{bf} in the rest frame of the initial particle. The conservation law corresponding to (11.66) is now

$$(p)_i = (p_a)_f + (p_b)_f, \tag{11.70}$$

where $(p)_i$ is the four-vector belonging to the single particle. In its own rest frame it has zero speed so that the energy conservation law corresponding to (11.68) simplifies, for small speeds in this frame, to

$$m_0 c = (m_a + m_b)c + \tfrac{1}{2}m_a v_{af}^2/c + \tfrac{1}{2}m_b v_{bf}^2/c.$$

Hence

$$m_0 - (m_a + m_b) = \frac{\Delta T}{c^2}, \tag{11.71}$$

where ΔT is the gain in kinetic energy,

$$\Delta T = \tfrac{1}{2}m_a v_{af}^2 + \tfrac{1}{2}m_b v_{bf}^2. \tag{11.72}$$

This seems to be quite wrong, for it is 'well known' that when an object breaks into two or more parts, the sum of their masses is equal to the mass of the original single object. Equation (11.71), by contrast, asserts that *it is less.* Does this not disprove the extension of the conservation law (11.61) to inelastic processes? Before abandoning it too hastily we should remember the warning given in Chapter 1: physical laws are only proved, or disproved, within the accuracy of the experiments that test them.

To give some sense of magnitude to equation (11.71) let us take the case of a rifle, of mass 5 kg (m_a), firing a bullet of mass 0·1 kg (m_b) at a speed of 1000 m s^{-1} (v_{bf}). The gain in kinetic energy comes almost entirely from that of the bullet, since the recoil of the rifle (v_{bf}) is negligible. Hence, according to equation (11.71),

$$m_0 - (m_a + m_b) \cong \frac{1}{2}m_b\frac{v_{bf}^2}{c^2} = \frac{0{\cdot}1}{2}\left(\frac{10^3}{3 \times 10^8}\right)^2 \cong 6 \times 10^{-13}\,\text{kg}. \tag{11.73}$$

Thus the initial mass of the stationary rifle and bullet, which we would expect to be 5·1 kg, should exceed this by approximately 6×10^{-13} kg. In order, therefore, to show that the masses satisfy the 'well-known' classical result

$$m_0 - (m_a + m_b) = 0,$$

rather than the prediction (11.73), we should need to measure the masses to an accuracy of one part in 10^{13}. It is clear from this that no normal laboratory experiment in mechanics, even one involving high supersonic speeds, could decide between the two relationships.

In a similar way, if two masses, m_a and m_b, moving with speeds small compared with that of light, collide and stick together to form a single mass m_0 with, in this case, a *loss* of kinetic energy ΔT, the conservation law (11.61) leads to

$$m_0 - (m_a + m_b) = \frac{\Delta T}{c^2}. \tag{11.74}$$

Equations (11.71) and (11.74) together show that a change in total kinetic energy ΔT, which may be positive or negative, should be accompanied by a change in total rest mass Δm_0, where

$$\Delta m_0 + \frac{\Delta T}{c^2} = 0. \tag{11.75}$$

This is equivalent to saying that energy can be created from rest mass and vice versa. The rifle and bullet example showed how difficult it would be to detect this process in a typical laboratory mechanics experiment. More generally, in an experiment involving masses with a typical value m_0 and speeds with a typical value u, the ΔT will have an order of magnitude $m_0 u^2$. The relationship (11.75) then gives an order of magnitude value for the change in mass as

$$\left(\frac{\Delta m_0}{m_0}\right) \sim \left(\frac{u}{c}\right)^2. \tag{11.76}$$

Thus, to attain 1 per cent changes in mass, speeds of at least $c/10$ would be needed. It is only in the field of nuclear physics that such speeds commonly occur, and it is here that a whole range of experiments linking the masses of the stable atomic nuclei and the disintegration schemes of the unstable nuclei have indeed demonstrated the validity of the relationship (11.75). The creation of energy from the destruction of nuclear mass (in nuclear power stations, and in fission and fusion bombs) has been one of the most dramatically verified predictions of special relativity.

Even so, when only 1 per cent or less of the mass is exchanged for kinetic energy, the phenomena can be thought of as small, but important, corrections to the classical law of mass conservation, and are not, by themselves, convincing proof of the truth of the conservation law (11.61) from which (11.75) is derived as a special case. However, in the study of cosmic rays, and more recently in the operation of large proton accelerators, we are able to observe collisions and disintegrations in the field of high-energy nuclear physics (or elementary particle physics), where the speeds are very near to that of light. Then the kinetic energies greatly exceed the rest energies of the nuclear particles taking part, and interchange between them can, and does, result in the creation and disappearance of quite new particles. The evidence from this study fully confirms the validity of equation (11.70), and when many particles are concerned, the more general result (11.61). (See Section 11.12.7, page 376.)

We may therefore consider the momentum-energy conservation law (equation 11.61),

$$(p)_i = (p)_f, \tag{11.77}$$

to be valid not only for all speeds up to that of light but for *inelastic* as well as elastic collisions. Extensive experimental verification of the predictions of this and the preceding section in reference frames connected by the Lorentz transformation provide the firmest evidence for the validity of that transformation and the assumed invariance law (11.14) on which it is based. (See Section 11.12.8, page 379.)

11.9. Addition of velocities

We saw in Section 11.6 that we cannot make a four-vector from velocity. The relationship between velocities in two frames is consequently rather complicated. It can be derived from the infinitesimal displacements (see equations 11.28 to 11.31):

$$\mathrm{d}x_1 = \frac{\mathrm{d}x_1' + U\,\mathrm{d}t'}{(1 - U^2/c^2)^{1/2}}, \qquad \mathrm{d}x_2 = \mathrm{d}x_2', \qquad \mathrm{d}x_3 = \mathrm{d}x_3',$$

$$\mathrm{d}t = \frac{\mathrm{d}t' + U\mathrm{d}x_1'/c^2}{(1 - U^2/c^2)^{1/2}}.$$

Dividing the first three equations by the last we obtain the required relationships,

$$v_1 = \frac{v_1^* + U}{1 + v_1^* U/c^2}, \tag{11.78}$$

$$v_2 = \frac{v_2^*(1 - U^2/c^2)^{1/2}}{1 + v_1^* U/c^2}, \tag{11.79}$$

$$v_3 = \frac{v_3^*(1 - U^2/c^2)^{1/2}}{1 + v_1^* U/c^2}. \tag{11.80}$$

When the speed of the particle in the S' frame and the speed of S' relative to S are both small compared with the speed of light, these results reduce to the classical values,

$$v_1 = v_1^* + U, \qquad v_2 = v_2^*, \qquad v_3 = v_3^*.$$

However, as the speeds increase the resultant velocity components in the S frame become less than this simple addition law would indicate. This difference is shown most clearly by the invariance of the four-vectors $(\mathrm{d}x)$, $(\mathrm{d}x')$, which, as we saw in Section 11.6, can be written as

$$(v^2 - c^2)\,\mathrm{d}t^2 = (v^{*2} - c^2)\,\mathrm{d}t'^2.$$

Since in this equation all the quantities are real and their squares positive, we can see from it that if the speed of a particle is less than that of light in one frame it will be so in the other. In its own rest frame the speed will be zero. Hence in all other frames it will be less than that of light. Thus the speed of light is an upper limit for the observed speed of any particle possessing rest mass. This was pointed out as an experimental fact in Section 11.3.

11.10. Space-like and time-like intervals; causality

The invariant length of the interval between two events may be written as

$$\Delta x^2 - c^2\,dt^2 = \Delta x'^2 - c^2\,dt'^2, \tag{11.81}$$

where

$$\Delta x^2 = \Delta x_1^2 + \Delta x_2^2 + \Delta x_3^2, \qquad \Delta x'^2 = \Delta x_1'^2 + \Delta x_2'^2 + \Delta x_3'^2.$$

When both sides of equation (11.81) are positive,

$$\left|\frac{\Delta x}{\Delta t}\right| > c \quad \text{and} \quad \left|\frac{\Delta x'}{\Delta t'}\right| > c,$$

and therefore no material particle, or even a light signal, could link the two events. Such an interval is called *space-like*, and we would conclude that in no sense could one event be caused by the other. We may also note that

$$\Delta t' = \Delta t \frac{1 - (\Delta x_1/\Delta t)U/c^2}{(1 - U^2/c^2)^{1/2}} \tag{11.82}$$

and therefore that, depending upon the magnitudes and signs of $\Delta x_1/\Delta t$ and of U, it is possible for Δt and $\Delta t'$ to have opposite signs. Thus S and S' may even disagree about which event occurred first!

However, when both sides of equation (11.81) are negative,

$$\left|\frac{\Delta x}{\Delta t}\right| < c \quad \text{and} \quad \left|\frac{\Delta x'}{\Delta t'}\right| < c,$$

and it *is* possible for a particle or light signal to link the two events. The interval is then called *time-like*. Equation (11.82) shows that in this case Δt and $\Delta t'$ necessarily have the same sign, and it is then possible for one event to be 'caused' by the other, with all observers agreeing about which event occurred first and which resulted from it.

11.11. Invariant mass

In Section 11.6.3 we saw that when a momentum-energy four-vector refers to a single particle of rest mass m_0 its 'length' is given by $-m_0^2c^2$, and that the invariance of this length is simply another way of stating that all observers agree about the rest mass of the particle. In the rest frame of the particle itself the four-vector may be designated $(p)_{cm}$ and reduces to

$$(p)_{cm} = \begin{pmatrix} 0 \\ 0 \\ 0 \\ m_0 c \end{pmatrix}.$$

The concept which links invariant length to rest mass is, however, more generally applicable. Suppose a group of particles with momentum-energy vectors (p_a), (p_b), ... have the total momentum-energy vector (p):

$$(p) = (p_a) + (p_b) + \ldots.$$

(p) will have an invariant length which may be used to define a rest mass M_0 for the whole group according to

$$p_1^2 + p_2^2 + p_3^2 - E^2/c^2 = -M_0^2 c^2. \tag{11.83}$$

Although, as it stands, this is simply a dimensionally correct definition of a mass M_0, this invariant 'group mass' has great physical significance if the particles should coalesce, even momentarily. For, since the total momentum-energy is conserved by any mutual interactions between the particles, M_0 will be the rest mass of any composite particle that they might form.

Analysis of invariant masses for groups of particles in high energy nuclear interactions has in recent years led to the discovery of a whole range of new nuclear particles or 'resonances'. Their existence is so transient that they became recognized as entities only by the concentration of invariant group masses at or near particular values. (See Section 11.12.9, page 381.)

11.12. Comments and worked examples

11.12.1. See page 343

Although we have discussed instaneous measurement and geometrical invariance separately, they are related problems. In both cases we have to decide how to define and measure simultaneous events. In the first, we should like to set up a perpendicular, one of whose ends coincides with a scale mark on an axis while the other coincides, *at the same time*, with the end of a vector. In the second, to compare a moving body with a stationary one of supposedly the same size and shape we need to know whether all the corresponding points coincide *at the same time.*

The basic problem, therefore, is how systematically to determine the simultaneity of events which are a finite distance apart when there is no experimental method of linking them that is faster than the speed of light. For more discussion on this point see, for example, C. Møller, *The Theory of Relativity*, Oxford University Press, London (1952).

11.12.2. See Section 11.5

When the form (11.17), with imaginary fourth component, is used for the four-vector, the corresponding form for the Lorentz transformation is then

$$(\Lambda) = \begin{pmatrix} \Gamma & 0 & 0 & jB\Gamma \\ 0 & 1 & 0 & 0 \\ 0 & 0 & 1 & 0 \\ -jB\Gamma & 0 & 0 & \Gamma \end{pmatrix}.$$

The inverse transformation which can be calculated either directly, or by the reversal of the velocity U ($B \to -B$), is

$$(\Lambda)^{-1} = \begin{pmatrix} \Gamma & 0 & 0 & -jB\Gamma \\ 0 & 1 & 0 & 0 \\ 0 & 0 & 1 & 0 \\ jB\Gamma & 0 & 0 & \Gamma \end{pmatrix}.$$

In this form (Λ) has properties similar to those of the three-dimensional transformation:

$$(\tilde{\Lambda}) = (\Lambda)^{-1} \quad \text{or} \quad (\Lambda)(\tilde{\Lambda}) = (\tilde{\Lambda})(\Lambda) = (1)$$

11.12.3. See Section 11.5

Example 36

A measuring rod, whose length in its own rest frame is l', is moving with a velocity $\mathbf{U}$ relative to an observer S. What is the length of the rod according to S when the velocity is (a) parallel with, (b) perpendicular to, the length of the rod?

To discuss length we must choose *two events* which describe the ends of the rod. Each event requires the measurement of three space coordinates *and* one time coordinate. Suppose S' is the rest frame of the rod. Since the rod is fixed in this frame it does not matter when the measurements of position are made. The coordinate intervals $\Delta x'_1$, $\Delta x'_2$, $\Delta x'_3$ obtained from the S' measurements will give the rest length l':

$$l'^2 = \Delta x_1'^2 + \Delta x_2'^2 + \Delta x_3'^2.$$

However, if S is to determine the length of a rod which is moving in his frame he must measure the position of its two ends *at what to him is the same instant*. Thus the events must be such that

$$\Delta t = 0$$

if S is to assert that the length of the rod, l, is given by

$$l^2 = \Delta x_1^2 + \Delta x_2^2 + \Delta x_3^2.$$

The Lorentz relationship for the appropriate space-time intervals is therefore

$$\begin{pmatrix} \Delta x'_1 \\ \Delta x'_2 \\ \Delta x'_3 \\ c\Delta t' \end{pmatrix} = \begin{pmatrix} \Gamma & 0 & 0 & -B\Gamma \\ 0 & 1 & 0 & 0 \\ 0 & 0 & 1 & 0 \\ -B\Gamma & 0 & 0 & \Gamma \end{pmatrix} \begin{pmatrix} \Delta x_1 \\ \Delta x_2 \\ \Delta x_3 \\ 0 \end{pmatrix}, \tag{11.84}$$

from which

$$\Delta x'_1 = \Gamma \Delta x_1, \qquad \Delta x'_2 = \Delta x_2, \qquad \Delta x'_3 = \Delta x_3.$$

The four-vectors in equation (11.84) represent the rod in an arbitrary orientation. The velocity is along the axes 1, 1′ because of the form of the Lorentz transformation. Hence if the velocity is to be parallel with the length of the rod we must choose

$$\Delta x'_2 = \Delta x_2 = \Delta x'_3 = \Delta x_3 = 0.$$

The length according to S is then

$$l = \Delta x_1 = \frac{\Delta x'_1}{\Gamma} = \frac{l'}{\Gamma}. \tag{11.85}$$

Thus an observer measuring the moving rod finds it apparently *shorter* than its rest length by the factor

$$1/\Gamma = (1 - U^2/c^2)^{1/2}.$$

This effect is known as the *Lorentz contraction.*

When the velocity is perpendicular to the length of the rod we must choose

$$\Delta x'_1 = \Delta x_1 = 0.$$

Then

$$\Delta x'_2 = \Delta x_2, \qquad \Delta x'_3 = \Delta x_3,$$

and the length according to S is unchanged:

$$l = (\Delta x_2^2 + \Delta x_3^2)^{1/2} = (\Delta x_2'^2 + \Delta x_3'^2)^{1/2} = l'.$$

11.12.4. See Section 11.5

Example 37

A time interval $\Delta t'$ is measured by a clock in its rest frame S'. If the clock is moving with a constant velocity $\mathbf{U}$ relative to S, what is the corresponding time interval according to S?

As in the case of the moving rod we must decide exactly what experiment must be carried out to determine the required measurements, and then describe these by the appropriate events. Here we must observe coincidences between the moving S' clock and two synchronized clocks fixed in the S frame. The positions and times of these two coincidences constitute the required events.

In the S' frame the clock is fixed in position. Hence the space intervals in the S' frame derived from the two events are zero:

$$\Delta x'_1 = \Delta x'_2 = \Delta x'_3 = 0,$$

while $\Delta t'$ is the given time interval. The corresponding space-time intervals in the S frame are then completely determined by the Lorentz transformation:

$$\begin{pmatrix} \Delta x_1 \\ \Delta x_2 \\ \Delta x_3 \\ c\,\Delta t \end{pmatrix} = \begin{pmatrix} \Gamma & 0 & 0 & B\Gamma \\ 0 & 1 & 0 & 0 \\ 0 & 0 & 1 & 0 \\ B\Gamma & 0 & 0 & \Gamma \end{pmatrix} \begin{pmatrix} 0 \\ 0 \\ 0 \\ c\,\Delta t' \end{pmatrix}.$$

Hence

$$\Delta t = \Gamma\,\Delta t', \tag{11.86}$$

and the corresponding time interval therefore appears larger to S. This result is often expressed by saying that the moving clock appears to S to be running slowly by the factor

$$1/\Gamma = (1 - U^2/c^2)^{1/2},$$

and the effect is known as *time dilatation.*

Note that both the Lorentz contraction and time dilatation are *symmetrical* phenomena, expressible in terms of U^2/c^2. S' would have exactly the same opinion of measuring rods and clocks fixed in the S frame.

11.12.5. See Section 11.5

Since the establishing of special relativity depends upon a very careful examination of the experimental methods of measuring the positions and times of physical events, it is worth outlining a procedure for setting up a frame of reference and making four-vector measurements in it.

(1) We start with a long rigid straight rod. Its straightness is determined by ensuring that it coincides with the path of a beam of light.

(2) A standard clock is chosen, equipped to time the emission and reception of light signals in terms of an arbitrary time unit.

(3) Two marks on the rod are chosen as an arbitrary unit of distance. By setting the clock on one of these and a mirror on the other the time taken for a

light signal to traverse twice the unit distance is determined. This measurement determines the velocity of light, c.

(4) By moving the mirror along the rod to give twice, thrice, ... the time for the light signal to traverse double the unit path, the rod may be marked out in multiples of the unit distance. Subdivisions of the unit distance may be marked out in a similar manner.

(5) Similar clocks are placed at regular spatial intervals along the rod. They are adjusted to run at the same rate, and to show the same time, by ensuring that light signals emitted by any pair of clocks at what, according to them, is the same instant arrive simultaneously at the mid-point between them.

(6) The marked rod and its associated clocks now constitute an axis with a distance scale and a method of measuring the *same* time at any point along it.

(7) Set up a second axis intersecting, and perpendicular to, the first. It will be perpendicular when a light signal sent from any point on it reaches simultaneously two clocks on the first axis which are equidistant from the point of intersection.

(8) Calibrate distances on the second axis by moving a mirror along it and finding the positions for which a light signal sent from the intersection of the axes takes the same time to travel there and back as it does along the known lengths of the first axis.

(9) Equip the second axis with similar clocks as in (5).

(10) Set up a third axis perpendicular to the other two, calibrate it and equip it with clocks as in (7), (8) and (9).

(11) We can now suppose that any event in space-time initiates a light signal. This spreads out in a spherical wave which will eventually reach the three axes. Suppose the first clock on axis 1 to receive the signal is at x_1 and does so at time t_1. Suppose x_2, t_2 and x_3, t_3 are the corresponding quantities for the axes 2 and 3.

(12) The space coordinates of the event are then x_1, x_2, x_3.

(13) The distance travelled by the light signal from the event to the axis 1 is $(x_2^2 + x_3^2)^{1/2}$. It does so at a speed c and arrives there at time t_1. Hence the time coordinate of the event itself is

$$t = t_1 - \frac{1}{c}(x_2^2 + x_3^2)^{1/2}.$$

(It is left to the reader to show that similar expressions derived from the time of arrival at the other two axes will give the same time for the event.)

The procedure outlined above is not the only, or necessarily the simplest, one to give a method of measuring the four coordinates of an event (see problem 11.3.).

11.12.6. See Section 11.6.3

When real particles are concerned the right-hand side of equation (11.53) will always be negative. A more 'natural' choice of the invariant length of the four-vector in problems involving particle collisions is therefore

$$\frac{E^2}{c^2} - p_1^2 - p_2^2 - p_3^2 = m_0^2 c^2.$$

In order to obtain this a convenient form for the energy-momentum four-vector is

$$\begin{pmatrix} p_0 \\ p_1 \\ p_2 \\ p_3 \end{pmatrix} = \begin{pmatrix} m_0\gamma c \\ m_0\gamma v_1 \\ m_0\gamma v_2 \\ m_0\gamma v_3 \end{pmatrix} = \begin{pmatrix} E/c \\ p_1 \\ p_2 \\ p_3 \end{pmatrix} = \begin{pmatrix} E/c \\ \mathbf{p} \end{pmatrix}, \tag{11.87}$$

and for the metric is

$$(G) = \begin{pmatrix} 1 & 0 & 0 & 0 \\ 0 & -1 & 0 & 0 \\ 0 & 0 & -1 & 0 \\ 0 & 0 & 0 & -1 \end{pmatrix}. \tag{11.88}$$

Note that p_1, p_2, p_3 are the three components of momentum and are just the same as the p_1, p_2, p_3 of the definition (11.47). We have simply relabelled the energy as p_0 and put it as the first element of the four-vector.

The corresponding matrices for the Lorentz transformation are

$$(\Lambda) = \begin{pmatrix} \Gamma & -B\Gamma & 0 & 0 \\ -B\Gamma & \Gamma & 0 & 0 \\ 0 & 0 & 1 & 0 \\ 0 & 0 & 0 & 1 \end{pmatrix}, \quad (\Lambda)^{-1} = \begin{pmatrix} \Gamma & B\Gamma & 0 & 0 \\ B\Gamma & 0 & 0 & 0 \\ 0 & 0 & 1 & 0 \\ 0 & 0 & 0 & 1 \end{pmatrix}. \tag{11.89}$$

Since p_2 and p_3 are unchanged by this Lorentz transformation, in most cases the calculations involved in collision problems require the manipulation only of the simple 2×1 and 2×2 matrices:

$$\begin{pmatrix} E/c \\ p_1 \end{pmatrix}, \begin{pmatrix} \Gamma & -B\Gamma \\ -B\Gamma & \Gamma \end{pmatrix} \text{ and } \begin{pmatrix} \Gamma & B\Gamma \\ B\Gamma & \Gamma \end{pmatrix}.$$

11.12.7. See Section 11.8

Example 38

A beam of pions interacts inelastically with protons at rest in the laboratory to create kaons and lambda-particles according to the reaction

$$\pi^- + \mathrm{p} \rightarrow \Lambda + \mathrm{K}^0.$$

What is the least pion momentum, in the laboratory, that will give such a reaction, and what is the corresponding momentum of the lambda-particle? If the mean life of the lambda is $2{\cdot}51 \times 10^{-10}$ s in its own rest frame, what is the mean distance it travels in the laboratory before decaying? (The rest energies of the π^-, p, Λ and K^0 are respectively 140, 938, 1115 and 498 MeV.)

This type of problem is usually best approached by considering the energy-momentum four-vectors of the particles and groups of particles concerned, both in the laboratory and in the centre-of-mass frames. The physics of the problem lies in the conservation of the total four-vector in either frame and in the invariance of the length of the four-vectors, and of their scalar products with each other, in the two frames. (p and Λ, like π^- and K^0, are used here as particle symbols and must not be confused with their connotation as momentum and Lorentz transformation.)

Let us take the (space) axis 1 of the laboratory frame and the axis 1′ of the centre-of-mass frame to be parallel with the direction of motion of the pion. At threshold the Λ and K^0 will, by definition, be created at rest in the centre-of-mass frame, and therefore have a velocity component only along axis 1 in the laboratory frame. Hence the second and third components of momentum will be zero for every particle in both frames, and this will, of course, also be true of the first component in a frame in which any particle is at rest.

Using the notation of the preceding section the energy-momentum four-vectors of the π^-, p, Λ and K^0 in the laboratory frame will therefore be of the form

$$(p_\pi) = \begin{pmatrix} E_\pi/c \\ p_\pi \\ 0 \\ 0 \end{pmatrix}, \quad (p_p) = \begin{pmatrix} m_p c \\ 0 \\ 0 \\ 0 \end{pmatrix}, \quad (p_\Lambda) = \begin{pmatrix} E_\Lambda/c \\ p_\Lambda \\ 0 \\ 0 \end{pmatrix}, \quad (p_K) = \begin{pmatrix} E_K/c \\ p_K \\ 0 \\ 0 \end{pmatrix}, \tag{11.90}$$

while in the centre-of-mass frame they will be

$$(p'_\pi) = \begin{pmatrix} E'_\pi/c \\ p' \\ 0 \\ 0 \end{pmatrix}, \quad (p'_p) = \begin{pmatrix} E'_p/c \\ -p' \\ 0 \\ 0 \end{pmatrix}, \quad (p'_\Lambda) = \begin{pmatrix} m_\Lambda c \\ 0 \\ 0 \\ 0 \end{pmatrix}, \quad (p'_K) = \begin{pmatrix} m_K c \\ 0 \\ 0 \\ 0 \end{pmatrix}, \tag{11.91}$$

where m_π, m_p, m_Λ and m_K are the rest masses of the particles. Note that since total momentum must, by definition, be zero in the centre-of-mass frame, this has been ensured by taking

$$p'_{\pi 1} = p' = -p'_{p1}.$$

There is conservation of energy-momentum in the laboratory frame,

$$(p_\pi) + (p_\mathrm{p}) = (p_\Lambda) + (p_\mathrm{K}),$$

and therefore the invariant lengths of $(p_\pi) + (p_\mathrm{p})$ and of $(p_\Lambda) + (p_\mathrm{K})$ are equal. But the length of $(p_\Lambda) + (p_\mathrm{K})$ must be the same as that of $(p'_\Lambda) + (p'_\mathrm{K})$ since they are linked by a Lorentz transformation. Thus, adding the first and last pairs of four-vectors,

$$\begin{pmatrix} E_\pi/c + m_\mathrm{p}c \\ p_\pi \\ 0 \\ 0 \end{pmatrix} \quad \text{and} \quad \begin{pmatrix} m_\Lambda c + m_\mathrm{K}c \\ 0 \\ 0 \\ 0 \end{pmatrix}$$

must have the same invariant length; that is,

$$(E_\pi/c + m_\mathrm{p}c)^2 - p_\pi^2 = (m_\Lambda c + m_\mathrm{K}c)^2$$

or

$$(E_\pi/c)^2 + 2E_\pi m_\mathrm{p} - p_\pi^2 = (m_\Lambda c + m_\mathrm{K}c)^2 - m_\mathrm{p}^2c^2. \tag{11.92}$$

Now the invariant length of (p_π) gives the rest mass of the pion:

$$(E_\pi/c)^2 - p_\pi^2 = m_\pi^2c^2. \tag{11.93}$$

Substituting this into equation (11.92) we find an expression for the total energy of the pion in the laboratory:

$$E_\pi = \frac{(m_\Lambda c^2 + m_\mathrm{K}c^2)^2 - m_\mathrm{p}^2c^4 - m_\pi^2c^4}{2m_\mathrm{p}c^2}.$$

The rest energies of all the particles on the right-hand side of this equation are known, and yield the result

$$E_\pi = 907\ \mathrm{MeV}.$$

Equation (11.93) may then be used again to give

$$p_\pi c = (E_\pi^2 - m_\pi^2c^4)^{1/2} = 897\ \mathrm{MeV}.$$

Hence the required pion momentum is 897 MeV/c.

To find the lambda momentum we use momentum conservation in the laboratory frame. From equations (11.90) this gives

$$p_\pi = p_\Lambda + p_\mathrm{K}. \tag{11.94}$$

In the laboratory the lambda and the kaon have the same velocity (that of their centre-of-mass), and their momenta will therefore be proportional to their rest masses. Hence equation (11.94) may be written

$$p_\pi = p_\Lambda\left(1 + \frac{p_K}{p_\Lambda}\right) = p_\Lambda\left(1 + \frac{m_K}{m_\Lambda}\right).$$

Hence

$$p_\Lambda = \frac{m_\Lambda}{m_\Lambda + m_K} p_\pi = \frac{m_\Lambda c^2}{m_\Lambda c^2 + m_K c^2} p_\pi,$$

from which, using the values given or calculated above, we find

$$p_\Lambda = 620 \text{ MeV}/c.$$

Let v be the velocity of the lambda in the laboratory. Then if τ_l is its mean life in the laboratory, the mean distance travelled is

$$l = \tau_l v. \tag{11.95}$$

Because of time dilatation τ_l is greater than the mean life, τ_0, in the lambda rest frame. As we know from Section 11.12.4, page 373,

$$\tau_l = \gamma \tau_0, \tag{11.96}$$

where

$$\gamma = \frac{1}{\sqrt{1 - v^2/c^2}}.$$

The momentum of the lambda in the laboratory is

$$p_\Lambda = m_\Lambda \gamma v. \tag{11.97}$$

Then, substituting from (11.96) and (11.97) into (11.95),

$$l = \frac{p_\Lambda \tau_0}{m_\Lambda} = \left(\frac{p_\Lambda c}{m_\Lambda c^2}\right)\tau_0 c.$$

With the values given and calculated above this expression yields

$$l = 4{\cdot}18 \text{ cm}.$$

11.12.8. See Section 11.8

Zero rest mass; photons and neutrinos. The relativistic concepts we have considered so far have given us an enormous extension of the application of energy and momentum conservation. We are no longer restricted to the low speeds of common experience or to processes in which the particles and their rest masses are the same before and after their interactions with each other.

However, the definitions (11.45) and (11.47) will hold only for $v < c$. Is there any way in which energy and momentum concepts can still be applied at the limiting speed c? Since γ becomes infinite for this speed, we could only obtain finite values for energy and momentum if the rest mass were zero. Now we know from experiment of a process which carries energy and momentum with velocity of light. That is light itself (an electromagnetic wave). The energy is carried in discrete units of $\hbar\omega$, and the momentum in discrete units of $\hbar\mathbf{k}$, where ω is the angular frequency, $\mathbf{k}$ is the wave vector, and $\hbar$ is (Planck's constant)/2π. (The wave propagates in the direction of $\mathbf{k}$, with wavelength $2\pi/k$ and frequency $\omega/(2\pi)$.)

Hence we can say that the energy and momentum content of an electromagnetic wave can be represented by some multiple of the four-vector

$$(p) = \begin{pmatrix} \hbar k_1 \\ \hbar k_2 \\ \hbar k_3 \\ \hbar\omega/c \end{pmatrix} \equiv \begin{pmatrix} \hbar\mathbf{k} \\ \hbar\omega/c \end{pmatrix}. \tag{11.98}$$

This leads to a picture of the wave as equivalent to a stream of particles, called *photons*, each of zero rest mass and possessing momentum described by (11.98).

If photons merely illustrated these dynamical properties of an electromagnetic wave, they could hardly be regarded as a serious alternative description. However, when we consider the interaction between electromagnetic radiation and material particles, an analysis in terms of a wave colliding with a particle is complicated and difficult. It turns out that the picture of a photon interacting with the particle, in a collision that conserves relativistic energy-momentum, gives a simple and accurate description of the process. Moreover, the changes of frequency and direction of a wave as they appear to different observers accords accurately with the Lorentz transformation of the four-vector (11.98) between their frames of reference (see problems 11.9 and 11.10). Thus we may add the photon to the field of physics that is covered by the relativistic description of energy-momentum conserving interactions.

However, the photon might still be regarded as a very untypical particle since it is linked to the frequency and wavelength of electromagnetic radiation; and the latter has a special role in the development of relativity, as we have seen. Even here the restriction is unnecessary. There are other particles of zero rest mass, called *neutrinos*, which are quite unconnected with electromagnetism. The four-vector description of their energy and momentum also fits beautifully into the transformation and conservation scheme of special relativity.

11.12.9. See Section 11.11

When a particle has a transient existence, with a mean lifetime δt, there is a spread, δE, in its rest energy m_0c^2 that is governed by the uncertainty relationship:

$$\delta t\, \delta E \sim \hbar = 1{\cdot}05 \times 10^{-34}\ \mathrm{J\,s} = 6{\cdot}6 \times 10^{-22}\ \mathrm{Me\,V\,s}.$$

This is why we observe *concentrations* about particular rest masses rather than unique values, and the particles are referred to as 'resonances'. For a typical lifetime $\sim 10^{-23}$ s, the spread or 'width of the resonance' would be

$$\delta E \sim 100\ \mathrm{MeV}.$$

11.13. Problems

In the following problems, unless it is stated otherwise, the observers S, S' may be assumed to have rest frames with corresponding axes parallel, S' having a velocity U relative to S directed along axis 1 ($B = U/c$, $\Gamma^{-2} = 1 - B^2$).

11.1. Write down the Lorentz transformations for the relative velocities $U = 0{\cdot}1c$, $U = 0{\cdot}8c$, $U = 0{\cdot}98c$.

11.2. Suppose the axes 1, 1′ used by S, S' are collinear, and that both S and S' take their time zeros as the time when the origins of their two frames coincide. For each of the transformations of problem 11.1 calculate:
(a) The event observed by S' which appears to S as the position (5, 7, 9) m at time −5 s,
(b) The event observed by S which appears to S' as the position (6, −5, 2) m at time 20 s.

11.3. Discuss critically the steps outlined in Section 11.12.5 for setting up a frame of reference. In particular comment on
(a) alternative definitions of 'straightness',
(b) alternative methods of establishing a distance scale on the axes,
(c) the minimum number of clocks required to establish a time scale. How could you test experimentally the equivalence of the alternatives?

11.4. Suppose each axis of a Cartesian frame is equipped with a sliding trolley from which projects a rigid rod perpendicular to the axis. As a particle moves in space these three rods are kept in coincidence with the particle. The positions of the trolleys should therefore give instantaneous values of the three coordinates of the particle. What are the practical problems that would arise in such a procedure? Which of these are purely technical and which are in principle insoluble? (At least one must be insoluble if the discussion at the beginning of Chapter 11 is not to be contradicted.)

11.5. Show that the mid-point of a rod at any orientation is the same for any two observers moving with constant velocity relative to each other.

11.6. The S' frame is fixed in a train of length $2l$ with origin O′ at its mid-point and axis 1′ along the length of the train. The train moves with constant velocity U relative to S along the axis 1. As O′ coincides with O the clocks of both observers are started and light signals are emitted from O′ forward and backward along the train. These are reflected by mirrors at the ends of the train and then move back towards each other. Determine the four-vectors for both S and S' which describe the following sequence of events:

(a) The emission of the light signals from O′.
(b) The reflexion of the light signal at the front of the train.
(c) The reflexion of the light signal at the rear of the train.
(d) The arrival at O of the first light signal to reach it.
(e) The meeting of the two light signals.
(f) The arrival at O of the second light signal to reach it.

Verify that, although S and S' measure different positions and times for all these events except the first, the velocity of light is the same throughout for both observers, and show that the light signals reach O with a time difference $4B\Gamma l/c$ according to S and $4B\Gamma^2 l/c$ according to S'.

11.7. The matrix (K) is defined by

$$(K) = \begin{pmatrix} A & \vdots & 0 \\ \cdots & \cdots & \cdots \\ 0 & \vdots & 1 \end{pmatrix} \equiv \begin{pmatrix} a_{11} & a_{12} & a_{13} & 0 \\ a_{21} & a_{22} & a_{23} & 0 \\ a_{31} & a_{32} & a_{33} & 0 \\ 0 & 0 & 0 & 1 \end{pmatrix}, \tag{11.99}$$

where (A) is the 3×3 rotation matrix of Chapter 8. Show that (K) has the Lorentz transformation property

$$(K)(G)(K)^{-1} = (G),$$

and that it relates two frames which are at rest with respect to each other, but whose axes are at different orientations. Hence show that in this case the invariance property (11.14) represents the geometrical invariance of classical physics.

11.8. If S' is at rest relative to S, but with reoriented axes, and S'' is moving with constant velocity U in the direction 1′ relative to S', and with axes parallel with those of S', show that the transformation relating the frames S'' and S is given by $(\Lambda)(K)$ where (Λ) and (K) represent the transformations (11.26) and (11.99). Show that the matrix

$$\begin{pmatrix} \Gamma\cos\phi & \Gamma\sin\phi & 0 & -B\Gamma \\ -\sin\phi & \cos\phi & 0 & 0 \\ 0 & 0 & 1 & 0 \\ -B\Gamma\cos\phi & -B\Gamma\sin\phi & 0 & \Gamma \end{pmatrix}$$

is of this form, and describe the transformation it represents.

11.9. A plane wave propagating in the direction $\mathbf{k}$ and with angular frequency ω in the S frame will have all its field quantities varying as $\exp(\mathbf{k}\,.\,\mathbf{x} - \omega t)$. Since the phase angle $\mathbf{k}\,.\,\mathbf{x} - \omega t$ must be the same in whichever frame it is calculated, it must be an invariant for transformations between frames with constant relative velocities. By expressing the phase angle as $(\tilde{k})(G)(x)$, where

$$(\tilde{k}) = (k_1, k_2, k_3, \omega/c),$$

show that (k) is a four-vector. What is the physical significance of the invariant 'length' of (k)?

A plane electromagnetic wave is propagating at right angles to the axis 3 and at an angle ϕ to the axis 1. Show that if ω' and ϕ' are the angular frequency and direction in the S' frame, then

$$\cos\phi' = \frac{\cos\phi + B}{1 + B\cos\phi}, \qquad \frac{\omega'}{\omega} = \Gamma(1 + B\cos\phi).$$

(These results express the relativistic *aberration* and *Doppler effect*.)

11.10. Light of wavelength λ is scattered through an angle ϕ by an electron of mass m which is initially at rest. Show, by considering this as a relativistic energy-momentum conserving collision between a photon and an electron, that the wavelength, λ', of the scattered light is given by

$$\lambda' - \lambda = \frac{2\hbar}{mc}\sin^2\frac{\phi}{2}.$$

(This change in wavelength is known as the *Compton effect*.)

Index

Page numbers in bold refer to chapter and section headings; those in italics to end-of-chapter problems.